International Association of Fire Chiefs

National NFPA

Fundamentals of Technical Rescue

JONES AND BARTLETT PUBLISHERS

Sudbury, Massachusetts

BOSTON   TORONTO   LONDON   SINGAPORE

Jones and Bartlett Publishers, LLC
World Headquarters
40 Tall Pine Drive
Sudbury, MA 01776
978-443-5000
info@jbpub.com
www.jbpub.com

Jones and Bartlett Publishers Canada
6339 Ormindale Way
Mississauga, Ontario L5V 1J2
Canada

Jones and Bartlett Publishers International
Barb House, Barb Mews
London W6 7PA
United Kingdom

National Fire Protection Association
1 Batterymarch Park
Quincy, MA 02169-7471
www.NFPA.org

International Association of Fire Chiefs
4025 Fair Ridge Drive
Fairfax, VA 22033
www.IAFC.org

Jones and Bartlett's books and products are available through most bookstores and online booksellers. To contact Jones and Bartlett Publishers directly, call 800-832-0034, fax 978-443-8000, or visit our website, www.jbpub.com.

The procedures and protocols in this book are based on the most current recommendations of responsible sources. The International Association of Fire Chiefs (IAFC), National Fire Protection Association (NFPA®), and the publisher, however, make no guarantee as to, and assume no responsibility for, the correctness, sufficiency, or completeness of such information or recommendations. Other or additional safety measures may be required under particular circumstances.

Additional photographic and illustration credits appear on page 276, which constitutes a continuation of the copyright page.

Notice: The individuals described in "You are the Rescue Responder" and "Rescue Responder in Action" throughout the text are fictitious.

Production Credits
Chief Executive Officer: Clayton Jones
Chief Operating Officer: Don W. Jones, Jr.
President, Higher Education and Professional Publishing:
   Robert W. Holland, Jr.
V.P., Sales: William J. Kane
V.P., Design and Production: Anne Spencer
V.P., Manufacturing and Inventory Control: Therese Connell
Publisher, Public Safety Group: Kimberly Brophy
Senior Acquisitions Editor, Fire: William Larkin
Associate Editor: Amanda Brandt

Production Manager: Jenny L. Corriveau
Associate Photo Researcher: Jessica Elias
Director of Sales: Matthew Maniscalco
Director of Marketing: Alisha Weisman
Marketing Manager, Fire: Brian Rooney
Cover Image: © Tose/Dreamstime.com
Text Design: Anne Spencer
Cover Design: Scott Moden
Composition: Publishers' Design and Production Services
Text Printing and Binding: Courier Kendallville
Cover Printing: Courier Kendallville

Library of Congress Cataloging-in-Publication Data
Rhea, Robert.
    Fundamentals of technical rescue / Robert Rhea, Brian Rousseau.
      p. cm.
    ISBN-13: 978-0-7637-3837-2 (pbk.)
    ISBN-10: 0-7637-3837-9 (pbk.)
    1. Emergency management—United States—Handbooks, manuals, etc.   2. Rescue work—United States—Planning—Handbooks, manuals, etc.   3. Search and rescue operations—United States—Handbooks, manuals, etc.   I. Rousseau, Brian.   II. Title.
    HV551.3.R5 2009
    363.34'81—dc22
                                                                2009014460

6048

Brief Contents

Contents

Skill Drills

Resource Preview

Fundamentals of Technical Rescue

Beginning with an introduction to technical rescue and progressing through discussions of tools and equipment, incident management, and conducting search operations, this text will introduce rescue organizations and their members to all aspects of the rescue process and the various environments in which they may be responding. *Fundamentals of Technical Rescue* presents in-depth coverage of structural collapse, confined space and trench rescue, vehicle rescue, and water and wilderness rescue to allow rescue organizations to approach any rescue situation safely and confidently.

    Fundamentals of Technical Rescue includes coverage of the awareness-level requirements found in the 2009 Edition of NFPA 1670, *Standard on Operations and Training for Technical Search and Rescue Incidents* and some of the base line job performance requirements found in the 2008 Edition of NFPA 1006, *Standard for Technical Rescuer Professional Qualifications.*

Chapter Resources

Fundamentals of Technical Rescue serves as the core of a highly effective teaching and learning system. The features reinforce and expand on essential information and make information retrieval a snap. These features include:

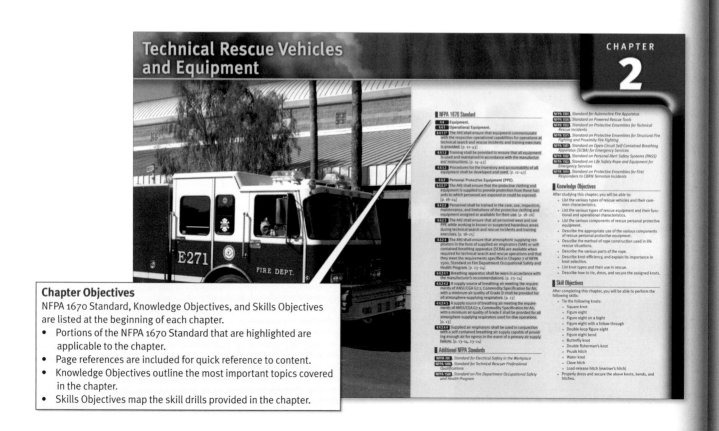

Chapter Objectives

NFPA 1670 Standard, Knowledge Objectives, and Skills Objectives are listed at the beginning of each chapter.

- Portions of the NFPA 1670 Standard that are highlighted are applicable to the chapter.
- Page references are included for quick reference to content.
- Knowledge Objectives outline the most important topics covered in the chapter.
- Skills Objectives map the skill drills provided in the chapter.

You Are the Rescue Responder

Each chapter opens with a case study intended to stimulate classroom discussion, capture students' attention, and provide an overview for the chapter. An additional case study is provided in the end-of-chapter Wrap-Up material.

You Are the Rescue Responder

You are a first responder assigned to an emergency medical services (EMS) response vehicle. Your unit is dispatched to a report of a building that has collapsed as a result of high wind conditions. A severe thunderstorm accompanied by the threat of tornadoes has been reported in your response area. Your response unit is dispatched to the collapse event along with two fire department engine companies, a ladder company, and a rescue company. The dispatch center transmits a situation update reporting that the building is a two-story wood-frame house that was under construction. Witnesses report that high winds from the thunderstorm have blown the house over and that two construction workers are trapped under the debris.

As you are responding to the event, you realize that your unit will be the first responders to arrive on the scene. You know that a structural collapse incident can be very dangerous for responders. You also know that specialized resources are required to conduct search and rescue operations at a structural collapse event.

You arrive to find a wood-frame structure, under construction, that has suffered a lateral rack-over collapse of the first floor. Two construction workers approach your unit and are bleeding from the head. Another construction worker is lying on the ground and appears to be in pain. One of the walking wounded tells you that two of his buddies are still trapped under the collapsed house, but he is not sure where they are located. As you start to perform a size-up of the event, you remember that a structural collapse incident is resource intensive and that specially trained responders and specialized tools and equipment are needed to operate effectively.

1. What should you be looking for during a structural collapse size-up?
2. Which hazards might rescuers find at a structural collapse event?
3. Which actions can first responders take at this incident to more effectively manage the event?
4. How should you search for victims at a structural collapse event?
5. Which victims should be treated first?

Introduction

Structural collapse operations cover a wide range of incident scenarios. These scenarios vary in difficulty based on the significance of the collapse and the size, type, and occupancy of the building involved. Given the inconsistency of their presentation, it is necessary to implement a standard but flexible plan of action that can be used for rescue operations at all structural collapse situations.

Difficult collapse rescue operations require a combination of specialized tools and equipment, as well as specially trained emergency responders who apply unique techniques, for their effective mitigation. Acquiring the appropriate resources as early as possible is an important step in the successful outcome of structural collapse rescue. For this reason, responders must recognize the need for search and rescue operations at a building collapse event quickly and to take the appropriate actions immediately.

Structural collapse emergencies have a variety of causes. Some collapses are a result of natural occurrences such as hurricanes, tornadoes, and other high wind conditions. In addition, earthquakes, mudslides or landslides, floods, and snow or ice loads on roofs are natural causes of building collapses **Figure 6-1 ▸**.

Buildings may also collapse due to fire damage when a fire weakens structural support components such as floors, roofs, walls, or columns. Interior or exterior blast loading on a structure can also result in a collapse situation if building support

Voices of Experience

As fire fighters, are we really prepared for every incident that we might encounter?

It happened one July night. At approximately 0200 hours, the fire department received a call for a natural gas leak inside a building. This is a common, everyday occurrence for most every fire department in the country. The following few minutes would change the response dramatically, however. While the duty shift was preparing for a response, there was a sudden and loud explosion. A few windows in the fire station were blown out from the force. Those of us who were off duty were awakened by the explosion as it resonated throughout the downtown area. The force activated the master box in the bank across the street from the building. The engine pulled out of headquarters and turned right to find the street blocked from the debris of the building lying across the street.

What happens now? Do we know how many victims might be home? Which hazards might be encountered upon our arrival? What was the actual cause of the explosion? As the first-arriving companies started gathering information, we soon found that there were approximately 12 people who lived in the residence. With the help of the local police, we identified and found all but four of these occupants. As the companies were completing a size-up and surveying the building for surface victims, we had two adult occupants crawl out of a small void space on the B side of the building. They looked up at us and stated that their two young children were still in there, and they were alive and talking to them just before they left to crawl out. We made a decision to make an entry and attempt to locate the two children.

A few department members were part of a trained district technical rescue team, and they were immediately activated for a response. We started setting up a staging area for the equipment to be parked near the scene. The ICS was in use, and early positions established were the IC, operations, rescue team leader, and safety officer. We found a need for early shoring operations, but also found a lack of available lumber to meet these needs.

Upon arrival of the technical rescue team, we set up our command staff positions. We went about shoring operations with what we had available and made entry into the collapse area. During entry operations, the team was evacuated from the building due to an active natural gas leak on the D side of the structure. What we found was that the collapse had landed the building on top of the shut-off valves; there would be a significant time lapse waiting for the leak to be shut off from a remote location.

We learned that most departments don't have a trained team available or know how to locate a team should one be needed. The names and contact numbers for after-hours acquisition of materials, such as lumber, were not active or up-to-date. We fell back on the five basic stages for structural collapse incidents. We had the initial response on scene; we could easily do a search for and safe removal of surface victims. Once we had a technical rescue team on scene, we could start an interior search of void spaces. We did not have to get to selected or general debris removal. The technical rescue team did make a recovery of both children as soon as the natural gas leak was shut down.

The final stage of our response was to have all the members operating on scene go through critical incident stress debriefing (CISD) prior to being demobilized. By establishing the command presence early, we kept our members safe and injury free, which made it a successful response and recovery for the team.

Francis M. Clark
Hopkinton Fire Department
Massachusetts Fire District 14 Technical Rescue Team
Hopkinton, Massachusetts

“Their two young children were still in there, and they were alive and talking to them just before they left to crawl out.”

Voices of Experience

In the Voices of Experience essays, veteran rescue responders share their accounts of memorable incidents while offering advice and encouragement. These essays highlight what it is truly like to be a rescue responder.

Skill Drills

Skill Drills provide written step-by-step explanations and visual summaries of important skills and procedures. This clear, concise format enhances student comprehension of important procedures.

Skill Drill **2-12**

Tying a Clove Hitch (Closed Object)

1. Loop the rope completely around the object, with the working end below the running end.

2. Loop the working end around the object a second time to form a second loop, slightly above the first. Pass the working end under the second loop, just above the point where the second loop crosses over the first loop.

3. Secure the knot by pulling on both ends.

4. Tie a safety knot on the working end of the rope.

---

58   Fundamentals of Technical Rescue

rescue skills and equipment. High angle rope rescue victims may be suffering from traumatic injuries and/or other medical emergencies. Advanced life support (ALS) capabilities are required to treat fall victims effectively once they are moved to a safe location.

Incident Management Requirements

The strategic objective when operating at an event that requires rope rescue equipment and personnel trained in rope rescue is to control and manage the event in such a way as to evaluate the scene and identify potential victims and their location, to initiate operations to minimize hazards to operating personnel and trapped victims, to search the area, and to rescue and remove trapped victims, and to minimize further injury to victims during search, rescue, and removal operations.

The Incident Command System (ICS) positions are staffed as determined by the scope of the event. In most cases, at a minimum, management personnel will be required to fill the positions of incident commander (IC), technical rescue group supervisor, and safety officer.

The technical rescue group supervisor is responsible for implementing tactical rescue decisions to support the strategy established by the IC **Figure 5-8**. These tactical benchmarks include hazard mitigation, entry team readiness, rapid intervention capabilities, and emergency medical care for the patient.

Incident management personnel are responsible for strategic and tactical management of the emergency response. One of their responsibilities is the development of an incident action plan (IAP), which identifies the overall control objectives for the emergency. Because an emergency requiring specialized rope rescue skills may include unique operations and hazards related to another causative factor (e.g., confined space, building collapse, water rescue), the event may require additional specialized resources. These specialized resources could include swiftwater rescue teams, building collapse assets, and/or trench rescue teams. The incident management personnel may need to gain input from other knowledgable people, including personnel who have been trained in specific areas of technical rescue, such as technical search personnel, rescue team engineers, or emergency medical professions

---

114   Fundamentals of Technical Rescue

Figure 5-7   The Larkin Frame may be used in wilderness rescues as substitutes for tripods.

and operate rope rescue systems safely. Rope rescue techniques include establishing anchor systems, establishing fall protection, building and operating lowering systems, building and operating mechanical advantage systems, building and using high-line systems, and **ascending** rope. Ascending is a means of safely traveling up a fixed rope with the use of ascent devices. Any of these skills may be required at a high-angle rope rescue incident.

Awareness level responders are not properly trained to operate rope rescue equipment or to use any of the previously listed rope rescue techniques. Instead, awareness level is the first level of rescue training provided to all responders. It emphasizes the ability to recognize hazards at an incident, secure the scene, and call for appropriate assistance and resources.

Rescue Tips

Rope rescue tools and techniques may be required to move patients effectively within a variety of rescue environments, including swift water, trench collapse, and confined spaces. Awareness level responders must recognize the type of rescue environment involved and ensure that rescue teams crossed-trained for both rope rescue and the rescue environment are requested and made available.

Emergency Medical Services Resources

Emergency medical services (EMS) resources are required so rescue teams can hand off patients for treatment and transport to medical facilities once the victims are retrieved utilizing rope

Figure 5-8   Team briefing at a technical rescue

Rescue Tips

Rescue Tips offer important and insightful information for rescue responders.

---

98   Fundamentals of Technical Rescue

should be established. During the assessment phase, rescuers also determine which resources will be required to extricate patients from any entrapment and which equipment and staffing will be required to package and move patients out of the hazard zone. For this reason, rescuers must have a good knowledge of the internal resources available from their response organization as well as the response times when requesting specialized resources from outside the organization, such as cranes, specialized rescue teams, or utility company assistance.

Patient assessment methods and patient treatment must always follow the medical protocols established by the authority having jurisdiction (AHJ), which in turn must comply with all applicable local, state, and federal laws. Protection from bloodborne pathogens must be ensured through the use of proper PPE and adherence to written protocols related to **body substance isolation (BSI)**. Body substance isolation is an infection control concept that treats all bodily fluids as potentially infectious.

Victim Stabilization

Initial medical care to a patient (BLS) includes establishing and maintaining an adequate airway, providing respiratory ventilation to maintain air flow, controlling severe bleeding, and maintaining the circulatory system. In addition, spinal immobilization must be assured, and the victim should be treated for shock if this situation is present or suspected. In these situations, local medical protocols must always be followed. If the rescuers are not trained in emergency medical care, they must know where to get these resources and be prepared to assist the emergency medical staff with patient packaging and movement. **Patient packaging** refers to the process of preparing the victim for movement as a unit, often accomplished with a long spine board or similar device.

Triage

Some technical rescue events, such as a building collapse or vehicle accident involving a bus, may result in multiple victims who need medical care **Figure 4-5**. Whenever rescuers find multiple patients at a technical rescue incident, the victim

assessment must include determining which patients are most severely injured, then prioritizing which patients receive treatment first and which patients can wait to receive delayed care. In general, this process (referred to as triage) sorts patients based on the severity of their injuries or medical conditions.

Safety Tips

Victim management may expose responders to bloodborne pathogens. For this reason, rescuers must maintain the proper level of PPE, including BSI, during all patient contact.

Triage is essential at all **mass-casualty incidents** because an increased number of patients places an increased demand on resources. At small-scale incidents, the first provider on scene with the highest level of training usually begins the triage process. Victims are ranked in order of severity of their injuries, and the victim with the most severe injury is given priority attention. After determining the total number of victims and requesting additional assistance, initial assessment of all victims should begin. As more resources arrive, the triage personnel should assign crews to give assistance to those patients in most need first.

At a large-scale mass-casualty incident, triage should occur in several steps. The following triage steps are accepted by the majority of larger-scale mass-casualty operations:

1. Life-saving care is rapidly administered to those in need.
2. Color coding is used to indicate priority for treatment and transportation at the scene. Red-tagged victims are the first priority, yellow-tagged victims are the second priority, and green- or black-tagged patients are the lowest priority **Figure 4-6**.
3. Rapid removal of red-tagged victims for field treatment and transportation occurs as ambulances become available.
4. A separate treatment area is set aside to care for red-tagged victims if transport is not immediately available. Yellow-tagged victims can also be monitored and cared for in a treatment area while waiting for transportation.

Figure 4-5   A rescue event involving multiple victims will require triage.

Figure 4-6   Triage tags are color coded to indicate treatment priority.

Safety Tips

Safety Tips reinforce safety-related concerns for rescue responders.

Wrap-Up
End-of-chapter activities reinforce important concepts and improve students' comprehension. Additional instructor support and answers for all questions are available on the Instructor's ToolKit CD-ROM.

Rescue Responder in Action
The Rescue Responder in Action feature promotes critical thinking through the use of case studies and provides instructors with discussion points for the classroom presentation.

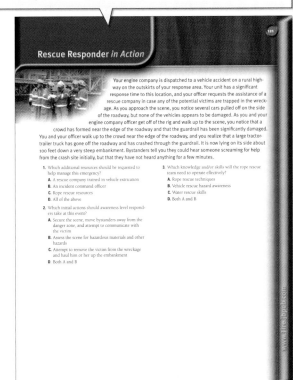

Wrap-Up

Chief Concepts

- Rescuers may encounter rope rescue situations where patients are located hanging on the side of a building from a fall protection harness, dangling from a ledge on the side of a rock face, located on the top of a building collapse rubble pile, situated at the bottom of an excavation, or trapped on an upper floor of a building under construction. Rescuers must have a clear understanding of how to recognize which type of rescue environment they are working in and which hazards may be present.
- High-angle rope rescue emergencies require the use of specialized resources, including rope rescue hardware and software and specially trained rope rescue teams. Responders should have a good understanding of where to acquire the proper resources to perform a successful rope rescue.
- Rope rescue hazards can include fall hazards, falling debris, energy hazards, environmental hazards, and crowd control issues. Hazard mitigation efforts are required before responders can initiate actions to access, treat, package, and remove a patient. Awareness level responders must have a thorough understanding of which high-angle hazards may be present and know how to identify them.
- Awareness level responders may be able to initiate some initial scene control activities such as crowd control, edge control, and fall protection. However, at no time should they attempt to gain access to a patient or to remove a patient in a situation where fall protection is required.

Hot Terms

Ascending A means of traveling up a fixed rope safely with the use of an ascent device.

Awareness level The first level of rescue training provided to all responders, which emphasizes recognizing hazards, securing the scene, and calling for appropriate assistance. There is no actual use of rescue skills at the awareness level.

Carabiner A piece of metal hardware used in rope rescue operations. It is generally an oval-shaped device with a spring-loaded gate that can be used to connect pieces of rope, webbing, and other rope hardware.

Descending A means of traveling down a fixed rope safely with the use of a descent control device.

Edge protection A device that prevents damage to rope or other rope software from sharp or jagged edges or from friction.

Fall protection A system of associated rope hardware and software used to protect workers from falling from an elevated position. Fall protection also refers to equipment to which workers may be attached while working at elevations or near a fall hazard, which is meant to capture the individuals if they fall.

Harness A piece of safety equipment that can be worn by a rescuer and attached to a rope rescue system, thereby allowing the rescuer to work safely in a high-angle environment during rope rescue operations.

High-angle rope rescue A rope rescue operation where the angle of the slope is greater than 45 degrees. Rescuers depend on the rope for the primary support mechanism rather than a fixed support surface such as the ground.

Low-angle rope rescue A rope rescue operation where the slope of the ground is less than 45 degrees. Rescuers depend on the ground for their primary support, and the rope system becomes the secondary means of support.

Mechanical advantage system A system that creates a leverage force using levers, pulleys, or gearing and is described in terms of a ratio of output force to input force.

Packaging The process of preparing a victim for movement as a unit, often accomplished with a long spine board or similar device.

Rope hardware Rigid mechanical auxiliary equipment that can include, but is not limited to, anchor plates, carabiners, and mechanical ascent and descent control devices.

Rope rescue incident A situation where rescuing an injured or trapped patient requires rope rescue equipment and techniques.

Rope rescue system A system consisting of rope rescue equipment and an appropriate anchor system intended for use in the rescue of a subject.

Rope software A flexible fabric component of a rope rescue equipment that can include, but is not limited to, anchor straps, pick-off straps, and rigging slings.

Rescue Responder in Action

Your engine company is dispatched to a vehicle accident on a rural highway on the outskirts of your response area. Your unit has a significant response time to this location, and your officer requests the assistance of a rescue company in case any of the potential victims are trapped in the wreckage. As you approach the scene, you notice several cars pulled off on the side of the roadway, but none of the vehicles appears to be damaged. As you and your engine company officer get off of the rig and walk up to the scene, you notice that a crowd has formed near the edge of the roadway and that the guardrail has been significantly damaged. You and your officer walk up to the crowd near the edge of the roadway, and you realize that a large tractor-trailer truck has gone off the roadway and has crashed through the guardrail. It is now lying on its side about 100 feet down a very steep embankment. Bystanders tell you they could hear someone screaming for help from the crash site initially, but that they have not heard anything for a few minutes.

1. Which additional resources should be requested to help manage this emergency?
   A. A rescue company trained in vehicle extrication
   B. An incident command officer
   C. Rope rescue resources
   D. All of the above

2. Which initial actions should awareness level responders take at this event?
   A. Secure the scene, move bystanders away from the danger zone, and attempt to communicate with the victim
   B. Assess the scene for hazardous materials and other hazards
   C. Attempt to remove the victim from the wreckage and haul him or her up the embankment
   D. Both A and B

3. Which knowledge and/or skills will the rope rescue team need to operate effectively?
   A. Rope rescue techniques
   B. Vehicle rescue hazard awareness
   C. Water rescue skills
   D. Both A and B

Hot Terms
Hot Terms are easily identifiable within the chapter and define key terms that the student must know. A comprehensive glossary of Hot Terms also appears in the Wrap-Up.

Chief Concepts
Chief Concepts highlight critical information from the chapter in a bulleted format to help students prepare for exams.

Instructor Resources

Instructor's ToolKit CD-ROM

ISBN: 978-0-7637-7695-4

Preparing for class is easy with the resources on this CD-ROM. The resources found in the Instructor's ToolKit CD-ROM have been formatted to be seamlessly integrated into popular course administration tools. The CD-ROM includes the following resources:

- **Adaptable PowerPoint Presentations.** Provides instructors with a powerful way to create presentations that are educational and engaging to their students. These slides can be modified and edited to meet instructors' specific needs.
- **Detailed Lesson Plans.** The lesson plans are keyed to the PowerPoint presentations with sample lectures, lesson quizzes, answers to all end-of-chapter student questions found in the text, and teaching strategies. Complete, ready-to-use lecture outlines include all of the topics covered in the text. The lecture outlines can be modified and customized to fit any course.
- **Electronic Test Bank.** The test bank contains multiple-choice and scenario-based questions and allows instructors to create tailor-made classroom tests and quizzes quickly and easily by selecting, editing, organizing, and printing a test along with an answer key that includes page references to the text.
- **Skill Sheets.** Provides instructors with a resource to track students' skills and conduct skill proficiency exams.
- **Image and Table Bank.** Instructors can use these graphics to incorporate more images into the PowerPoint presentations, make handouts, or enlarge a specific image for further discussion.

Student Resources

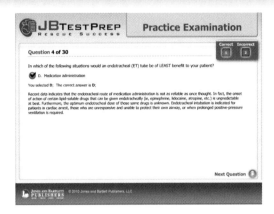

JBTest Prep: Rescue Success

ISBN-13: 978-0-7637-7696-1

JBTest Prep: Rescue Success is a dynamic program designed to prepare students to sit for rescue certification examinations by including the same type of questions they will likely see on the actual examination.

It provides a series of self-study modules organized by chapter offering practice examinations and simulated certification examinations using multiple-choice questions. All questions are page referenced to *Fundamentals of Technical Rescue* for remediation to help students hone their knowledge of the subject matter.

Students can begin the task of studying for certification examinations by concentrating on those subject areas where they need the most help. Upon completion, students will feel confident and prepared to complete the final step in the certification process--passing the examination.

www.Fire.jbpub.com

This site has been specifically designed to complement *Fundamentals of Technical Rescue* and is regularly updated. Resources available include:

- **Chapter Pretests** that prepare students for training. Each chapter has a pretest and provides instant results, feedback on incorrect answers, and page references for further study.

- **Interactivities** that allow students to reinforce their understanding of the most important concepts in each chapter.
- **Hot Term Explorer**, a virtual dictionary, allowing students to review key terms, test their knowledge of key terms through quizzes and flashcards, and complete crossword puzzles.

Acknowledgments

Authors

Robert Rhea

Battalion Chief (Ret.) Robert Rhea is a 29-year veteran of the Fairfax County (Va.) Fire and Rescue Department and served the last 13 years as a Battalion Commander assigned to the Operations Division. Robert served for 16 years with the department's Technical Rescue and Urban Search and Rescue (USAR) Team, including serving as Task Force Leader and USAR Program Manager. Robert served three years in the Training Division as the Director of Fire and Rescue Training Programs, and also served as the Senior Technical Officer in charge of the departments' Technical Rescue Operations Team where he was instrumental in designing, implementing, and evaluating confined-space, structural collapse, and trench rescue training programs and operational procedures. Robert is a principal committee member on the National Fire Protection Association committees: NFPA 1026, *Standard for Incident Management Personnel Professional Qualifications*, NFPA 1670, *Standard on Operations and Training for Technical Search and Rescue Incidents*, and NFPA 1006, *Standard for Technical Rescuer Professional Qualifications*. Chief Rhea served as a content expert for the FEMA USAR System development team, and as an adjunct instructor for the National Fire Academy. Robert has a BS in Business Management and has completed Masters Degree studies with a focus in training and human resource development. Robert is co-owner of ARK Technical Rescue Training Services, Inc.

Brian Rousseau

Deputy Chief Brian Rousseau has 34 years of fire/rescue experience, the last 24 years with the New York State Office of Fire Prevention and Control. During the last 19 years, he has been primarily involved in technical rescue training, program development, and consulting, ranging from very small organizations to some of the largest fire departments in the world as well as local, regional, and national governments and rescue teams. He has been involved in training, consulting, and exercise evaluation throughout the US, Canada, the Caribbean, and Europe. He has authored and co-authored training programs in basic rescue, rope rescue, confined space rescue, structural collapse, and trench/excavation rescue and has been actively involved for over 13 years in the development of the NFPA international standards for rescue training, operations, and professional certification.

Contributing Authors

Jerry McGhee

Chapter 9: Vehicle and Machinery Search and Rescue

Chief Jerry McGhee is a Training and Exercise Analyst for EWA Information & Infrastructure Technologies (EWA IIT) and is responsible for the training and readiness of Weapons of Mass Destruction Civil Support Teams (WMD-CST), CBRNE Enhanced Response Force Package teams, Joint Operations Centers, State Offices of Emergency Services, State Health, States Joint Force Headquarters (JFHQ), Fire Departments, Emergency Medical Service, Law Enforcement, etc. Prior to his international/domestic work, Jerry has worked in fire/EMS for over thirty years; he worked as a field fire fighter/paramedic with the City of Charleston Fire Department before becoming Deputy Director of EMS division. He later accepted a position as the Occupational Safety & Health safety manager for two Aluminum Industrial facilities, one located in West Virginia, the other in Vernon, California. Jerry has designed innovative programs in fire science, hazardous materials, both light and heavy vehicular extraction, technical rescue, confined space entry and rescue, weapons of mass destruction (WMD) for both fire fighter/rescuers and EMS personnel, and many other programs in association with West Virginia University Fire Service Extension, Louisiana State University, Department of Education, NFPA/OSHA, National Fire Academy and State Fire Commission. Jerry is also a member of the International Association of Fire Chiefs, National Fire Protection Association (NFPA), and International Society of Fire Service Instructors. Jerry has trained and worked with all agencies across the United States in scenario development and planning for small, minimum and large scale training exercises involving the US and International Military, Federal and State Task Force teams, United States Coast Guard, Federal Bureau of Investigation, Explosive Ordinance Division, SWAT teams, US Postal Service, Secret Service, Technical Escort, FEMA, Fire Departments, Law Enforcement, Hazardous Material Teams, and Emergency Medical Services. Jerry currently serves as Chief of the Marmet Fire Department.

Ben Waller

Chapter 10: Water Search and Rescue

Ben Waller is a Battalion Chief with Hilton Head Island Fire & Rescue, currently assigned as the chief of training. He has been involved in fire, rescue, and emergency medical services since 1975. He is a fire fighter, paramedic, hazardous materials technician, and USAR rescue specialist. Ben is a swiftwater rescue instructor and raft guide, was a member of the 1996 Olympic Games swift-water rescue team, and was the swiftwater rescue team leader for the 2000 U.S. Olympic Team Trials. He is an adjunct instructor with the South Carolina Fire Academy, where he is also a member of the Water Rescue and Rope Rescue Technical Development Committees. He has responded to numerous flood and swiftwater rescues and has been a swiftwater rescue team member for several international-level whitewater competitions. Ben's education includes a Master's of Public Administration degree from Clemson University and undergraduate degrees in Fire Administration and Paramedic/Allied Health. Ben is also a member of the Hilton Head Island-Bluffton Hazardous Materials/COBRA Team and is Safety Officer for the Hilton Head Island-Bluffton USAR Team, SC-RRT-4. Ben was a contributor to the Jones and Bartlett *Swiftwater and Flood Rescue Field Guide* and is the Water Rescue Consulting Editor for *Technical Rescue Magazine.*

Jon Trapp

Chapter 11: Wilderness Search and Rescue

Captain Jon Trapp is a fire fighter/medic for Red Lodge Fire Rescue in Montana. He heads the technical rope rescue team and is the coordinator for Carbon County Search and Rescue. He runs the Wildland Urban Interface (WUI) home assessment and mitigation program for Red Lodge Fire. He is also the Geographic Information System (GIS) specialist for Red Lodge. Prior to joining the fire service, Jon served as an intelligence officer and taught combat survival skills in the Air Force. He is certified in wilderness first aid, wilderness-EMT, Advanced Cardiac Life Support, Pre-Hospital Trauma Life Support, and Pediatric Advanced Life Support. He is a Graduate of the USAF Survival School. He received a Bachelor's degree in Political Science from Northern Arizona University and a Master's in Conservation Biology from Prescott College in Prescott, Arizona.

Editorial Board

Frank Florence
National Fire Protection Association
Quincy, Massachusetts

Robert Rhea
Fairfax County (Va.) Fire and Rescue Department (retired)
Fairfax, Virginia

Brian Rousseau
New York State Office of Fire Prevention and Control
Albany, New York

John Woulfe
International Association of Fire Chiefs
Fairfax, Virginia

Contributors

Francis M. Clark
Fire Fighter/EMT
Hopkinton Fire Department
Massachusetts Fire District 14 Technical Rescue Team
Hopkinton, Massachusetts

Steve Crandall
Heavy Rescue Captain
Salt Lake City Fire Department
Owner, HeavyRescueTraining.com
Salt Lake City, Utah

Aaron McDowell, BS, REMT-Paramedic
Deputy Chief
Red Lodge Fire Rescue
Red Lodge, Montana

Edward J. Meiman III
Captain, Special Unit Coordinator
City of Louisville Division of Fire
Louisville, Kentucky

Derek J. Peterson
Fire Fighter/Paramedic
Saint Paul Fire Department
Saint Paul, Minnesota

Thomas J. Rinoldo
Lieutenant
Town of Framingham, Massachusetts Fire Department
Framingham, Massachusetts

Alan Tresemer
Battalion Chief
Painted Rocks Fire & Rescue Company
Darby, Montana

Reviewers

J. Michael Albee
Assistant Fire Chief
Greater Prudhoe Bay Fire Department
Anchorage, Alaska

Michael Allard
Lieutenant
Manchester, New Hampshire Fire Department
Manchester, New Hampshire

Clarence S. Bennett
Curriculum Developer & Rescue Instructor
South Carolina Fire Academy
Columbia, South Carolina

Mark Bosse
Chief
Poland Fire Rescue
Poland, Maine

Jeff Boushee
Fire Technology Instructor
Northland College
East Grand Forks, Minnesota

Randy Brown
Fire Fighter/Paramedic/Instructor
Webmaster IndianaFirefighter.com
Former Co-Chair Indiana Firefighter Special Training Task Force
Fire Service Instructor, EMS Instructor, Training Officer Angola Fire Department (Retired)
Angola, Indiana

David O. Couvelha, MS
Program Manager
Kansas Fire and Rescue Training Institute
Lawrence, Kansas

Steve Crandall
Heavy Rescue Captain
Salt Lake City Fire Department
Owner, HeavyRescueTraining.com
Salt Lake City, Utah

J. David Feichtner
Lieutenant
Farmington Hills Fire Department
Farmington Hills, Michigan

Bob Fitzgerald
Special Operations Chief
West Valley City Fire Department
West Valley City, Utah

Michael Fox
Chief of Special Operations
Chicago Fire Department
Chicago, Illinois

Mike Gagliano
Captain
Seattle Fire Department
Seattle, Washington

Chris W. Gibson, NREMT-Paramedic
Captain
East Side Fire Department
Baton Rouge, Louisiana

Danny Gillum, Kansas EMT-IC
Associate Dean of Technical Education
Dodge City Community College
Wilderness Medicine Outfitters
Dodge City, Kansas

Jeff Hatcher
Fire Captain
Safety Officer, Lincoln Fire Rescue
NETF-1 FEMA Urban Search and Rescue Team
Lincoln, Nebraska

Clarence E. Hawkins
Training Officer
St. Paul Fire Department
St. Paul, Minnesota

Les Hawthorne, BA
Instructional Designer
Maryland Fire and Rescue Institute
College Park, Maryland

Jack Holliday, Jr., EFO
Lieutenant
Marion Township Fire Department
Marion, Ohio

Captain Eric Jones
Lincoln Fire and Rescue
FEMA/DHS Nebraska Task Force 1
Search Team Manager
Lincoln, Nebraska

Kevin L. Jump
Division Chief, Training Division
Worthington Fire Department
Louisville, Kentucky

Mark Larson
Idaho State Fire Marshal
Boise, Idaho

Mike Lee
Battalion Chief
Cunningham Fire Protection District
Denver, Colorado

Nick LoCicero
Rescue Division Chief
City of Tampa Fire Rescue
Tampa, Florida

Jason Louthan
Rescue Program Coordinator
Oklahoma Fire Service Training at Oklahoma State University
Stillwater, Oklahoma

Tommy E. MacPherson
Instructor
Alabama Fire College and National Cave Rescue Commission
Brindlee Mountain Fire Department
Union Grove, Alabama

Steve McKenna
Lieutenant/Paramedic
Salem, New Hampshire Fire Department
Technical Rescue Programs Coordinator, New Hampshire
   Department of Safety
Division of Fire Standards and Training
Concord, New Hampshire

Edward J. Meiman III
Captain, Louisville Fire Department
Special Unit Coordinator, Louisville Division of Fire
Louisville, Kentucky

Mark A. Morton
HTR Chief
Virginia Department of Fire Programs
Glen Allen, Virginia

Rick Nelson
Lieutenant
Reading Fire Department
Reading, Massachusetts

M.B. "Ollie" Oliver
Assistant Professor, Program Director Fire Protection
   Technology
Midland College
Midland, Texas

Derek J. Peterson
Fire Fighter/Paramedic
Saint Paul Fire Department/Rescue Squad 3
Saint Paul, Minnesota

Kenneth C. Rhodes
North Carolina Fire and Rescue Instructor
Hugo Fire & Rescue and Sand Hill VFD
Hugo, North Carolina

Todd Seitz
Deputy Fire Chief and Fire Training Coordinator
Brooklyn Park Fire Department
Hennepin Technical College
Eden Prairie, Minnesota

Chris Simpson
Training Officer
Choctaw Fire Department
Choctaw, Mississippi

William "Smokey" Simpson
Captain
Seattle Fire Department
Seattle, Washington

Richard Stone
Lieutenant
Westport Fire Department
Westport, Connecticut

Jeff Travers
Director of Public Safety
Great Oaks Institute
Cincinnati, Ohio

Alan Tresemer
Battalion Chief
Painted Rocks Fire & Rescue Company
Darby, Montana

Jerry Wade
ConocoPhillips Company
Oklahoma

Robert M. Wisneski
Perth Amboy Fire Department
Middlesex County Fire Academy
Perth Amboy, New Jersey

Photographic Contributors

We would like to extend a huge "thank you" to Glen E. Ellman, the photographer for this project. Glen is a commercial photographer and fire fighter based in Fort Worth, Texas. His expertise and professionalism are unmatched.

Thank you to the following organizations that opened up their facilities for this photo shoot:

Butler County Community College, Butler, Pennsylvania

Mel Bliss, Public Safety, Training Facility Coordinator, Certification Site Coordinator

Steven V. Nickell, Director, Industrial Safety Training Programs

Ernest J. Early, Instructor, Fire Training Programs

Introduction to Technical Rescue

NFPA 1670 Standard

4.1.2 At a minimum, all technical search and rescue organizations shall meet the awareness level for each type of search and rescue incident for which the AHJ has identified a potential hazard (see 4.2.1). [p. 7]

4.1.3* In jurisdictions where identified hazards might require a search and rescue capability at a level higher than awareness, a plan to address this situation shall be written. [p. 7]

4.1.3.1 The AHJ shall determine distribution of roles and responsibilities in order to focus training and resources at the designated level to maintain proficiency. [p. 2]

4.1.3.2 Where an advanced level of search and rescue capability is required in a given area, organizations shall have a system in place to utilize the most appropriate resource(s) available, through the use of local experts, agreements with specialized resources, and mutual aid. [p. 2]

4.1.5* It is not the intent of this document to have an organization deem itself capable of an advanced skill level in any of the disciplines defined herein simply by training or adhering to the requirements set forth. Maintaining an operations- or technician-level capability in any discipline shall require a combination of study, training, skill, and frequency of operations in that discipline. [p. 2]

Additional NFPA Standards

NFPA 1006, *Standard for Technical Rescuer Professional Qualifications*

Knowledge Objectives

After studying this chapter, you will be able to:

- List the major technical rescue specialties and common subspecialties.
- Describe how different specialties require the use of other specialties.
- Describe the difference between regulations and standards.
- List those regulations and standards affecting the rescue community.
- Describe the importance of compliance with the various regulations and standards.

Skill Objectives

There are no skill objectives for this chapter.

*y*ou are a fire fighter in a small rural community that has experienced a significant amount of residential, commercial, and industrial growth within the last few years. Recognizing this fact, your chief has asked you to be part of a committee that is to make recommendations on the future rescue needs of your department.

1. What are some of the things the committee needs to consider to make sound recommendations to the chief?

Introduction

The purpose of this program is to provide you, the emergency responder, with basic knowledge and skills necessary for further advancement into specific areas of **technical rescue**—that is, the process of applying special knowledge, skills, and equipment to resolve difficult rescue situations. These core skills and knowledge form the backbone of basic rescue response and prepare you for further training in the more specialized and advanced skills necessary in various technical rescue disciplines. No matter what the specialty considered, basic training in site operations, search, helicopter ground support operations, victim management, equipment maintenance, and ropes/rigging is necessary. If a more advanced level of search and rescue capability is needed in a particular area, the rescue organization must have a system in place to use the most appropriate resource(s) available (local experts, agreements with specialized resources, mutual aid, etc.).

Optional modules also allow for awareness level training in a number of disciplines. These modules are designed to provide for awareness level organizational capabilities as specified in NFPA 1670 in those specialties deemed necessary by your local organization or as required in any applicable rules, regulations, or laws. It is the responsibility of the **authority having jurisdiction (AHJ)** to determine how roles and responsibilities should be distributed to focus training and resources in an effort to maintain proficiency. However, NFPA 1670 warns that meeting the requirements of the standard is not sufficient for an organization to consider itself at the advanced level. "Maintaining an operations- or technician-level capability in any discipline shall require a combination of study, training, skill, and frequency of operations in that discipline."

History

Rescue, the act of saving another person's life, has long been considered an honorable and heroic deed. Many have chosen to make such deeds their life's work, either as a profession or as a volunteer. Whichever approach they choose, the need for people and organizations capable of performing rescue service has gradually grown over time. Advancements in the field of search and rescue have also gained momentum as technology has allowed our societies to build bigger cities, move into isolated areas, and have enough free time to pursue leisure activities such as hiking, mountain climbing, sailing, or exploring caves. Technology has also allowed for widespread and easy movement of people and information. This ease of movement has allowed for national and international rapid rescue capabilities but has also led to an increased clash of cultural beliefs, which in turn has fostered the development and deployment of weapons of terror of a type and scale previously unimagined. Whether because of natural disaster, accident, or terrorism, the end result is often the same—the need for specialized rescue capabilities.

Some aspects of the technical rescue field have been in existence as an organized effort for hundreds—if not thousands—of years. In 46 B.C.E., Roman law established **regulations** (rules, ordinances, or laws) requiring that assistance be provided to shipwrecked seamen. In 1708, the earliest Chinese life-saving services were documented with the formation of the Chinkiang Society for the Saving of Life. Detailed rules were established for rescues, including a scale of funds to be awarded for types of rescues performed. There was even punishment set forth for poor performance and/or malfeasance.

Numerous cases report the formation of specialized rescue services in North America and Europe in the eighteenth, nineteenth, and early twentieth centuries, although these efforts were typically a result of a specific need:

- In 1807, the Massachusetts Humane Society built the first surf boats and huts of refuge to be used by volunteer crews when performing rescues.
- In the 1930s, mountain climbing clubs in the British Isles joined together to design a rescue stretcher and first aid equipment list, with the first civilian teams being formed in 1947.
- Modern **urban search and rescue (USAR)** techniques, used to locate and rescue victims in collapses, were documented in the United States and United Kingdom in the

Figure 1-1 Worker safety considerations contributed to the establishment of mine rescue teams.

1930s; they were complemented by significant knowledge gained in the United Kingdom during the German army's Blitz of World War II. Much of the training material produced during this time was later used to develop "high and heavy rescue" training in the United States during the Cold War era of the 1950s and 1960s.

Prior to the early 1900s, mine rescue was typically performed by the miners themselves. With the advent of breathing apparatus and in the face of growing concerns about worker safety, however, specialized mine rescue teams were formed in countries around the world **Figure 1-1 ▲**.

Modern technical rescue did not become a truly widespread, organized effort until the last half of the twentieth century. Attempts were made by the U.S. Congress in 1950 to create a national response capability with the passage of the Federal Civil Defense Act. Although organizations and regions have long formed specialty teams to meet their specific needs, with few exceptions these efforts have rarely been organized and managed on a large scale. These specialized organizations have sometimes been part of emergency management, fire service, police, EMS, or industrial organizations. Many more, however, are separate, not-for-profit groups formed for the sole purpose of providing specialized rescue services—especially in the wilderness and cave environments.

More recently, Congress passed the Robert T. Stafford Disaster Relief and Emergency Assistance Act, establishing a national USAR system. Congress also passed legislation supporting the formation of the **Emergency Management Assistance Compact (EMAC)**, allowing for the establishment of a formal state-to-state mutual assistance system on a national level. In a similar vein, various state and local mutual aid pacts have provided for a more efficient and coordinated response to these specialty incidents.

This development of organized rescue capabilities has become a truly global movement. Rescue teams from all over the world, whether from the Americas, Europe, Africa, Asia, or Australia, are now commonplace and able to react on a moment's notice to local, national, or international disasters. These teams bring with them specialized knowledge, skills, and equipment to deal with the situation at hand.

Rescue Specialties

Although the list of technical rescue disciplines provided in this chapter is by no means a complete accounting of all possible specialties, it covers those disciplines chosen by the members of the NFPA committees for rescue as meeting the majority of the fire service's needs. Additionally, other training and operations **standards** (accepted models for conduct) exist to meet specific needs, such as those promulgated by the U.S. Department of Labor and administered by its Mine Safety and Health Administration and Occupational Safety and Health Administration (OSHA).

Oftentimes, skills and knowledge from a different discipline may be required to help effect a rescue involving another discipline. For example, rope rescue skills might be needed during a trench rescue incident **Figure 1-2 ▼**. Likewise, confined-space rescue skills might be needed to deal with a structural collapse incident. Because of this "cross-pollination," it is common practice for organizations to equip and train for multiple capabilities.

Each of the following specialties may very well have subspecialties that, while requiring most or all of the skills in the main specialty, also require other training and equipment to properly perform a rescue in that environment. Examples of these subspecialties and possible interdependencies among them follow.

■ Rope Rescue

Rope rescue is used in every other type of technical rescue situation. Many different types of rope rescue environments require specific skills and equipment—for example, low-angle rescue, wilderness rescue, and urban rescue. Subspecialties include tower rescue, helicopter rescue, and cliff rescue, to name but a few possibilities.

■ Structural Collapse Rescue

Fortunately, structural collapse incidents are rare events for most organizations. They can, however, totally overwhelm local resources and sometimes even regional and national resources.

Figure 1-2 Many incidents require multiple rescue skills—for example, rope rescue and trench rescue skills.

Voices of Experience

Technical rescue incidents will happen in your community. Whether your department is prepared to handle them is another issue.

Whether or not your department is prepared, these incidents will happen and the level of participation you are prepared to give will be critical to the outcome. If your department is trained and prepared only to the awareness level, then you must be the best at that level and know how to recognize the needs of the incident, activate the needed resources, and assist in the overall operation. If your department chooses to train and function at higher levels, then again you must be the best at those levels and know your role. Technical rescue is not something you can "kind of do;" there must be a commitment by all involved. Because they are low frequency occurrences, there is the mindset in some cases to prepare for and fund those aspects of our operations that occur more frequently. The needs of those operations we respond to less frequently get shuffled to the back burner. Because they are high risk situations, if you do not adequately plan for them, your crew will do what they think needs to be done, often with disastrous results.

It is rare that an agency will be prepared to respond to and handle every possible technical rescue incident that can occur. However, by doing a needs assessment and finding out what is unique to your community, your department can focus on those particular disciplines that you would most likely be encountering and budget for the needed resources. By entering into mutual aid agreements or regionalizing your responses, you may be able to share the costs and resources needed and still provide the level of protection required for any incident.

Every department and all personnel should be trained to the awareness level at minimum. However there is absolutely nothing wrong with a department realizing that they cannot adequately train for or fund a technical rescue response team; that is one of those many tough decisions that must be made. Knowing how to determine the situation, what response is needed, where to get the needed resources, and how to interact with them, however, must be addressed and planned out well in advance.

Developing an understanding of the unique aspects of technical rescue incidents will help your department in providing a professional response, but most importantly, in helping keep your members as safe as possible while operating.

Thomas J. Rinoldo
Lieutenant
Town of Framingham, Massachusetts Fire Department
Framingham, Massachusetts

> **"Every department and all personnel should be trained to the awareness level at minimum."**

Subspecialties of this type of rescue typically focus on a particular construction type, ranging from light-frame wood to heavy steel and concrete. The training and equipment necessary to effect rescues in this environment can vary widely depending on the construction type and type of damage incurred. Additionally, it is not unusual for rope, confined-space, and extrication skills to be needed for this type of incident.

Confined Space Rescue

Confined spaces are all around us, from sewer systems to agricultural and industrial spaces to mechanical areas in local buildings. A victim could be trapped in any of these types of spaces, and rescue could possibly require rope, machinery, structural collapse, or rescue skills and equipment.

Trench Rescue

Most areas have the potential for trench rescue incidents. A trench is a relatively narrow excavation made below the surface of the earth. When an excavation is made in the earth, there is almost always a possibility of its collapse. The promulgation of regulations for the construction industry has lessened the likelihood of collapse, but some people will always ignore the rules, whether from ignorance or from intentional disregard. Rescuers should always view a trench collapse as happening in the worst type of soil. Nevertheless, differences in knowledge and skills may arise based on the depth of the excavation and any other special circumstances, such as the presence of water, involvement of toxic environments, or location. As with structural collapse, it is not unusual for rope, confined-space, and extrication skills to be needed to mitigate this type of incident.

Subterranean Rescue

Typically considered a specialty involving mines, tunnels, shafts, and caves, subterranean rescue can require different knowledge, skills, and equipment depending on the environment and conditions. Equipment and knowledge used in structural collapse shoring, confined-space, extrication, and rope rescue can be necessary.

Vehicle and Machinery Rescue

In addition to using rope skills, a rescuer might find machinery in a confined space or collapsed trench, necessitating vehicle and machinery rescue capabilities. Subspecialty examples include rescue from passenger vehicles, buses, trains, elevators, industrial machinery, and farm machinery.

Surface Water Rescue

Surface water rescue skills can vary depending on the environment. For example, these incidents may occur in still, swift, surf, and ice environments. Rope rescue skills are often used in this type of rescue, but it is also possible that machinery extrication skills might be required if the victim is entangled. Likewise, water rescue skills might be used in trench or structural collapse situations where water mains have burst and the rescue area is filling with water.

Dive Rescue

While dive rescues often have a recovery mission, many documented cases describe victims who have survived by finding air pockets or have benefited from the phenomenon of mammalian diving reflex (MDR). Rope, extrication, and surface water knowledge, skills, and equipment are occasionally used in this type of incident.

Wilderness Rescue

There are tremendous variations in the skills needed to perform wilderness rescue based on the location, terrain, and weather conditions. These skills can include search, rope rescue, survival, helicopter operations, subterranean rescue, and water rescue.

Regulations and Standards

It is important to understand the difference between regulations and standards, and to appreciate how both affect the requirements placed on your organization. Additionally, understanding which regulations and standards apply can save a tremendous amount of time and money—both in the development of your program and in potential fines for noncompliance. Essentially, regulations and standards applying to emergency services serve to ensure that you and your organization are properly equipped, trained, and organized to perform the services rendered in a safe and expeditious manner.

Rescue Tips

While some standards-setting organizations develop international standards, many countries have their own agencies that develop standards specific to that country. A list of these countries can be obtained by contacting the American National Standards Institute (ANSI).

Regulations are developed by various government or government-authorized organizations. As such, these regulations carry the force of law. Regulations can be federal, state, or local in nature. They are commonly enforced by the AHJ, the organization, office, or individual responsible for approval of such regulations. As a consequence, it is quite possible that neighboring organizations might have different rules. An example of one set of regulations that applies nationwide is the OSHA standards, which apply to both government (including emergency services) and private industry in the United States.

Standards are typically developed to provide guidance on the performance of processes, products, individuals, or organizations. It is a common practice for these standards to be adopted by a government body, allowing the standard to take on the force of law. An example of such a standard is NFPA 1001, *Standard for Fire Fighter Professional Qualifications*. Many states have adopted this standard as a requirement for some or all fire fighters within their state. Also, it is not uncommon for regulations to specifically reference a standard, thereby making the standard part of the regulation.

Even if they are not formally adopted by government, standards may nevertheless take on the force of law when they are viewed as a generally accepted practice. For example, a standard may be cited in a civil lawsuit, where the standard is considered something about which the plaintiff or defendant, through their actions, would have readily known and would be expected to act within whose bounds.

Other types of standards exist on a local level; they are typically referred to as policies and **standard operating procedures (SOPs)**. Policies are developed to provide definite guidelines for present and future actions and may, in some cases, be considered regulations. For example, certain personnel policies might be adopted by a city or county, which then requires all its employees—including emergency services personnel—to follow those policies. Policies are typically considered administrative in nature and may or may not have emergency incident applications.

SOPs provide specific information on the actions that should be taken to perform specific tasks. Some organizations prefer the use of the term **standard operating guidelines (SOGs)**, as it implies more flexibility in the decision-making process while still requiring accountability. SOGs are usually operational in nature and may or may not have administrative applications.

Regulations

Regulations are adopted to provide a specific set of rules with which an organization must comply. As stated earlier, they hold the force of law, and an organization found not to be in compliance can face sanctions (e.g., corrective actions or fines). The regulations most commonly affecting emergency services are those promulgated by OSHA. Although not considered a federal regulatory agency, the Federal Emergency Management Agency (FEMA) also establishes emergency management plans and procedures that influence how state and local response agencies manage large-scale emergency incidents. Response agencies that use federal funding sources are often required to follow federal response guidelines as a stipulation for receiving federal funds.

Federal Emergency Management Agency

Currently, all federal agencies have adopted the National Incident Management System (NIMS). All jurisdictions were required to adopt and meet the NIMS implementation requirements by September 30, 2006, as a condition of receiving federal preparedness funding assistance in fiscal year 2007. This directive, in effect, has the force of regulation unless the states and local jurisdictions do not want federal preparedness funds.

OSHA Regulations and State Plans

OSHA issues regulations that are intended to create a safer working environment for U.S. residents. OSHA standards are currently in place for general, maritime, construction, and agriculture industries. In general, this federal agency is responsible for enforcement of its standards across private industry. States are allowed to have plans under 29 CFR Part 1952 whereby the state enforces the regulations among public employers. Those states without approved plans fall under 29 CFR Part 1956.

Those regulations affecting emergency services are generally contained in 29 CFR Part 1910. Regulations of specific interest to emergency service providers include the following:

- OSHA 29 CFR 1910.120: Hazardous Waste Operations and Emergency Response
- OSHA 29 CFR 1910.133: Eye and Face Protection
- OSHA 29 CFR 1910.134: Respiratory Protection
- OSHA 29 CFR 1910.135: Head Protection
- OSHA 29 CFR 1910.136: Occupational Foot Protection
- OSHA 29 CFR 1910.146: Permit-Required Confined Spaces
- OSHA 29 CFR 1910.147: The Control of Hazardous Energy (Lockout/Tagout)
- OSHA 29 CFR 1910.1030: Bloodborne Pathogens

Many other OSHA regulations may not necessarily have direct implications for emergency services. They are considered "generally accepted practices" and, therefore, are worth incorporating into your training and operations plans. Examples of such regulations include 29 CFR 1926, Subpart P—Excavations, and 29 CFR 1926.800—Underground Construction. While not designed specifically to guide emergency response to excavation and underground incidents, the information contained in these regulations may nevertheless be extremely beneficial to those developing rescue teams for these situations.

Standards

There are many standards-setting agencies and organizations in the world today. Most are specific to an industry, including some of those listed in this section. Because standards are considered voluntary, it is not uncommon to have more than one standards organization creating a standard on the same subject. This proliferation of guidelines can also happen with regulations, especially when the regulation does not provide the depth of information necessary to provide the required background to comply with the regulation. An example is the confined-space segments of NFPA 1006 and 1670, which were developed even though an OSHA regulation governs confined-space rescue.

National Fire Protection Association (NFPA)

NFPA 1006, *Standard for Technical Rescuer Professional Qualifications,* lists the minimum job performance requirements for personnel who perform the following technical rescue operations:

- Rope rescue
- Confined space rescue
- Trench rescue
- Structural collapse
- Vehicle and machinery rescue
- Surface water rescue
- Swiftwater rescue
- Dive rescue
- Ice rescue
- Surf rescue
- Wilderness rescue
- Mine and tunnel rescue
- Cave rescue

As mentioned earlier, this standard contains a core set of requirements common to all forms of rescue—specifically, the knowledge and skills used in site operations, search, helicopter ground support operations, victim management, equipment maintenance, and basic ropes/rigging. Beyond this core set, each discipline has its own set of knowledge and skills unique to that discipline. These specific requirements are outlined in detail in the appropriate chapter later in this textbook.

NFPA 1670, *Standard on Operations and Training for Technical Search and Rescue Incidents,* is designed to assist organizations in developing a technical rescue capability in their community. It is commonly referred to as an *organizational standard,* meaning that the organization as a whole (as compared to individual members) must comply with the requirements of the standard.

Like NFPA 1006, the NFPA 1670 standard contains a core set of requirements that are common to all forms of rescue, including requirements related to medical care, hazard analysis and risk assessment, incident response planning, equipment, safety, incident management system, and fitness. Specific specialty requirements are also included for the following scenarios:

- Rope rescue
- Structural collapse search and rescue
- Confined space search and rescue
- Vehicle search and rescue
- Water (dive, ice, surf, surface, and swift water) search and rescue
- Wilderness search and rescue
- Trench and excavation search and rescue
- Machinery search and rescue
- Cave search and rescue
- Mine and tunnel search and rescue
- Helicopter search and rescue

Each of the specialties in NFPA 1670 includes three response levels: awareness, operations, and technician. For each of the areas with a potential hazard, the rescue organization must be trained to at least the awareness level (and to higher levels if the hazards dictate). These specific requirements are outlined in more detail in the appropriate chapter later in this textbook.

The following NFPA standards also have implications for the rescue community:

- NFPA 328, *Recommended Practice for the Control of Flammable and Combustible Liquids and Gasses in Manholes, Sewers, and Similar Underground Structures*
- NFPA 472, *Standard for Competence of Responders to Hazardous Materials/Weapons of Mass Destruction Incidents*
- NFPA 1001, *Standard for Fire Fighter Professional Qualifications*
- NFPA 1404, *Standard for Fire Service Respiratory Protection Training*
- NFPA 1500, *Standard on Fire Department Occupational Safety and Health Program*
- NFPA 1521, *Standard for Fire Department Safety Officer*
- NFPA 1561, *Standard on Emergency Services Incident Management System*
- NFPA 1581, *Standard on Fire Department Infection Control Program*
- NFPA 1981, *Standard on Open-Circuit Self-Contained Breathing Apparatus (SCBA) for Emergency Services*
- NFPA 1982, *Standard on Personal Alert Safety Systems (PASS)*
- NFPA 1983, *Standard on Life Safety Rope and Equipment for Emergency Services*

ASTM International

ASTM International, formerly known as the American Society for Testing and Materials, is primarily a developer of product stan-

dards. It also has four working committees actively developing standards in the following emergency service areas:

- Technical rescue (Committee F32)
- Protective clothing (Committee F23)
- Homeland security applications (Committee E54)
- Emergency medical services (Committee F30)

American National Standards Institute

The American National Standards Institute (ANSI) does not develop standards, but rather helps facilitate the development of standards and lists them for purchase and use. Additionally, ANSI promotes the use of U.S. standards in the international community. Each standard developed by another organization that is accredited by ANSI is assigned an ANSI standard number. ANSI's listing includes a number of standards developed by NFPA as well as many international organizations.

U.S. Department of Homeland Security/FEMA

In addition to the NIMS requirement mentioned earlier, the Department of Homeland Security (DHS), which includes FEMA, has established a standardized typing system for emergency services. The FEMA/NIMS Integration Center Resource Typing Definitions facilitate communication among emergency responders by standardizing the terminology used to request resources during an emergency.

An additional effort currently under way is a certification program called the National Emergency Responder Credentialing System. This system, which documents the professional qualifications, certifications, training, and education of emergency responders, is meant to assist in verifying the identity and qualifications of emergency responders at an incident.

National Institute of Standards and Technology

The National Institute of Standards and Technology (NIST) has implemented an initiative called Technologies for Public Safety and Security that is intended to foster the development and standardization of technologies for the emergency services community. Some of the more interesting emergency responder initiatives include USAR robots; detection devices for chemical, biological, and other threats; and communications improvements.

In cooperation with other organizations, NIST has developed a series of performance and compliance standards for hazardous materials suits and other equipment that is designed to protect first responders from chemical, biological, radiological, and nuclear threats. In February 2004, the DHS adopted eight of these standards. As a consequence, state and local governments receiving DHS equipment grants will use the standards recommended through the NIST-led effort for technical guidance when making purchasing decisions.

International Organization for Standardization

While perceived primarily as a standards development organization for industry, the International Organization for Standardization (ISO) includes several working committees that are responsible for promulgating standards with direct and indirect implications for the rescue service. One such committee is TC 94/SC14/WG5, Firefighter Personal Equipment/Non-Fire Rescue Incidents.

Wrap-Up

Chief Concepts

- While the mission of rescue services has always been to save lives, the available equipment and expertise have improved greatly over the last 100 years.
- Technical rescue comprises many specialties requiring different training and equipment. More than one of these specialties is often required at a rescue scene.
- Basic rescue knowledge and skills should be part of every emergency responder's training.
- Regulations and standards are an important part of the rescue service and are designed to help the rescue responder do the job better and more safely.

Hot Terms

Authority having jurisdiction (AHJ) The organization, office, or individual responsible for approving equipment, materials, installation, or a procedure.

Confined space A space large enough and so configured that a person can enter and perform assigned work, that has limited or restricted means for entry or exit (e.g., tanks, vessels, silos, storage bins, hoppers, vaults, and pits), and that is not designed for continuous human occupancy.

Emergency Management Assistance Compact (EMAC) An organization established by the U.S. Congress that provides form and structure to interstate mutual aid. Through EMAC, a state that has experienced a disaster can request and receive assistance from other member states.

Regulation A rule, ordinance, or law requiring specific conduct of individuals and organizations.

Rescue Those activities directed at locating endangered persons at an emergency incident, removing those persons from danger, treating the injured, and providing for transport to an appropriate healthcare facility.

Standard A model or example commonly or generally accepted for the conduct of individuals and organizations.

Standard operating guideline (SOG) An organizational directive that is similar to a standard operating procedure but allows more flexibility in the decision-making process.

Standard operating procedure (SOP) A written organizational directive that establishes or prescribes specific operational or administrative methods to be followed routinely for the performance of designated operations or actions.

Technical rescue The application of special knowledge, skills, and equipment to safely resolve unique and/or complex rescue situations.

Trench An excavation that is relatively narrow in comparison to its length. In general, the depth of the trench is greater than its width, but the width of a trench is not greater than 15 feet.

Urban search and rescue (USAR) Those activities directed at locating and rescuing victims in the collapse of a structure.

Rescue Responder *in Action*

You have been in the fire service for a number of years and pride yourself on your knowledge of the history of the fire service. You have been asked to write a brief history of the rescue service for your department's newsletter.

1. In 1807, the Massachusetts Humane Society established huts of refuge for which type of rescue?

A. Mine rescue

B. Surf rescue

C. Wilderness rescue

D. Trench rescue

2. Which war had a great impact on the knowledge gained in USAR techniques?

A. Cold War

B. World War I

C. World War II

D. Korean War

You have recently been assigned to a rescue company and are anxious to start your training. A number of different courses are available to you, but you don't know where to start. When you meet with your new captain, she advises you that you must start with the basics. This makes sense, so you sign up for an Introduction to Technical Rescue course.

1. According to NFPA 1006, *Standard for Technical Rescuer Professional Qualifications*, what are some of the core requirements for all areas of technical rescue?

A. Basic rope rescue, helicopter operations, awareness training in the various disciplines

B. Rescue management, equipment purchasing and maintenance

C. Helicopter ground support operations, search and victim management

D. Equipment maintenance, rescue management, awareness training in rope rescue

2. You are thinking of specializing in confined space rescue. Which agency's consensus standard has established regulations that must be followed for this type of rescue?

A. NFPA

B. ASTM

C. FEMA

D. OSHA

Technical Rescue Vehicles and Equipment

CHAPTER 2

NFPA 1670 Standard

4.4 Equipment.

4.4.1 Operational Equipment.

4.4.1.1* The AHJ shall ensure that equipment commensurate with the respective operational capabilities for operations at technical search and rescue incidents and training exercises is provided. [p. 12–43]

4.4.1.2 Training shall be provided to ensure that all equipment is used and maintained in accordance with the manufacturers' instructions. [p. 14–43]

4.4.1.3 Procedures for the inventory and accountability of all equipment shall be developed and used. [p. 12–43]

4.4.2 Personal Protective Equipment (PPE).

4.4.2.1* The AHJ shall ensure that the protective clothing and equipment is supplied to provide protection from those hazards to which personnel are exposed or could be exposed. [p. 18–24]

4.4.2.2 Personnel shall be trained in the care, use, inspection, maintenance, and limitations of the protective clothing and equipment assigned or available for their use. [p. 18–26]

4.4.2.3 The AHJ shall ensure that all personnel wear and use PPE while working in known or suspected hazardous areas during technical search and rescue incidents and training exercises. [p. 18–25]

4.4.2.4 The AHJ shall ensure that atmospheric supplying respirators in the form of supplied air respirators (SAR) or self-contained breathing apparatus (SCBA) are available when required for technical search and rescue operations and that they meet the requirements specified in Chapter 7 of NFPA 1500, Standard on Fire Department Occupational Safety and Health Program. [p. 23–24]

4.4.2.4.1 Breathing apparatus shall be worn in accordance with the manufacturer's recommendations. [p. 23–24]

4.4.2.4.2 A supply source of breathing air meeting the requirements of ANSI/CGA G7.1, Commodity Specification for Air, with a minimum air quality of Grade D shall be provided for all atmosphere-supplying respirators. [p. 13]

4.4.2.4.3 A supply source of breathing air meeting the requirements of ANSI/CGA G7.1, Commodity Specification for Air, with a minimum air quality of Grade E shall be provided for all atmosphere-supplying respirators used for dive operations. [p. 13]

4.4.2.4.4 Supplied air respirators shall be used in conjunction with a self-contained breathing air supply capable of providing enough air for egress in the event of a primary air supply failure. [p. 13–14, 23–24]

Additional NFPA Standards

NFPA 70E, *Standard for Electrical Safety in the Workplace*

NFPA 1006, *Standard for Technical Rescuer Professional Qualifications*

NFPA 1500, *Standard on Fire Department Occupational Safety and Health Program*

NFPA 1901, *Standard for Automotive Fire Apparatus*

NFPA 1936, *Standard on Powered Rescue Tools*

NFPA 1951, *Standard on Protective Ensembles for Technical Rescue Incidents*

NFPA 1971, *Standard on Protective Ensembles for Structural Fire Fighting and Proximity Fire Fighting*

NFPA 1981, *Standard on Open-Circuit Self-Contained Breathing Apparatus (SCBA) for Emergency Services*

NFPA 1982, *Standard on Personal Alert Safety Systems (PASS)*

NFPA 1983, *Standard on Life Safety Rope and Equipment for Emergency Services*

NFPA 1994, *Standard on Protective Ensembles for First Responders to CBRN Terrorism Incidents*

Knowledge Objectives

After studying this chapter, you will be able to:

- List the various types of rescue vehicles and their common characteristics.
- List the various types of rescue equipment and their functional and operational characteristics.
- List the various components of rescue personal protective equipment.
- Describe the appropriate use of the various components of rescue personal protective equipment.
- Describe the method of rope construction used in life rescue situations.
- Describe the various parts of the rope.
- Describe knot efficiency, and explain its importance in knot selection.
- List knot types and their use in rescue.
- Describe how to tie, dress, and secure the assigned knots.

Skill Objectives

After completing this chapter, you will be able to perform the following skills:

- Tie the following knots:
  - Square knot
  - Figure eight
  - Figure eight on a bight
  - Figure eight with a follow-through
  - Double-loop figure eight
  - Figure eight bend
  - Butterfly knot
  - Double fisherman's knot
  - Prusik hitch
  - Water knot
  - Clove hitch
  - Load-release hitch (mariner's hitch)
- Properly dress and secure the above knots, bends, and hitches.

our organization has a "heavy rescue." This vehicle is quite large and is equipped for fire rescue and extrication of individuals from automobile accidents. A neighboring organization will not call for this unit when they get a rescue call, stating that the vehicle is not a real heavy rescue.

1. Do you think that this vehicle is a real heavy rescue, and why?
2. If not, which equipment do you think should be carried on a heavy rescue, and why?

Introduction

When responding to rescue incidents, it is the combination of personnel, knowledge, skills, and equipment that gets the job done. This chapter focuses on the equipment that rescuers may use in the performance of their duties. It is not enough, however, to simply know that these tools exist. To ensure that the rescue operation is as safe and efficient as possible, you must choose the proper equipment and be skilled in its use. It is equally important that these tools be properly maintained so that they will function as intended during an actual emergency situation.

Rescue Vehicles

Many different types of emergency vehicles may respond to rescue incidents. Some are single, self-contained units; others comprise truck/trailer combinations. Some are specialized vehicles dedicated to a single purpose; others are designed to support multiple specialties. In addition, some types of units are designed to perform both firefighting and rescue functions.

Specialized rescue vehicles are commonly classified into four categories: light, medium, heavy, and special purpose/multipurpose **Figure 2-1 ▼**. This distinction has more to do with the function of the vehicle than with its size. For example, a large

vehicle might carry equipment and support items only for basic vehicle **extrication** (removal of trapped victims from entrapment). Because of its minimal rescue capability, this vehicle may meet the criteria of only a light rescue. The design and equipping of any rescue vehicle should comply with NFPA 1901 in addition to the needs determined by your organization.

Light Rescue Vehicles

Light rescue vehicles are equipped for basic rescue tasks, so they typically carry only hand tools and basic extrication and medical care equipment. Because these vehicles tend to be smaller in size, they are ideal for quick response—they can get to a scene quickly, handle small incidents, or stabilize the scene until additional equipment and personnel arrive. Typically built on a 1- to 1½-ton chassis, these vehicles can have a standard or crew cab, so they may be able to carry as many as five responders. Additionally, their light weight and small size make them ideal for going into areas where larger apparatus will not fit.

One of the most common uses for light rescue vehicles is for response to extrication incidents. The equipment typically carried on such units includes pry bars, bolt cutters, air chisel, jacks, stabilization equipment and cribbing, hand and power saws, lighting, and portable hydraulic rescue tools. Most engines and ladder trucks that carry some rescue equipment fall into this category.

A.

B.   C.

Figure 2-1 **A.** Light rescue vehicle. **B.** Medium rescue vehicle. **C.** Heavy rescue vehicle.

Medium Rescue Vehicles

Medium rescue vehicles are designed to handle most rescue situations likely to be encountered by the responding department. They may carry basic to advanced equipment applicable to a variety of specialties, but normally are not equipped to the advanced level in more than one or two areas. For example, such units might provide advanced capabilities for confined-space and rope rescue, along with basic equipment for trench and structural collapse rescue. Other capabilities may include built-in electrical power generation, a hydraulic rescue tool pump, and an air compressor (breathing or regular air).

Many medium rescue units are designed to serve a single purpose, however. This purpose could relate to any of the specialty areas, although the most commonly encountered are units devoted to trench and excavation, confined-space, rope, or ice/water rescue. Other specialty units, though not as common, focus on structural collapse, heavy extrication, and mine/tunnel rescue.

Heavy Rescue Vehicles

Heavy rescue vehicles are, by definition, the most heavily equipped vehicles in the rescue vehicle class. They are designed to handle almost any rescue incident that the responding organization might encounter. These vehicles tend to be capable of advanced capabilities in multiple areas, and they carry a wide variety of specialized rescue tools. While many of these units are supplied by municipalities, regional and national heavy rescue organizations tend to own units that are on the high end of this capability spectrum. These teams, which are often referred to as urban search and rescue (USAR) teams, specialize in structural collapse. Nevertheless, many also have advanced capabilities in trench, confined-space, rope, heavy extrication, and water rescue.

Special-Purpose/Multipurpose Vehicles

Special-purpose and multipurpose vehicles are also quite common. As mentioned earlier, many organizations have engines, ladder trucks, and vans that carry a limited amount of technical rescue equipment while also supporting fire scene and rescue operations. Special-purpose vehicles include small, off-road vehicles and boom trucks Figure 2-2 ▶ . The off-road vehicles are growing in popularity because of their small size, all-wheel-drive capability for accessing difficult locations, ability to carry both victims and equipment, and basic firefighting capability when equipped with an optional skid pump/tank unit.

Rescue Vehicle Accessories

Regardless of which type of rescue vehicle your organization uses, many accessories can be added to the unit to meet the organization's specific needs. These accessories may be ordered when the vehicle is manufactured or added later as the need arises. Among the more commonly encountered accessories are compressed air capabilities, hydraulic pumps, electrical generators, lighting, winches, booms/gins, and stabilizers.

A.

B.

Figure 2-2   Special-purpose vehicles include off-road vehicles (**A**) and boom trucks (**B**).

Compressed Air (Breathing, Tool)

Compressed air capabilities on rescue vehicles take two forms: stored and generated. Stored compressed air consists of installed bottles that are almost always used for breathing air. Often referred to as **cascade systems**, these systems are used for refilling SCBA bottles and/or regulating for direct supply to supplied-air respirators Figure 2-3 ▶ . For example, this type of direct supply is often found on aerial platform vehicles, where a long-term supply is desired for the occupants in the bucket.

Air compressors either provide breathing air or are intended for general use; those found on rescue units can be either portable or fixed Figure 2-4 ▶ . Breathing air compressors are generally larger because of the filtration required to meet ANSI/CGA G7.1, *Commodity Specification for Air*. These compressors may feed cascade units or supply air directly. A cascade connection is preferred when filling supplied-air respirators; with this approach, if the compressor fails there is still a significant

Figure 2-3 Cascade systems are used for refilling SCBA bottles and/or regulating for direct supply to supplied-air respirators.

A.

B.

Figure 2-4 Air compressors can be used for breathing (**A**) or for general use (**B**).

quantity of air left for continued operations or evacuation. Care needs to be taken to ensure that the proper fittings are provided, as there are several styles from which to choose.

Like breathing air compressors, general-use compressors may be either mounted on the vehicle or portable; these units can vary in size from a single tool (for low-volume use) to multiple-tool equipment (for high-volume use). The larger units are frequently trailer mounted and use gasoline or diesel as fuel. With this kind of air compressor, it is critical to ensure that the proper air fittings are provided, as there is a wide variety from which to choose.

Maintenance tasks for air compressors include draining water from the filters and tank, filter replacement, and, in the case of breathing air compressors, periodic testing as required. With compressors used for nonbreathing purposes, the oil levels in the compressor (if applicable) and engine fluid levels should be checked after each use. In addition to these general guidelines, you should be familiar with any specific recommendations or field maintenance guidance provided by the manufacturer of the equipment.

Hydraulic Pumps

Many rescue vehicles carry hydraulic pumps that are used to power a variety of rescue tools. These hydraulic pumps may be built into the vehicle or portable; likewise, they may be powered by the vehicle engine or have their own engine Figure 2-5 ▶ . These tools are those primarily used for vehicle extrication, although some other tools are appropriate for structural collapse incidents. Examples of hydraulic rescue tools include spreaders, shears/cutters, rams, saws, jackhammers, and pumps.

Maintenance of hydraulic pumps includes checking the hydraulic oil after every use for proper level and evidence of any leaks. It is also important to inspect the equipment for any damaged or improperly functioning parts. Check the fittings for proper operation and to ensure that they are clean and free of debris. For units powered by an attached engine, engine fluids and proper operation of the engine should also be checked. In addition to these general guidelines, you should be familiar with

A.

B.

Figure 2-5 Hydraulic pumps can be built into the rescue vehicle (**A**) or portable (**B**).

A.

B.

Figure 2-6 Electrical generators may be either portable (**A**) or fixed (**B**).

any specific recommendations or field maintenance guidance provided by the manufacturer of the equipment.

Electrical Power

Electrical generators are an important and very common accessory on any rescue apparatus. These generators, which may be either portable or fixed units, are used primarily to provide scene lighting and to run power tools and equipment Figure 2-6 ▶. They range in capacity from less than 1000 watts (1 kW) to 75,000 watt (75 kW) and larger. Depending on their size, they provide output as 120 or 240 volts. When selecting generators for a rescue, remember that their sizes reflect the maximum power output—the rated output can be 10 to 20 percent less

than the maximum output. Because of the nature of the power generated, it may be necessary to use a line conditioner when operating some electronic equipment.

Rescue Tips

Different countries and regions use different voltages to run electrical equipment. Despite these variations, the types of equipment and concepts of operation remain the same from region to region.

Four types of vehicle-mounted generators are distinguished: inverters, power take-off (PTO), hydraulic, and internal combustion engine generators. Inverters convert 12-volt DC vehicle power to 120-volt AC power; their output ranges from 125 watts to 5000 watts, which is typically appropriate for low-power needs. PTO generators use the vehicle's engine to drive the generator by means of a shaft (similar to a fire pump) or belt (similar to a vehicle alternator). By contrast, hydraulic generators use the vehicle's engine to power a hydraulic pump, which then drives the generator. Internal combustion engine generators use a separate engine to run the generator, using either the vehicle's fuel source or the generator's own fuel source.

Portable generators come in a variety of sizes and styles. Units may use either a pull-start or battery-start model, and the output may range from 100 watts to 12,000 watts. They are run by their own internal combustion engine and range from 30 pounds to several hundred pounds in weight. Some heavier models have wheels mounted on the bottom of the unit, ensuring that they can be rolled to where they are needed. Others have lights attached directly to the generator or are mounted on specially designed carts that include the generator, lights, and power outlets. Recent advances have even allowed for some generators to be linked in parallel to increase available wattage through the generator outlets.

Maintenance tasks for electrical power generators include checking for evidence of any leaks or damaged parts. For those units powered by an attached engine, check the engine fluids and ensure proper operation of the engine. Familiarize yourself with any specific recommendations or field maintenance guidance provided by the manufacturer of the equipment.

While the generator supplies the electrical power, auxiliary lead cords are required to get the power to where it is needed. These auxiliary lead cords, which are stored coiled on reels, should consist of heavy-duty, exterior-grade equipment. They should meet OSHA and Bureau of Mines (MSHA) standards and be resistant to oil, moisture, and abrasion. Some organizations use a standard three-prong household plug at the end of these lead cords, but many prefer to use twist lock connections to avoid disconnecting the plug. No matter which type of lead cord is used, be sure to have adapters available that will work with the equipment provided by mutual aid departments, all types of generators, and building outlets.

Junction boxes are commonly employed when multiple outlets are required Figure 2-7 ▶ . These boxes should be heavy-duty models that are supplied with ground fault circuit interrupters (GFCI) and that comply with *NFPA 70E*. For long-term incidents where the generator is large and does not have plug and breaker distribution capabilities, portable distribution panels may be used. These panels accept full power from the generator and provide circuits to parcel this power out to individual outlets. These assemblies should be certified as meeting Underwriters Laboratories (UL) Standard 1640, be weatherproof, and provide protection from accidental contact with live electrical connections.

After use, wipe down all electrical equipment to eliminate any contamination from abrasive materials such as sand or dirt

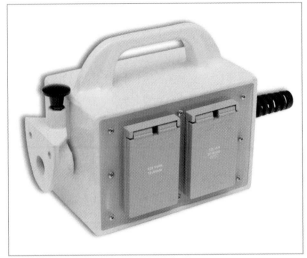

Figure 2-7 Heavy-duty junction boxes are used at incidents where multiple outlets are needed.

and petroleum products, which might otherwise damage the unit's protective coating. During this process, also check the electrical equipment for any physical damage, such as cuts, cracks, wire pulled from sockets, or disfigurement caused by overloading or other injury.

■ Lighting

Most rescue vehicles carry an assortment of portable and mounted lights.

Typically ranging from 300 watts to 1000 watts or more per light, portable lights come in a variety of styles. They are meant to provide light where fixed lights cannot. Portable lights Figure 2-8 ▶ are usually adjustable for elevation and have a broad base to prevent them from tipping over. Some portable light stands also contain multiple lights. These stands are adjustable in height and are very useful for providing large-area or higher-height lighting.

Fixed lighting is mounted on the rescue vehicle and, like portable lights, comes in a variety of styles. Some units are mounted into the body of the vehicle and intended to provide perimeter lighting; such lights are not adjustable. Other models are mounted so they can be adjusted for direction and elevation, with some units being of fixed height and other units being attached to adjustable poles. More elaborate systems, referred to as light towers Figure 2-9 ▶ , consist of telescoping masts with a bank of lights that provide up to 6000 watts of lighting. These masts, which can be vehicle or trailer mounted, can be 30 to 40 feet in height and provide a significant amount of overhead lighting. Because of the height of these towers, stabilizers are often installed on the vehicle or trailer to prevent it from tipping over during high-wind situations.

Take care to ensure that the lights you are using do not exceed the rated capacity of the generator supplying the power. Consider not only the possible power usage of the lights, but also any other needs you may have. Overloading the generator

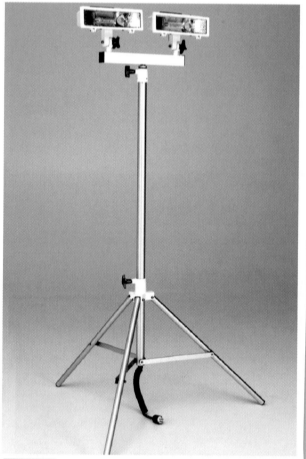

Figure 2-8 Portable lights are adjustable and have a broad base to prevent them from tipping over.

Figure 2-9 Light towers mounted to vehicles offer a significant amount of overhead lighting.

could cause damage to the generator, the lights, or other attached equipment.

After use of a lighting system, check all lights for physical damage, such as wires pulled from sockets or disfigurement caused by overloading or other injury. Bulbs are susceptible to shock damage and are usually easy to change, so keep a stock for replacement as needed. Never replace a bulb with one that has a higher wattage than the light's rating. Also, never touch a bulb with your bare hands, as the residue left may make the bulb fail prematurely. Keep the lens and reflector clean and the lens protector in place to ensure maximum light output. Additionally, familiarize yourself with any specific recommendations or field maintenance guidance provided by the manufacturer of the equipment.

Safety Tips

Lights must not exceed the rated capacity of the generator supplying the power for them. Overloading the generator could damage the generator, the lights, or other attached equipment.

Power Winches

Power winches are used by many rescue organizations for a variety of lifting, pulling, and holding operations. Non-person-rated winches are found in two varieties: permanent and attachable. In both cases, the winch is typically driven by an electric motor and uses steel cable. Various models come in pulling capacities ranging from 2000 to 12,000 lb or more. NFPA 1901, *Standard for Automotive Fire Apparatus*, requires chassis-mounted winches to be rated with at least a 6000-lb pulling capacity and to be remotely operated with at least a 25-foot cord. Any winch is only as strong as what it is attached to, so this equipment is normally connected in some fashion to the frame of the rescue vehicle **Figure 2-10 ▸**. Bumper-type winches are mounted permanently in or on the bumper of the vehicle. By contrast, tow hitch-type winches are attached to the tow hitch receiver of a vehicle and are designed to be removed when not in use.

When inspecting a winch, examine the cable, hook, and gears closely for indications of damage or overloading. Periodically inspect the connections to the vehicle frame for rust, bolt tightness, or weld cracks. Regular lubricating and other inspection/maintenance procedures should be performed in accordance with any recommendations or field maintenance guidance provided by the manufacturer of the equipment.

Booms/Gins

Booms, gin poles, and A-frames are some of the more uncommon accessories found on rescue vehicles and are designed to provide vertical lifting capabilities at a point away from the vehicle. A **gin pole** consists of a single pole that attaches to the vehicle, with stabilizing cables attached to the vehicle on either side. An **A-frame** consists of two poles that are connected at the working end, with the base of each pole being connected to the vehicle, thereby creating the shape of the letter A. An A-frame is stabilized by one or more cables attached to the vehicle. Neither a gin pole nor an A-frame is designed to withstand lateral stress. If necessary, however, additional rope or cable can be attached to help reduce this stress.

A.

B.

Figure 2-10 Winches can be attached to the bumper of a vehicle (**A**) or to a tow hitch receiver (**B**).

Booms can be hydraulically or electrically driven and consist of a telescoping beam that can be rotated 360 degrees. Booms are frequently used to lift heavy objects stored on top of the rescue vehicle, such as boats. They are permanently attached to the vehicle and can sometimes be remotely operated.

Take care never to lift more than the rated load of the boom, gin, or A-frame device or exceed the chassis capacity of the rescue vehicle. Doing so could lead to catastrophic failure of the device or damage to the device or vehicle. Because of the stress placed on the vehicle, stabilizer jacks or outriggers are frequently used for this purpose. These units reduce the possibility of tipping the vehicle and reduce stress on the vehicle suspension.

When inspecting these devices, check for any signs of damage, including disfigurement from overloading. Check all connection points for damage. Pulleys should turn freely with no off-center movement, loose bolts, or cracked welds. Examine ropes and cables for wear or damage, and check hydraulic units for proper fluid level. In addition to these guidelines,

familiarize yourself with any specific recommendations or field maintenance guidance provided by the manufacturer of the equipment.

Personal Protective Equipment

When performing any type of rescue, the safety of the rescuer is the most important consideration. Just as in structural firefighting, the use of **personal protective equipment (PPE)** to create a total protective envelope provides for the safety of the rescue personnel. This total protective envelope consists of body, respiratory, head, foot, hand, and eye protection. The selection of the various components will depend on the hazards present, operating environment, weather, and available equipment, as well as the type of incident and its duration. For example, standard structural firefighting gear may be appropriate for automobile accidents but should not be worn while performing a shore-based water rescue. Additionally, standards such as NFPA 1951, *Standard on Protective Ensembles for Technical Rescue Incidents*, and your organization's own standard operating procedures (SOPs) will dictate what you wear or use for personal protection.

Body Protection

The selection of body protection, as well as any other PPE, will be made based on the known and potential hazards. Consideration also needs to be given to the operating environment and incident needs. Protection from fire, bloodborne pathogens, temperature extremes, and other physical hazards must be taken into account. The equipment chosen must also provide for maneuverability and comfort. As a general rule, body protection should cover all parts of the body to the neck, wrists, and ankles. While in some situations it may seem appropriate to wear shorts, such as during a wilderness rescue in a hot environment, consider the potential for cuts, scrapes, infection, and exposure to allergens such as poison ivy. Body protection should also provide for vis-ibility appropriate to the situation, including the use of reflective striping and lettering.

Typical body protection used by rescue services includes fire fighter clothing; coveralls; work uniforms; and specialty garments such as ice, wet, and dry suits Figure 2-11 ▼ . Structural fire fighter's clothing and USAR ensembles include a coat and pants and protect against short-duration extremes of temperature. This clothing also offers some protection against cuts, abrasions, certain chemicals, and bloodborne pathogens. On occasion, the need for a higher level of flash/flame protection may require the use of a proximity suit. The bulky construction of these garments may present mobility, heat, or weight problems resulting in significant physical stress on the wearer.

Coveralls and work uniforms/battle dress uniforms (BDUs) may or may not be flame/flash resistant. Some of these items are available with insulation to protect the wearer against cold temperatures. Such clothing is typically not bulky, so it provides for easy mobility. On the downside, it does not provide chemical, bloodborne pathogen, or significant fire/flash protection.

Chemical-protective clothing provides protection from direct chemical contact and is used in situations where chemical exposure is an issue. It can be very bulky, however, and may not work in some situations, such as in environments strewn with sharp objects that might tear the garment.

Wet, dry, and ice suits are used in water rescue situations. They provide thermal and abrasion protection and, in the case of dry suits, some protection from contaminants in the water.

A.

B.

C.

D.

Figure 2-11 **A.** Structural fire fighting clothing. **B.** USAR clothing. **C.** Battle dress uniform. **D.** Dry suit.

A.

B.

C.

Figure 2-12 **A.** Fire fighter's helmet. **B.** Rescue helmet. **C.** Water helmet.

Keep all body protection items clean to facilitate their inspection and to avoid damage to the material from dirt and chemicals. Regularly inspect clothing for all types of damage, including rips, tears, and broken or missing components. Test dry and ice suits to ensure seam integrity against leakage. In addition to these general guidelines, familiarize yourself with any specific recommendations or field maintenance guidance or procedures provided by the manufacturer of the equipment.

Head Protection

The most important piece of protective clothing for the rescuer is head protection. A variety of helmets are used by rescue services, where the various models are designed differently to serve various purposes **Figure 2-12** . Nevertheless, all types of helmets are designed to do one thing: protect the wearer from the impact of objects on this sensitive part of your body. Any helmet chosen for rescue work should meet the requirements of OSHA 29 CFR 1910.135, *Head Protection*, and NFPA 1951 or 1971, *Standard on Protective Ensembles for Structural Fire Fighting and Proximity Fire Fighting*. In particular, the helmet should have a substantial chin strap and suspension system that is adjustable (unless it is sized specifically for an individual).

Fire fighters' helmets are frequently used in general rescue work because they afford maximum impact protection and their larger brim size does not pose a problem for the task at hand. For those situations where a smaller profile is desired, specially designed helmets are used. These models include rescue helmets, hard hat–type helmets, and climbing helmets. Another type of specialty helmet used by rescuers is the water helmet. In addition to providing impact resistance, the water helmet is designed to let water drain out and float should it become dislodged.

Hoods are often used in conjunction with helmets. Flame-resistant hoods are used to provide flame/flash protection as well as warmth. Other specialized hoods include wet suit hoods for water rescue and proximity suit hoods.

The inspection of helmets includes looking for any signs of damage to the shell such as cracks, major chips, or gouges. Such flaws could compromise the integrity of the shell, so the helmet should be removed from service if these defects are evident. The suspension system and chin strap should also be inspected for any signs of damage. Inspect helmet liners and hoods for unusual wear, tears, or other damage. Keep all head protection clean to facilitate its inspection. In addition to these general guidelines, familiarize yourself with any specific recommendations or field maintenance guidance or procedures provided by the manufacturer of the equipment.

Eye and Face Protection

One of the quickest and easiest ways to take rescuers out of service is for them to get something in their eyes. Exposure to dust, dirt, fumes, or bright light can not only cause temporary problems in terms of vision, but may, in fact, lead to permanent damage. To protect against temporary and permanent damage, several types of eye and face protection are available. Any items used by technical rescuers should conform to the requirements contained in ANSI Z87.1 and OSHA 29 CFR 1910.133, *Eye and Face Protection*.

Figure 2-13 A respirator face piece provides maximum protection for both the eyes and the face.

Figure 2-14 Rescuers may need specialty eye protection such as a face shield and goggles.

A.

B.

Figure 2-15 Ear plugs (**A**) and ear muffs (**B**) are types of hearing protection available to fire fighters.

A respirator face piece provides maximum protection for both the eyes and the face Figure 2-13 ▲ . Because of its totally contained nature and supplied air, this type of protection works well against fumes. Goggles are good choices for areas where respirators are not needed; they form-fit against the skin, providing for a good level of protection against debris, dust, and dirt. Safety glasses are commonly used where the primary hazard is larger debris. They should include a retainer strap (chums) and side shields. Tinting is available for bright-light situations. Helmet face shields provide protection against larger debris but can entrap dust and dirt-filled air. Specialty eye protection includes welding/cutting face shields and goggles when rescuers are using cutting torches, as well as diving masks for use during water rescue Figure 2-14 ▶ .

The inspection of eye protection should focus on looking for scratches or gouges that will impair the vision of the user, as well as broken or missing pieces. In addition to these general guidelines, familiarize yourself with any specific recommendations or field maintenance guidance provided by the manufacturer of the equipment.

Hearing Protection

Because rescuers are often exposed to high levels of noise from the equipment used to perform a rescue, hearing protection should be used routinely to prevent permanent hearing damage from loud noises and echo. Some hearing protection will actually filter out certain noises so that conversations can still be carried

out. The two types of hearing protection available are ear plugs and ear muffs (headsets) Figure 2-15 ▲ .

Ear plugs come in a variety of styles and are designed to be inserted into the outer ear canal. They conform to the shape of the canal and are disposed of after use. Ear plugs are inexpensive and are easily put into a pocket so that they are available when needed.

Ear muffs are designed to fit over the ears and can be more comfortable than ear plugs. Some are designed to fit over the head; others are designed to be mounted on the helmet.

No matter which kind of hearing protection you choose, make sure that it complies with ANSI S3.19 and has a noise reduction rating of at least 20 decibels. Inspection of hearing protection includes ensuring the equipment's cleanliness and checking for damage that would reduce the effectiveness of the plugs or muffs.

Foot Protection

When selecting an appropriate boot for rescue, remember that the footwear must protect and support your feet, provide traction on poor surfaces, and meet the appropriate NFPA and ANSI standards as well as OSHA 29 CFR 1910.136, *Foot Protection*. Steel toe and shank work boots are good choices for most rescue situations, although different designs are available for a variety of specialized situations. In any event, the boot must provide good support for the foot and ankle, and it should fit its wearer well. The sole should provide good traction. Most boots used in rescue work are either fire fighter's boots or safety work boots Figure 2-16 ▾.

A.

B.

Figure 2-16 Boots used in rescues are typically either fire fighters' boots (**A**) or safety work boots (**B**).

Fire fighters' boots are primarily used in conjunction with turnout gear. The two major types are bunker boots and 8- to 10-inch fire/rescue safety boots. Both come in varieties specifically suited to different needs. Although bunker boots provide a high level of protection, their bulk makes them difficult to use in many rescue situations. Steel toe and shank safety work boots provide impact and puncture resistance, and their design provides some protection for the ankles. Their light weight and flexibility make them a good option in many rescue situations. By contrast, safety shoes lack ankle support or protection and, therefore, are not the best choices for rescue work.

When inspecting footwear, look for wear, tears, or holes in the leather or rubber, as well as damage or excessive wear to the soles, laces, or zipper. Waterproof boots in accordance with the manufacturer's directions, and familiarize yourself with any specific recommendations or field maintenance guidance provided by the manufacturer of the equipment.

Hand Protection

A rescuer's hands need protection from the variety of hazards found in a rescue situation. Any durable leather gloves will work in most cases, but there are a variety of specialty gloves designed for rescue work. Rescue gloves come in full leather and as a synthetic body with leather components Figure 2-17 ▾. This type of glove has some puncture and cut resistance and provides a good "feel" for the rope and tools. Additionally, the reinforced palm provides good protection for rope work. Leather firefighting gloves, while designed to be puncture and cut resistant, should

Figure 2-17 Rescue gloves.

not be used for rope work because of potential contaminants from firefighting use, bulkiness that reduces dexterity, and a lack of palm reinforcement. Latex gloves are used to provide protection from contamination of the hands from body fluids and some chemicals. When handling victims, gloves should be worn in accordance with your organization's **infection control plan** (i.e., its written infection control system).

Inspect gloves for rips, tears, weak or missing stitching, and exposure to contaminants. Gloves should be cleaned, decontaminated, sanitized, and dried according to the manufacturer's instructions. You should also be familiar with any specific recommendations or field maintenance guidance provided by the manufacturer of the equipment.

■ Respiratory Protection

Because atmospheric hazards can pose one of the greatest risks to rescuers, the selection and use of **respiratory protection** is of the utmost importance during any incident. Hazards may take the form of chemicals, vapors, fumes, dust, or oxygen deficiency. Your organization's **respiratory protection program** should ensure that you are trained in identifying situations in which some form of respiratory protection is required. Additionally, this program should provide guidance in the selection of appropriate respirators and ensure that you are physically capable and adequately trained in the use of respirators available to you. Physical capabilities include facial features and/or lack of facial hair (which may compromise the face piece seal) and a determination of respiratory fitness (passing of a doctor's examination as specified in OSHA 29 CFR 1910.134, *Respiratory Protection*).

Five types of respirators are generally used by rescue personnel: air-purifying respirator (APR), self-contained breathing apparatus (SCBA), supplied air-respirator (SAR), rebreathers, and self-contained underwater breathing apparatus (SCUBA).

Air-Purifying Respirator

Air-purifying respirators rely on filtration to remove particulates, gases, or vapors from the atmosphere **Figure 2-18 ▶**. Take care when choosing these types of respirators because the filter medium may not handle all of the hazards present. An APR is not used when there is a possibility of an oxygen-deficient or -enriched atmosphere, or when multiple potential hazards require the use of several different filters. Medical and disposable dust masks fall into this category, as do canister respirators. Canister respirators are available in full- and half-mask varieties, with the full-mask type providing eye and face protection.

Self-Contained Breathing Apparatus

The self-contained breathing apparatus's air and full face piece provide a much greater level of protection than APRs do **Figure 2-19 ▶**. Available in 30- to 60-minute versions, these models are good choices for protecting against almost all airborne contaminants. Some units have a supplied-air connection that allows for extended work time; however, these units will then be limited by the amount of hose available from the air source (for further detail, see the discussion of supplied-air respirators). SCBA may be a disadvantage in some rescue situations because of the limited air supply (non-air-line types). Also, the bulk of

Figure 2-18 Air-purifying respirator.

Figure 2-19 Self-contained breathing apparatus.

the back frame and bottle may make working in confined or restricted spaces difficult, and the weight of the unit may cause excessive fatigue during long operations.

Supplied-Air Respirator/Breathing Apparatus

In a supplied-air respirator/breathing apparatus (SAR/SABA), breathing air is supplied by an air line from either a compressor or stored-air (bottle) system located outside the work area **Figure 2-20 ▶**. This type of apparatus has the same advantages as

Figure 2-20 Escape pack for a supplied-air respirator.

Figure 2-22 Self-contained underwater breathing apparatus.

SCBA, in terms of the face piece and self-contained air supply. System components of a SAR include the positive-pressure–type respirator, escape bottle (5–10 minutes; 10 minutes recommended), the air line (300 ft maximum, depending on the flow required and the manufacturer), and a compressor or stored-air system.

A SAR has several distinct advantages: The air supply is not limited to what you take with you, its small size allows the wearer to access smaller spaces and provides for more maneuverability, and its lighter weight helps reduce fatigue during long operations. Disadvantages to using SARs include the potential for air-line entanglement or damage and the possibility that the small escape bottle size may not allow much time in an emergency to evacuate to a safe place.

Rebreather Apparatus

Often used in mine and underwater rescue, positive-pressure rebreathers are long-duration units that recirculate the user's exhaled breath **Figure 2-21 ▼** . The exhaled breath passes through a carbon dioxide absorbent and back into a breathing chamber,

where fresh oxygen is added from an attached small oxygen cylinder. The replenished gas then becomes available for the next inhalation. These units have ratings for as long as 4 hours and are very beneficial when distance or time restrictions make the use of other, more common types of respirators impractical.

Self-Contained Underwater Breathing Apparatus

Self-contained underwater breathing apparatus units are similar to SCBA in their function; unlike with SCBA, however, the SCUBA regulators are designed to work in an underwater environment, which is characterized by increased pressures **Figure 2-22 ▲** . Personnel performing underwater rescue operations must receive extensive training in specialized operational skills and the use of this apparatus. They must also be informed of the associated hazards involved in this type of rescue.

Inspection and maintenance of respiratory protection equipment should, at a minimum, include replacement of air cylinders, inspection of system components for signs of excessive wear or damage, an air leak check, and proper cleaning and sanitation of the face and/or mouth piece. Check compressed air cylinders for damage to the cylinder shell and regulator, and ensure that the cylinder has been hydrostatically tested at the appropriate intervals. In addition to these general guidelines, familiarize yourself with any specific recommendations or field maintenance guidance provided by the manufacturer of the equipment.

■ Other Protective Equipment

Other types of PPE include elbow and knee protection (internal/external), personal lighting, and personal alert safety system (PASS) devices. Take care to choose the right type of personal lighting. Consider its ease of use, duration, and intensity. The most common types in use include hand-held and helmet-mounted flashlights and chemical light sticks. Wear a PASS device while working in areas or spaces where you cannot be

Figure 2-21 Rebreather apparatus.

Voices of Experience

When I first stumbled into the world of technical rescue in the early 1990s, I had a hard time distinguishing between a ladder truck and an engine, much less a rescue rope and a piece of webbing.

But I wasn't alone. At about the same time, the fire service was beginning to understand the need for local and regional technical rescue teams. With that understanding—and some significant funding—came the purchase of specialized rescue equipment: rescue rope, harnesses, helmets, carabiners, tripods, litters, and more. This is some unique and often expensive stuff, and it didn't take long to figure out that we needed a system for organizing, labeling, and storing the equipment. In fact, setting up a good inventory system is just as important as buying the right equipment in the first place.

Here are a few ideas I've picked up over the years. When you put your own plan in place, you'll find the tracking, stock rotation, and general condition of your gear is much easier to manage.

The first step is to sort your gear into categories: hardware, software, personal protective equipment (PPE), patient packaging, and specialized equipment. Then divide it again into areas of use: rope rescue, confined space, water, trench, collapse, and general use.

Next, go to work on each category. For example, your rope should be every different color under the sun that you can find. Don't use color to identify length. Why? Because if you're hanging over the edge of a cliff and tell the haul team, "Up on red," what's going to happen if your main line and belay are both red? Likewise, choose one type of webbing (flat or tubular) and some standard lengths, and then assign a color to each length. For example, get 12-footers in yellow and 20-footers in orange. Now take it one step further: Mark the middle of the webbing and write the length on both sides and both ends. That might sound excessive, but when it's four in the morning and you *need* a 20-footer, you're going to thank me.

Now that you've figured out the basics, it's time to bag up your gear. I recommend storing different types of equipment for different types of rescues in separate places on the rig, or even on different rigs. Bag color and compartment separation can make a world of difference. When you tell a rookie, "Go to the first driver's-side compartment and bring me the red bag," the chances he'll come back with the right gear are better than if you say, "It's on the big red truck somewhere in one of the cabinets."

Keep in mind that these are just recommendations. Identify whatever works best for your team and go with it. But be sure to keep it simple—if you come up with some kind of complicated Dewey Decimal System, all you'll do is drive your personnel crazy.

Derek J. Peterson
Fire fighter/paramedic
Saint Paul Fire Department
Saint Paul, Minnesota

> **"[Specialized rescue equipment] is some unique and often expensive stuff, and it didn't take long to figure out that we needed a system for organizing, labeling, and storing the equipment."**

Figure 2-23 PASS devices and other protective equipment may offer benefits at a rescue scene.

seen **Figure 2-23 ▲**. One exception to this rule would be when other communications devices are available and the device poses a greater danger to the rescuer than it provides a benefit, such as in confined-space situations where the device could get caught on something.

PPE Maintenance

In addition to the general guidelines mentioned for each specific type of PPE, it is important to be familiar with any additional, specific recommendations or field maintenance guidance provided by the manufacturer of the equipment. Follow the manufacturer's directions and department SOPs regarding frequency of inspection. At the very least, however, PPE should be inspected before and after each use to check for any damage that may have occurred. Additionally, inspect PPE periodically for any damage that may have occurred during storage. Follow the manufacturer's directions in terms of inspection, cleaning, and repair methods. Document all inspection and maintenance activities performed in accordance with your organization's SOPs.

Rescue Equipment

After your organization has determined which types of rescue it will perform and at which level, two critical components need to be in place to achieve those goals—properly trained and protected personnel and tools appropriate for the required task. Once these tools are selected and purchased, rescuers need to know how and why they work, what their uses and limitations are, and which operating procedures and techniques should be followed. These tools may be commonplace, simple-to-use items, or they may be highly specialized equipment, requiring significant training for the rescuer to become proficient in their use. Because of the variety of tools, uses, and types of rescues, rescuers must become proficient through training and experience in the safe use and maintenance of all of the tools at their disposal.

Thousands of tools are classified as appropriate for use in four basic rescue activities: search, scene stabilization, victim access/extrication, and victim packaging. Rescue tools can be categorized by type, function, or use in a given specialty. In any event, all tools that technical rescuers use will fall into one of these four categories.

Hand Tools

Hand tools are the simplest and sometimes most effective tools rescuers carry. They tend to be lightweight and are easily carried to where they are needed; typically, these tools need little or no setup time. While power tools may be able to do the job faster once they are operating, the extraordinary mobility and quick setup of hand tools means that a rescue can sometimes be performed more quickly after a rescuer's arrival with the tool. Because of the different uses and methods of operation, hand tools are categorized as striking, leverage, cutting, and lifting/pulling items.

Striking Tools

Striking tools are used to apply an impact force to an object. Examples include driving a nail in a shoring system, forcing the end of a prying tool into a small opening, or breaking a vehicle window with a punch. Included in this category are hammers and punches.

Hammer-type striking tools have a weighted head with a long handle. They include the hammer, mallet, sledgehammer, maul, flat-head axe, and battering ram **Figure 2-24 ▼**.

Punches are used either to push on an object to move it or to impact an object to break it. Simple punches are normally used in conjunction with a hammer to push an object, such as a bolt or pin, while dismantling an object. The spring-loaded center punch is commonly used by rescuers to break tempered automobile glass **Figure 2-25 ▶**. This tool can exert a large amount of force on a pinpoint-size portion of tempered automobile glass, thereby disrupting the integrity of the glass and causing the window to shatter into small, uniform-sized pieces.

Leverage Tools

Two types of leverage tools are distinguished: rotating and prying **Figure 2-26 ▶**. Rotating tools are designed to turn objects; they include wrenches, pliers, and screwdrivers. These types of tools are commonly used for disassembly, such as in vehicle and machinery rescue.

Simple prying and spreading tools act as a lever to multiply the force that a person can exert to bend or pry objects apart. As

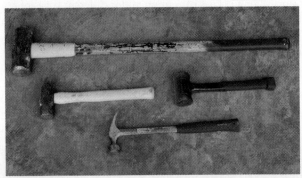

Figure 2-24 Striking tools (from bottom): hammer, mallet, maul, sledgehammer.

their name implies, they are used to pry apart objects; in addition, they are used to lift heavy objects. Common simple hand prying tools include the following equipment:

- Claw bar
- Crowbar
- Flat bar
- Halligan tool

Figure 2-25 A spring-loaded center punch can be used to break a car window safely.

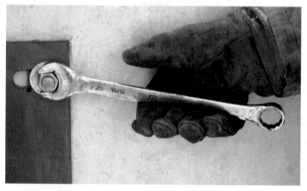

A.

B.

Figure 2-26 Leverage tools can be used for rotating (**A**) or prying (**B**).

- Kelly tool
- Pry bar

When using leverage tools, make sure you pick the right tool for the job. You should never have to modify a tool to make it work. For example, a pipe is sometimes (albeit inappropriately) used to extend the length of a tool handle. This modification is supposed to extend the distance from the fulcrum of the tool to where the force is applied, thereby increasing the amount of leverage exerted. In doing so, however, you may easily exceed the design strength of the tool and cause its damage or failure. Take leverage tools out of service if they show any signs of damage, such as cracks or bends.

Cutting Tools

Cutting tools have a sharp edge designed to sever an object. They range from tools placed in the pockets of turnout coats to larger tools that must be carried to where they are needed. Cutting tools fall into one of the following categories: saws, chopping, snipping/shears, knives, and chisels. Each type is designed to work on certain types of materials and to cut in a different manner. Because of their sharp edges and the force needed to make this equipment work, rescuers can be injured and cutting tools can be ruined if they are used incorrectly.

Common handsaws used in rescue include hacksaws, carpenter's handsaws, and jab, compass, keyhole, folding, and bow saws. Hacksaws are designed to cut metal, with different blades being applied based on the type of metal being cut **Figure 2-27**. These saws are useful when metal needs to be cut under closely controlled conditions or when the amount of metal to be cut is small. Blades are normally inserted so that the cutting action happens while the saw is being pushed away from the user. In some situations, rescuers may install two blades, one facing each way, to allow for cutting in both directions.

The carpenter's handsaw and the jab, compass, keyhole, folding, and bow saws are designed for cutting wood. Those with large teeth cut more quickly because they remove more material on each pass. Large-tooth saws, such as the bow saw, are effective tools for cutting large timbers or tree branches; as a consequence, they can prove especially useful at motor vehicle crashes where tree limbs might otherwise hamper the rescue

Figure 2-27 A fire fighter using a hacksaw to cut through metal.

A.

B.

Figure 2-28 Two types of axes: **A.** Flat-head axe. **B.** Pick-head axe.

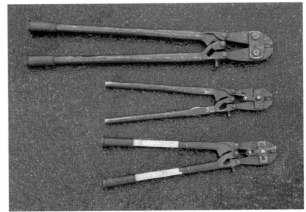

Figure 2-29 Bolt cutters are used to cut through items like chains and padlocks.

Figure 2-30 Folding knives with a serrated blade work well for cutting a variety of fabric materials, such as seat belts, during rescue work.

effort. The keyhole, jab, and compass saws are narrow, slender models that can be used to cut wood, plastic, and sheetrock quickly. Saw blades with finer teeth are designed for cutting finished lumber and tend to cut the material more slowly. For this reason, they are rarely used in rescue work.

Chopping tools include the axe and the pick **Figure 2-28 ▲**. These tools work by striking the object with a weighted head on a long handle, where the cutting edge of the head has been sharpened. These tools should have a semi-sharpened edge so that the edge will not chip if it hits an object such as a nail or a stone. Shovels may also be used as a chopping tool, and offer the added function of prying and material removal.

Snipping tools/shears include bolt cutters, cable cutters, insulated wire cutters, sheet metal snips, and EMS shears **Figure 2-29 ▶**. These types of tools operate on a leverage concept, with the fulcrum being located just behind the cutting edges. This concept allows for a concentration of the cutting force on a small area. Care must always be taken when cutting wires, including ensuring that they are not electrically energized. Special equipment operated by trained individuals is a must to avoid possible injury or death.

Rescuers also use many different types of knives **Figure 2-30 ▶**. Examples include folding knives, seat belt knives, linoleum knives, sheetrock knives, and box cutters. Knives should have a retractable or folding blade that locks when open. Keep knives sharp or replace them after each use to provide for maximum readiness the next time these tools are needed.

Chisels, which are used to cut wood and metal, are typically operated by striking the end of the chisel with a hammer or mallet **Figure 2-31 ▶**. They come in a variety of widths and styles and should be used only for cleaving the material for which they were designed. Wood chisels are used to remove small pieces of wood, whereas metal chisels (sometimes called cold chisels) are used to cut sheet metal or to cut off bolts and other objects.

Lifting/Pushing Tools

Manual lifting and pushing tools include mechanical jacks and hand-operated hydraulic jacks **Figure 2-32 ▶**. Mechanical jacks can be of the screw, ratchet lever, or cam type.

Screw-type jacks include bar screw jacks and house or trench screw jacks. These types of jacks operate on the inclined plane concept (screw) and may be capable of lifting very heavy loads. In most cases, however, these types of jacks are used to hold an object in place—for example, to hold trench panels in place or to support a floor beam.

Ratchet lever jacks (commonly known as high-lift or farmer's jacks) consist of a beam and a ratcheted lifting mechanism;

the lifting mechanism rides on the beam and can be controlled in either direction. Take care to ensure that the load being carried by the jack does not shift, as uneven distribution of the weight can cause the jack to tip and fall over. This consideration, which applies to all types of jacks, is especially critical when a ratchet lever jack is used, however.

Safety Tips

Ensure that the load being carried by the jack does not shift, as uneven distribution of the weight can cause the jack to tip and fall over.

Figure 2-31 Manual chisel.

Figure 2-32 Hand-operated jack.

Cam-type jacks (sometimes referred to as Ellis jacks or jack wrenches) use 2 × 2-inch, 4 × 4-inch, or 6 × 6-inch timbers placed side by side; these timbers are joined together by steel clamps. The cam mechanism is used to lengthen the timbers by grabbing one timber and lifting or pushing the other. These types of jacks can support a maximum safe working load as large as 16,000 pounds and can be as much as 18 feet long, depending on the size of the lumber used.

When using any type of jack, make sure it is located on a flat, level, hard surface that will not give way once weight is placed on the jack. If the surface underneath the jack is soft, place a flat board or steel plate under the jack to distribute the weight evenly. Inspect all jacks on a regular basis for any damage that may have occurred to them, including any signs of overloading, cracks, damaged parts, or improper operation.

Pulling Tools

Pulling tools can extend the reach of the fire fighter and increase the power that can be exerted upon an object. A variety of poles and hooks are used in rescue, including the pike pole. Another type of pulling tool is the manual winch and hoist, which uses either a chain or a cable (wire rope) and is used for pulling or lifting heavy objects **Figure 2-33 ▼**.

Figure 2-33 The manual winch and hoist uses either a rope, a chain, or a cable (wire rope) and is used for pulling or lifting heavy objects.

Chain winches come in two varieties: chain hoist and lever hoist. The chain hoist is used primarily for lifting; its lifting capacity ranges from 500 lb to 40,000 lb. The lever hoist is a lifting or pulling tool whose lifting capacity ranges from 250 lb to 12,000 lb.

Cable hoists use a wire rope instead of a chain and are used primarily for pulling. The two varieties of cable hoist are the integrated cable type (come along) and the pass-through cable type. Both of these styles can have a lifting capacity as high as 8000 lb.

When inspecting manual winches and hoists, look for any damage caused by overloading or friction. Check the winch for any warping, bending, or cracks. Examine cables and chains for signs of overloading, such as metal distention, cracks, or fraying. Lubricate moving parts properly, and keep the cables and chains free of rust.

Hydraulic Tools

Because of their power, versatility, and ease of setup, hydraulically powered tools have become very commonplace in the rescue service. Attachments that can pry, spread, lift, cut, and even pump water are available for use. Whether manual or powered, these tools operate by the use of compressed hydraulic fluid. Power units can be either portable or vehicle mounted. Whatever their type, the pressurized hydraulic fluid is sent to the tool by way of high-pressure hoses **Figure 2-34 ▾**. As a consequence, these power units are typically run by electric motors or gasoline engines, although some rely on compressed air for operation.

Powered Hydraulic Tools

The first powered hydraulic rescue tool was developed in 1972 to extricate racecar drivers after a crash. This unit consisted of a 32-inch spreader that was operated by a gasoline engine power unit. Today's spreaders are capable of both pushing and pulling actions and can produce as much as 22,000 psi of force at their tips.

Hydraulic shears/cutters can cut quickly through metal posts, bars, wood, and plastic, although some models cannot cut case-hardened steel. Because they may exert as much as 30,000 psi of cutting force, care should be taken to place the tool at a 90-degree angle to the object being cut. For those organizations with budget or space limitations, combination spreader/cutter units are available. These units tend not to have as much power as the single-use units have.

Hydraulic rams are used for pushing operations when the required distance to be covered is greater than the capacity of the spreaders. Available in models with a closed length of as much as 36 inches, the largest models can extend to more than 60 inches. While occasionally used for pulling operations, hydraulic rams pull with only half the strength of their pushing force (approximately 15,000 lb).

Hydraulic tools in use by the construction industry are now being employed by rescue units involved in structural collapse rescue **Figure 2-35 ▾**. These units may be equipped with a variety of additional tools, including concrete-cutting chainsaws, cutoff saws, jackhammers, drills, fans, and water pumps.

Nonpowered Hydraulic Tools

Hand-powered hydraulic tools are also used for prying, spreading, and lifting tasks. The use of hydraulic power enables rescuers to apply several tons of force on a very small area. One hand-powered hydraulic tool, called a rabbet tool, is designed for quickly opening doors. Hydraulic ram systems, sometimes referred to as "porta-power" units, are commonly used for this purpose as well. These kits contain a hand-operated hydraulic pump and a variety of accessories, including rams and spreaders. Because of the significant force applied by these devices, special training is required to operate the machines safely.

Hand-operated hydraulic jacks are used to lift heavy loads over a small distance. Sometimes referred to as bottle jacks, these jacks come in a variety of capacities, up to 35 tons. Air-over-hydraulic jacks are also becoming more popular, as they allow for either manual operation or the use of an air line to operate the jack. This flexibility can be especially helpful in tight areas where it would be very difficult to operate the pump handle.

Check hydraulic tools often for any indications of fluid leakage. Also check the fluid reservoir to ensure that it is at the proper level. Look for any signs of overloading, cracks, damaged parts, or improper operation.

Figure 2-34 Components of a hydraulic rescue tool.

Figure 2-35 Hydraulic construction tools may be useful in structural collapse rescue.

Pneumatic Tools

Pneumatic powered tools are operated by compressed air, which can be supplied from air compressors, SCBA cylinders, cascade systems, or vehicle-mounted compressors. A wide variety of pneumatic tools are available that can nail, drill, bolt, cut, vacuum, lift heavy loads, and stabilize trenches and structures. Most of these tools operate at forces between 90 and 250 psi and use adjustable regulators to provide the proper operating pressure. Because of this wide range of operating pressures, rescuers must take care to ensure that the proper regulators are used and adjusted to the right pressure so as to not damage the tool.

Pneumatic Cutting and Nailing Tools

Pneumatic cutting tools include chisels, shears, and saws. Pneumatic chisels (sometimes called air chisels, air hammers, or impact hammers) are used to cut sheet metal and shear off bolts Figure 2-36 ▾ . They normally create no sparks when operating. For these reasons, pneumatic chisels are often used in vehicle extrication incidents. Pneumatic shears are also used for cutting sheet metal; because of their limited application, however, they are rarely used in rescue work.

The pneumatic saws used in rescue work are of either the reciprocating or rotating type. Pneumatic reciprocating saws operate just like the electric variety. Models are available with saw speeds that range from 1600 to 10,000 revolutions per minute (rpm). Rotating pneumatic saws (also known as wizzer saws and die grinders) use a circular, composite cutoff blade that rotates at speeds as high as 25,000 rpm. Such saws are normally used when there is a space limitation, but can sometimes present a problem because of the sparks they generate.

Pneumatic nailers are used to drive nails or staples into wood or masonry. These nailers are frequently used by structural collapse and trench rescue teams when they are constructing shoring systems. They are just as dangerous as loaded guns.

Pneumatic Lifting Tools

Air bags come in three different types based on operating pressure. Low-pressure air bags operate at 7.25 psi. Medium-pressure air bags operate at 14.5 psi. Low- and medium-pressure bags are often referred to as air cushions. High-pressure air bags operate between 115 to 120 psi.

The capacity of any air bag is based on the number of square inches of bag surface area that come in contact with the object to be lifted Figure 2-37 ▾ . For example, an air bag with 20 square inches of contact operating at 119 psi has a lifting capacity of 2380 lb. (square inches × operating psi = lifting capacity). Air cushions are designed to have their entire surface in contact; therefore, they can operate at a lower pressure. Air cushions also have an advantage in that they can lift much higher than a high-pressure air bag can. Their major disadvantage is the size of the bag, which must be large to have plenty of contact area. High-pressure air bags, by contrast, form a pillow that reduces the surface contact area as it inflates. As a consequence, the lifting capacity of the bag decreases as it inflates.

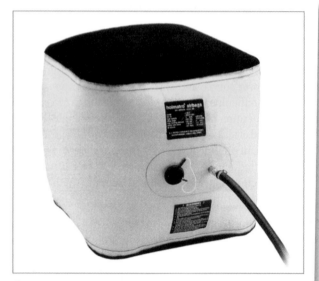

A.

B.

Figure 2-37   **A.** Air cushions are designed to have their entire surface in contact and can, therefore, operate at a lower pressure. **B.** Air bags require higher pressure.

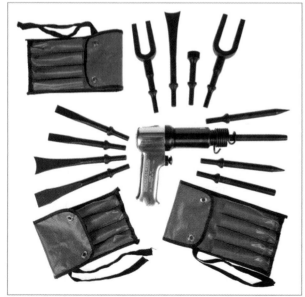

Figure 2-36   Pneumatic chisels are used to cut metal.

Other pneumatically operated tools include jackhammers, impact tools, hammer drills, and air knives and air vacuums in trench rescue. Impact tools are commonly used in vehicle and machinery extrication for removing bolts. In contrast, the air knife and the air vacuum are commonly used in trench rescue. The air knife, with its high air-exit velocity, is used to break up dirt so that the material can be more easily removed by hand or by an air vacuum. Rescue air vacuums use the same type of compressor as air knives and are capable of picking up loose soil and rocks to 2¾ inches in diameter. Commercial air vacuums, like those commonly used by highway maintenance organizations, can be used to pick up larger debris.

Inspection of all pneumatic tools includes checking for physical damage such as damaged air line connections, air leaks, improper operation, damaged high-pressure relief valves, and cracked and worn hoses. Some tools also require special lubrication prior to each use.

Figure 2-38 Electric and battery-powered tools are frequently used by rescue organizations.

Electric Tools

Electrically operated tools are used by many rescue organizations because of their light weight and low purchase cost **Figure 2-38 ▶**. These tools, which typically come in corded and battery-operated versions, primarily fall into one of the following categories: cutting, lifting/pulling, and drilling/breaking. While the cord versions tend to be more powerful, their requirement of a nearby power source and extension cords can present some operational difficulties. Additionally, exposure to moisture may create electrocution hazards. Cordless tools eliminate the extension cord and electrocution problems, but battery life, charging, and storage can present challenges. Because these tools are used relatively infrequently, the batteries tend to be left in chargers for a long period of time, which can degrade the battery and shorten its usage time.

Electric Cutting Tools

Electrically operated cutting tools include the circular saw, chainsaw, reciprocating saw, snipping devices such as the rebar cutter, and specialty items such as the plasma cutter.

Circular saws come in a variety of sizes and are used primarily for cutting wood, although special blades are available that will cut metal or masonry. Most rescue organizations use circular saws for cutting shoring lumber.

The electrically operated chainsaw is used by some organizations because of its light weight and low noise. It is especially advantageous for use in enclosed areas where high noise levels and carbon monoxide (CO) from internal combustion engines need to be avoided.

The reciprocating saw is very popular with rescue organizations because of its versatility. While there are different blades for metal and wood, demolition blades that will cut through just about anything are available.

Rebar cutters are used for cutting not only reinforcing bar, but also any other round metal that is ⅝ inch or less in diameter. They operate by placing the metal to be cut between a cutting edge and a ram, which then pinches the metal and sheers it.

The plasma cutter is a device that uses an electric arc to cut through metal. Rescue services use this tool when fine cutting and minimal sparks are required.

Electric Lifting/Pulling Tools

Electrically operated lifting and pulling devices include winches and hoists. Vehicle-mounted winches have been common on rescue vehicles for many years. Portable winches and hoists require only a secure place to attach them and are now being used in a variety of rescue situations. Always take care when using hoists or winches so the pulling force exerted by these devices does not injure rescuers or victims.

Electric Drilling/Breaking Tools

A variety of drilling and breaking tools are available for use by rescue services. Variable-speed, reversible drills come in both corded and battery-operated versions and are used for drilling holes and driving screws. There is also a right-angle version available for use in tight spaces. A hammer drill is used for drilling holes in concrete using special masonry bits. For those rescue organizations that need to be able to break up concrete and stone, the rotary hammer and the demolition hammer are popular choices.

One of the most important things to notice when inspecting electrical tools is any indication of damage to the cord or plug. Tools with damage to the plug or cord should be taken out of service immediately and repaired to avoid any possibility of electrocution. Also inspect tools for cracked or damaged housings, improperly operating trigger mechanisms, or improperly operating or damaged chucks.

Fuel-Powered Tools

The two major uses for fuel-powered tools are cutting and nailing **Figure 2-39 ▶**. Cutting tools include chainsaws, rotary saws, cutting torches, and exothermic torches. One of the major advantages of chainsaws and rotary saws is the high power they generate, which enables them to readily cut through materials. Disadvantages are that they are heavy to carry and sometimes difficult to start.

Fuel-Powered Cutting Tools

Fuel-powered chainsaws are available that cut wood, concrete, and even light-gauge steel. Standard steel chains are used to cut wood, carbide-tipped chains can cut wood and light-gauge metal, and diamond chains are used for cutting concrete. All chainsaws used by rescue personnel should be equipped with a chain break.

These tools should never be operated in enclosed spaces. Always wear appropriate PPE (including eye protection, hearing protection, and gloves) when working with these tools.

Fuel-powered rotary saws are also available that cut wood, concrete, and metal. Two types of blades are used on rotary saws: a round metal blade with teeth and an abrasive disk. These disks are made of composite materials and are designed to wear down as they are used. These disks come in different styles for concrete, asphalt, and metal. It is important to match the appropriate saw blade or saw disk to the material being cut.

Cutting torches produce an extremely high-temperature flame and are capable of heating steel until it melts, thereby cutting through the object. These tools are sometimes used in rescue situations to cut through heavy steel objects. Because these torches produce such high temperatures (5700°F or more), operators must be specially trained before using them. The most common type of cutting torch uses oxygen and acetylene to create the flame, but many rescue services have begun to use oxygen/gasoline torches. The latter torches can cut through as much as 14 inches of steel, and the sparks generated with their use produce little heat and weight, although there can be molten metal. Other advantages of the oxygen/gasoline models are that gasoline is readily available and is not as volatile as acetylene, and that these devices use much less fuel by weight.

Another type of cutting torch is the exothermic torch. It operates by igniting a combustible metal contained within a tube where oxygen has been forced down the center of the tube. While these torches can cut very heavy steel (even underwater), they also produce a tremendous amount of sparks and slag.

Fuel-Powered Nailing Tools

Fuel-powered nailing guns have found favor among many rescue services because of their light weight and portability. These tools work by igniting a combustible gas in a chamber inside the gun, which then drives the nail. The one major disadvantage to this type of gun is that it does not work very well in cold weather.

Monitoring/Detection Equipment

Rescuers are often exposed to hazards they cannot see, taste, smell, or hear. To deal with this situation, they rely on specialized monitoring and detection equipment to enhance their senses and to identify hazards before they become a danger. The most common hazards for which detection equipment is applied are atmospheric contaminants and electrically energized wires and equipment. Certain types of rescue situations, such as structural collapse, also require that rescuers monitor for any movement of the structure.

Atmospheric Monitoring

Multi-gas meters are commonly used by emergency services personnel Figure 2-40 ▾ . At a minimum, these meters should measure oxygen, combustible gases, and toxic gases such as

Figure 2-39 Gasoline-powered rotary disc saws can be used for a variety of forcible entry tasks.

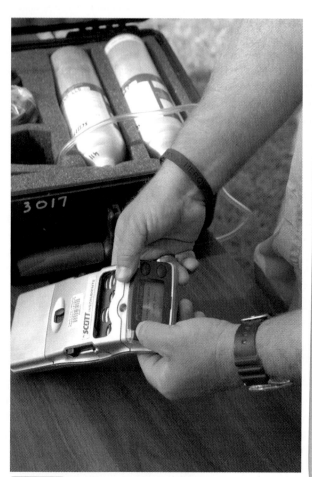

Figure 2-40 Multi-gas meters measure oxygen, combustible gases, and toxic gases.

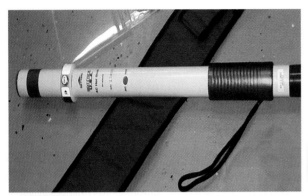

Figure 2-41 An AC power locator detects the presence of electrical frequencies below 100 Hz.

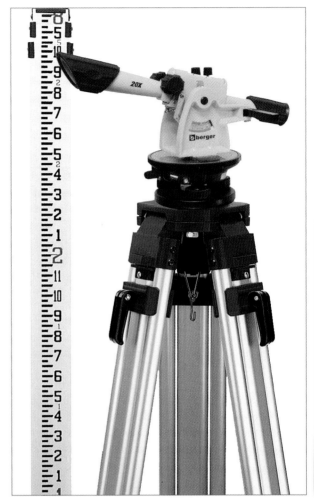

Figure 2-42 Transits and theodolites are sensitive surveying equipment that can identify even small structural movements.

carbon monoxide and hydrogen sulfide. If necessary, sensors can be installed to detect other toxic gases. A variety of other specialized meters are available, such as one for dust that can be used in structural collapse incidents.

Power Detection

Electrical hazards on a rescue scene are always a concern, and efforts should always be made to identify and isolate any hazards. Because it is often impossible to tell visually whether a wire or machinery is energized, every rescue organization should have an AC power locator available **Figure 2-41 ▲**. This tool is very good at detecting the presence of electrical frequencies below 100 Hz. It cannot, however, detect DC power or AC power contained in solid metal enclosures such as grounded metal conduits. Other meters can detect DC power or provide voltage readings if necessary. These models are less suitable for emergency response, however, because they require probes to be placed directly on the wire or object to be tested.

Movement Detection

At structural collapse incidents, it is critical that the structure be monitored for any movement. By the time this movement is recognized by the naked eye, it may already be too late to evacuate the rescuers inside the structure. To help identify even the smallest motion within the structure, sensitive surveying equipment such as the transit and theodolite are used **Figure 2-42 ▶**.

Search Equipment

Often the victims at a rescue scene are not easily located. Sometimes this difficulty arises because the area is dark and you cannot see until you get adequate lighting. In a collapsed structure, victims may be buried under rubble, with no visible indication where you should search. A variety of search tools and equipment are available not only to light up a scene, but also to help rescuers see things that are not visible with the naked eye and hear things that are not audible with normal hearing. Other equipment, such as GPS receivers, can tell rescuers exactly

where they are so they do not get lost themselves, requiring other searchers to then have to look for their colleagues.

Lighting

Effective and reliable lighting is critical for the safety of rescuers and to provide for a better search environment. Area lights, hand lights, and helmet lights are available for use by rescuers **Figure 2-43 ▶**. If there is the possibility of a flammable atmosphere, only intrinsically safe lights should be used. Because most AC-powered area lights are not intrinsically safe, the use of chemical light sticks has become common to provide minimal area lighting.

Visual Aids

Often rescuers cannot see the victims they are searching for. To overcome this problem, specialized cameras are available to the rescue services to assist them in finding victims. The most commonly encountered camera is the thermal imaging camera **Figure 2-44 ▶**. Many fire departments carry such cameras to help them detect hidden fire or hotspots, but these tools are equally

Figure 2-43 Hand lights provide a better search environment for rescuers.

Figure 2-44 Thermal imaging cameras are the most commonly encountered specialized cameras in rescue situations.

Figure 2-45 Acoustic listening devices can help rescuers find victims who cannot be seen.

useful in finding victims who are lightly covered in debris or dust or who cannot be seen because of darkness. Also available are specialized search cameras that incorporate a remotely controlled lens and fiber-optic cables, thereby allowing the operator to see into spaces where rescuers cannot gain entrance. Some of these models incorporate still and video cameras as well as have the ability to transmit video to the command post.

Audio Aids

Two types of acoustic listening devices are frequently used by rescue services **Figure 2-45 ▶**. The first type of device uses a microphone that is attached by an intrinsically safe wire that can be pushed or lowered into a space to listen for victims. Variations of this device allow for two-way communications when desired. The second type of device uses highly sensitive probes that, when the readings are evaluated by a trained technician at the receiving station, may actually be able to triangulate the location of the victim.

Other Search Equipment

One additional piece of equipment that is becoming more popular in rescue services is the GPS receiver. These tools are used for establishing the precise locations of rescuers who are performing grid searches and for marking the locations of victims. Although GPS receivers were once used primarily for wilderness search and rescue, structural collapse and other rescue teams are now using them for everything from obtaining travel directions to the incident to establishing precise locations for helicopter landing zones.

Victim Packaging and Removal Equipment

When victims are found, it is important that they not be further injured while rescuers are removing them to definitive medical care. This goal is accomplished by the use of stretchers and immobilization devices appropriate to the situation and, when necessary, high-point devices that will allow for the safe movement of the stretcher in a vertical environment. Collectively, these tools are known as **victim packaging and removal equipment**.

Stretchers/Litters

The stretchers most commonly used by rescue services when removing victims are collapsible, scoop, and basket stretchers, with each type having its own set of advantages and disadvantages **Figure 2-46 ▶**. Collapsible and basket stretchers allow for the incorporation of a backboard and may be rigged for vertical lifting. Basket stretchers can also be used in any water rescue environment when flotation devices are attached to them to provide buoyancy. The scoop structure is designed to lift a victim without moving the individual to place the stretcher under him

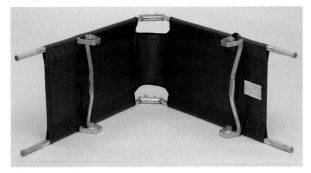

A.

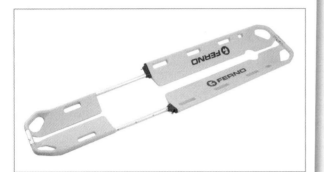

B.

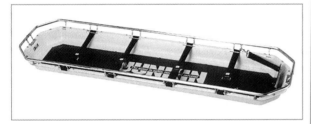

C.

Figure 2-46 Collapsible (**A**), scoop (**B**), and basket (**C**) stretchers are the types most often used by rescue teams.

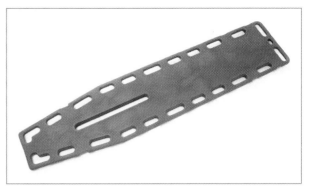

A.

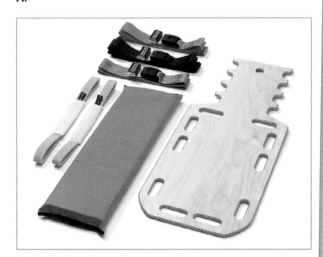

B.

Figure 2-47 Full (**A**) and half (**B**) backboards are the simplest immobilization devices.

or her. This type of stretcher is not designed to be used for vertical lifting by itself.

When inspecting stretchers, ensure that there are no bent rails or supports, cracked welds, rips, tears, damaged buckles, or missing pieces.

Immobilization and Combination Devices

The simplest immobilization devices are the full and half backboards, which are often used in conjunction with stretchers **Figure 2-47 ▶**. Other specialized devices incorporate a half board into a vest-style arrangement that can be used, depending on the type of equipment, either alone or in conjunction with a harness for lifting a victim vertically.

When inspecting immobilization devices, ensure that there are no cracks, splinters, gouges, worn straps, damaged buckles, or missing pieces.

Tripods and Offset High Points

Tripods are used in confined spaces and in other situations where a high point is needed to lift the victim out of a hole **Figure 2-48 ▶**.

Figure 2-48 Tripods are used in confined-space situations to lift a victim from a hole.

Other variations on this concept allow for the high point to be off-center for specific rescue situations. When choosing tripods or other offset high points, ensure that the device is rated for human use and that it is rated for a minimum of two people (800 lb).

Stabilization and Shoring Tools

Many rescue scenes are characterized by unstable vehicles, structures, soil, or other materials that create a dangerous environment for the rescuers and victims. Cribbing and shoring equipment is used to keep heavy loads from moving. This process of stabilization can employ the use of blocks of wood, specialized shoring jacks, and/or sheets of wood or metal to hold back dirt.

Cribbing

The simplest type of stabilization is referred to as cribbing Figure 2-49 ▾. Cribbing consists of wood, composite, or plastic pieces stacked across one another to build up a solid base for a load to sit on. Wedges are used to make up any difference between the cribbing and any object being supported. Cribbing is also used to create a solid surface for air bags or jacks when there is concern about their stability on soft ground.

Most cribbing is between 18 and 24 inches long and comes in various sizes, including 2 × 4 inch, 4 × 4 inch, and 6 × 6 inch. Often the ends are color coded to show the pieces' length, and handles are attached to aid in carrying and placing the cribbing. Plastic cribbing is often designed so that the pieces interlock and provide maximum stability.

Another type of cribbing that is frequently used is the step crib or chock. It allows for the use of cribbing at multiple fixed heights without having to add more lumber.

Cribbing is frequently stored in compartments with the handles facing out or in crates for easy portability. It should be inspected periodically for any major gouges, cracks, or contamination. Plastic cribbing is preferred by some organizations because any contamination (such as oil) can be cleaned off easily and will not be absorbed into the cribbing material.

Shoring

Shoring is used where the vertical distances are too great to use cribbing or the load must be supported horizontally, such as in a trench, or diagonally, such as in a wall shore Figure 2-50 ▾. Three types of shoring are used by rescue services: wood, pneumatic, and manual screw. These types of shoring should be used only when the rescuer is properly trained in the selection, use, and construction of shoring for the specific applications intended.

Wood shoring used by rescue services is usually 4 × 4 inches or 6 × 6 inches in dimension, although larger sizes may be required. It is typically made of Douglas fir or hemlock. Pneumatic shoring comes in a variety of lengths and may be extended by the use of compressed air. The shoring is then locked in place by pins, notches, or screw collars. Manual screw-type jacks include bar screw jacks and house or trench screw jacks. These devices are used in horizontal, vertical, and diagonal environments.

Sheeting

The sheeting used by rescue services is typically made of wood or metal and may be commercially made Figure 2-51 ▾. Sheeting is placed against the wall of a trench to stabilize it. For this

Figure 2-50 Shoring is used to support the walls in trench rescue.

Figure 2-49 Cribbing in the front and back of a vehicle's wheels stabilizes the vehicle.

Figure 2-51 Sheeting holds back running material in trench rescue.

application, it must be strong and able to withstand the pressures exerted by the wall of the trench as well as the shoring used to hold the sheeting in place. Wood used to create sheeting includes planks or trench panels of 1⅛-inch-thick CDX or white birch or Norwegian fir plywood. Commercial sheeting may be steel or aluminum and may incorporate integrated shoring.

Other Specialty Tools

Some rescue specialties require the use of tools that are used only in incidents involving that type of rescue. Examples include specialty tools for elevator rescue, water rescue, ice rescue, and structural collapse.

■ Elevator Tools

The two types of specialized equipment for use in elevator rescue are emergency keys and interlock release tools **Figure 2-52 ▾**. Emergency keys are used to open elevator doors where a key way is provided. To avoid the potential of anyone falling a great distance, these key ways are typically provided only on the first two floors of a building. To open doors at levels higher than the second floor, interlock release tools are used to slide between the door and the casing and release the mechanism holding the door shut. These keys and interlock release tools are specific to the manufacturer and door type, so they are often kept at the location of the elevator.

■ Rigging Equipment

Some structural collapse response teams carry their own rigging equipment for use with cranes and other heavy equipment. This

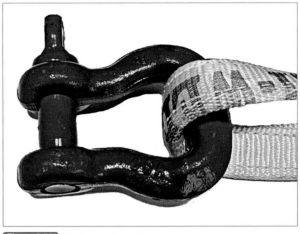
Figure 2-53 Rigging equipment.

equipment may include slings, clamps, shackles, and clips that are used to lift heavy objects **Figure 2-53 ▴**. Only individuals who have received special training in the rigging and hoisting of heavy objects should use such equipment.

■ Ice and Water Specialty Equipment

Specialized equipment used in ice and water rescue includes personal protection suits, breathing devices, boats, sleds, collars, personal flotation devices, beacons, throw bags, ice awls, and boot studs. Because of the specialized nature of this equipment, only trained individuals should be allowed to use it in a rescue environment. When inspecting ice and water specialty equipment, look for any holes, rips, or tears in suits, flotation devices, boats, and sleds. Also ensure that all other equipment has all the necessary parts and is in good working order.

■ Safety Equipment

Specialized safety equipment is designed to warn individuals against a hazardous situation or environment. Personal safety equipment used for this purpose includes reflective vests, cones, fire line tape, or barriers. Lockout/tagout kits are used to ensure electrical, gas, and liquid systems are rendered safe. They work by locking or tagging switches and valves so they cannot be turned on inadvertently.

■ Tool Inspection Maintenance

In addition to the general guidelines mentioned for each specific type of tool, familiarize yourself with any other recommendations or field maintenance guidance provided by the manufacturer of the equipment. Follow the manufacturer's directions and department SOPs on frequency of inspection. At the very least, inspect tools before and after each use to check for any damage that may have occurred during storage or at a rescue scene. Follow the manufacturer's directions regarding inspection, cleaning, and repair methods. Perform and document all inspection and maintenance activities in accordance with your organization's SOPs.

Figure 2-52 Elevator tools include emergency keys and interlock release tools.

Rope Rescue Equipment

In emergency services, ropes and rope rescue equipment are widely used to hoist or lower tools, appliances, or people; to pull a person to safety; or to serve as a lifeline in an emergency. Rope rescue equipment is categorized as either software or hardware. Rope rescue software encompasses life safety rope, accessory rope, webbing, and harnesses. Rope hardware includes carabiners, pulleys, friction devices, edge rollers, and other rope accessories made of steel or aluminum.

A rope may be your only means of accessing a trapped victim and bringing that individual to safety. For this reason, learning about ropes and knots is an important part of your training as a rescuer. You must be able to tie a variety of knots accurately without hesitation or delay. Tying knots is an acquired skill and usually requires a great deal of practice. This section on basic ropes and knots is placed early on during this book precisely for this reason. To become proficient in the use of **rope rescue systems**, you must first become proficient in the knots used in these systems.

Rope

There are two primary types of rope used in emergency services, each dedicated to a distinct function:

- **Life safety rope** is used solely for supporting people. It must be used whenever a rope is needed to support a person, whether during training or during firefighting, rescue, or other emergency operations.
- **Utility rope** is used in most other cases, when it is not necessary to support the weight of a person, such as when hoisting or lowering tools or equipment.

Life safety rope is a critical tool used only for life-saving purposes—it must never be used for utility purposes. Life safety rope must be used in every situation where the rope must support the weight of one or more persons. In these situations, rope failure could result in serious injury or death. Because a fire fighter's equipment must be extremely reliable, the criteria for design, construction, and performance of life safety rope and related equipment are specified in NFPA 1983, *Standard on Life Safety Rope and Equipment for Emergency Services*.

Life safety ropes are rated for either one person or two persons. A two-person rope must be used in rescue operations where both the rescued individual and the rescuer require support. NFPA 1983 lists very specific standards for the construction of life safety rope. NFPA 1983 also requires the rope manufacturer to include detailed instructions for the proper use, maintenance, and inspection of the life safety rope, including the conditions for removing the rope from service. In addition, the manufacturer must supply a list of criteria that must be reviewed before a life safety rope that has been used in the field can be used again. If the rope does not meet all of the criteria, it must be retired from service.

The two primary types of life safety ropes included in NFPA 1983 are light-use life safety rope and general-use life safety rope. A light-use life safety rope is designed to have a breaking strength of at least 20 kN (4496 lbf). A general-use life safety rope is designed to have a breaking strength of at least 40 kN (8992 lbf).

Rope Materials

Ropes can be made from many different types of materials. Because ropes have many different uses, different materials may work better than others in various situations. Since nylon was first manufactured in 1938, synthetic fibers have been used to make ropes. In addition to nylon, several newer synthetic materials such as polyester, polypropylene, and polyethylene are used in rope construction.

Synthetic fibers have several advantages over natural fibers. For instance, they are generally stronger than natural fibers, so it may be possible to use a smaller-diameter rope without sacrificing strength. Synthetic materials can also produce very long fibers that run the full length of a rope, thereby providing greater strength and added safety. Synthetic ropes are more resistant to rotting and mildew than natural-fiber ropes and do not age or degrade as quickly. Depending on the material, they may also provide more resistance to melting and burning than their natural-fiber counterparts. They also absorb much less water when wet and can be washed and dried. Some types of synthetic ropes can float on water, which is a major advantage in water rescue situations.

At the same time, ropes made from synthetic fibers have some drawbacks. Prolonged exposure to ultraviolet light or exposure to strong acids or alkalis can damage a synthetic rope and decrease its life expectancy. In addition, synthetic materials may be highly susceptible to abrasion or cutting.

Life safety ropes are always made of synthetic fibers. Before any rope can be used for life safety purposes, it must meet the requirements outlined in the most current version of NFPA 1983. These standards specify that life safety rope must be made of continuous filament virgin fiber and woven of **block creel construction** (without knots or splices in the yarns, ply yarns, strands, or braids of the rope). Rope of any other material or construction may not be used as a life safety rope.

The synthetic fiber most commonly used in life safety ropes is nylon. It has a high melting temperature, offers good abrasion resistance, and is both strong and lightweight. Nylon ropes are also resistant to most acids and alkalis. Polyester is the second most commonly used synthetic fiber in life safety ropes. Some life safety ropes are made of a combination of nylon and polyester or other synthetic fibers. Polypropylene is the lightest of the synthetic fibers. Because it floats and does not absorb water, polypropylene rope is often used for water rescue situations. However, polypropylene rope is not as suitable as nylon for fire department life safety uses because it is not as strong as nylon rope, is hard to knot, and has a low melting point.

Rope Construction

Several types of rope construction are possible. Because abrasion can damage the rope fibers and may reduce rope strength, rescue rope is made of kernmantle construction. Kernmantle rope consists of two distinct parts: the **kern** and the **mantle** `Figure 2-54 ▶`. The kern is the center or core of the rope; it provides approximately 70 percent of the strength of the rope. The mantle, or sheath, is a braided covering that protects the core of the rope from dirt and abrasion. Only 30 percent of the strength

Figure 2-54 Kernmantle rope consists of the kern (core) and the mantle (sheath).

Figure 2-55 Webbing has a variety of rescue applications.

of the rope comes from the mantle. Both parts of a kernmantle rope are made with synthetic fibers, although different fibers may be used for the kern and the mantle. Each fiber in the kern extends for the entire length of the rope without knots or splices; this block creel construction is required under NFPA 1983 for all life safety ropes. The continuous filaments produce a core that is stronger than one constructed of shorter fibers that are twisted or braided together. Kernmantle construction produces a very strong and flexible rope that is relatively thin and lightweight. This construction is well suited for rescue work and is very popular for life safety rope.

A rope can be either dynamic or static, depending on how it reacts to an applied load. A **dynamic rope** is designed to be elastic and will stretch when it is loaded. A **static rope** will not stretch as much as dynamic rope will under load. The differences between dynamic and static ropes result from both the fibers used in the rope and the construction method. Dynamic rope is typically used in safety lines for mountain climbing because it will stretch and cushion the shock if a climber falls for a long distance. In contrast, a static rope is more suitable for most technical rescue situations, where falls from great heights are not anticipated. Teams that specialize in rope rescue often carry both static and dynamic ropes for use in different situations.

Kernmantle ropes can be either dynamic or static. A dynamic kernmantle rope is constructed with overlapping or woven fibers in the core. When the rope is loaded, the core fibers are pulled tighter, which gives the rope its elasticity. The core of a static kernmantle rope has all its fibers laid parallel to one another. This type of rope has very little elasticity and demonstrates limited elongation under an applied load.

Most fire department life safety ropes use static kernmantle construction. Such ropes are well suited for lowering a person and can be used with a pulley system for lifting individuals. They can also be used to create a bridge between two structures.

Rope Strength

Life safety ropes are rated to endure a specific amount of force under the minimum requirements of NFPA 1983. The required minimum breaking strength for a life safety rope is based on an assumed loading of 300 lb per person with a safety factor of 15:1. The safety factor allows for reductions in strength as a result of knots, twists, abrasion, or any other cause. It also takes into account the possibility of shock loading if a weight is applied very suddenly. For example, shock loading could occur if the person who is tied to the rope falls and then is stopped by the rope. A personal escape rope is also expected to support a force of 13.5 kN (3034 lbf), representing one 300 lb person, with a safety factor of 10:1.

■ Webbing

Webbing is used for a number of applications in rescue, from creating anchors to tying victims into stretchers to creating site-made harnesses. Webbing comes in flat and tubular forms, varying from 1 to 3 inches in width **Figure 2-55 ▲**. Tubular webbing should be constructed of a continuous spiral weave (a continuous strand weaved around horizontal strands). Typically constructed of nylon, 1-inch tubular webbing has a strength of 4000 lb and is highly abrasion resistant.

■ Other Software

Other rope rescue equipment typically referred to as **rope rescue software** includes slings, looped straps, accessory cord, and specialty straps. Slings are normally used to wrap around an object to create an anchor. Looped straps include the etrier, foot loop, and multiloop strap. Looped straps are normally designed to be used by the individual and should never be incorporated into a system. Specialty straps include pick-off straps, adjustable straps, and load-release straps.

■ Hardware

Rope rescue hardware is defined as rigid mechanical auxiliary equipment that can include, but is not limited to, carabiners and other linking devices, mechanical ascent and descent control

Figure 2-56 Carabiners attach pieces of equipment together.

A.

devices, pulleys, and anchor plates. The use of these devices, in combination with rope and other software, constitutes a rope rescue system.

Linking Hardware

Commonly encountered linking hardware consists of carabiners, triangular screwlinks, rings, and rigging plates.

A carabiner attaches pieces of equipment together, such as joining ropes and anchors **Figure 2-56 ▲**. It is designed to take the load on the long axis, so it is rated only with the gate closed and secured. Carabiners are constructed of either steel or aluminum, with steel primarily being used in rescue work because of its high strength. Aluminum should be used only for a one-person load—never in a system.

Two types of carabiners are available: locking and nonlocking. The locking type should be used in rescue applications. The pin lock type has a lock configuration that prevents the gate side from opening in a high-force situation and loses 10 to 20 percent of its strength when left unlocked. The machined lock type has a gate-matching mechanism that holds the gate in line with the latch for alignment of the gate lock. It loses 50 to 90 percent of its strength when left unlocked. The specifications for carabiners include a minimum breaking strength:

- Personal use device: 6000 lbf (pounds of force)
- General use device: 9000 lbf

Triangular screwlinks are used where multiple-direction loading is expected. These devices are less expensive than carabiners and, like carabiners, must always be locked for use. Rings are used mostly for constructing multipoint bridles for stretchers. Rigging plates are primarily used for attaching multiple items to an anchor.

Care and maintenance of rope rescue hardware includes keeping the hardware clean and in good condition. Inspect the hardware for dents/burrs, rust, bending, or distention of the metal; proper gate function; and proper lock function. Clean hardware by wiping it with a cloth and using a small file or emery cloth for metal burr removal. Do not drop or throw hardware (a drop of more than 5 ft requires the hardware to be retired). Do not attach hardware to hard-edged metal anchor points if possible, as this can dent or burr the hardware.

B.

Figure 2-57 The figure eight plate (**A**) and the brake bar rack (**B**) are the most commonly used types of lowering hardware.

Lowering Devices

The two primary devices used for lowering in rope rescue are the figure eight plate and the brake bar rack **Figure 2-57 ▲**. The figure eight plate can be constructed of high-strength plate aluminum for low overall weight or, alternatively, can be all steel. The latter construction is more expensive and heavier, but nearly impossible to wear out. The typical breaking strength for an aluminum plate is 12,000 lb, while that for steel is 45,000 lb.

The figure eight plate is designed to create friction and is typically used for rappelling where the load is not expected to exceed one person. It is considered to be height limited because it imparts spin to the rope (an effect that is most noticeable at distances exceeding 200 ft). In addition, excessive weight will act as a brake on the plate, not allowing movement, such as a long length of rope hanging below creating a lot of weight.

A brake bar rack can be of aluminum or steel construction. Aluminum bars allow for a slower rappel but wear faster than steel. Racks work by generating friction based on the number of bars used and the space between the bars. The types available include five- or six-bar styles with a straight or twisted frame. The top bar is 1 inch in diameter and has a "training groove" that guides the rope. The second bar is ¾ inch in diameter, with a straight slot that allows the bar to fall out if improperly rigged. The third through sixth bars are ¾ inch in diameter, with an angled slot that snaps into place.

As mentioned earlier, the brake bar rack works by generating friction. The amount of friction generated can be varied over a wide range, which allows the operator to reduce the friction for long rappels or lower loads with less twisting. The rate of descent can be controlled by the spread of the bars as well as by the amount of rope running through the brake hand of the operator. Exercise caution when lacing the rope through the angled bars: Even when improperly laced, they may hold while you test the rigging, only to pop loose after the person on the line is committed over the edge. The training groove on the first bar (or the first and third bars) helps keep the rope in the center of the rack. Racks will accept two ropes, even if some of the bars have training grooves.

Do not drop or throw a rack or plate. Keep these devices clean and inspect them regularly for dents, burrs, cracks, sharp edges caused by rope wear, distortion of holes, wear of bars, integrity of the weld eye, tightness of the frame nut, and distortion of the frame. Clean the equipment by wiping it with a cloth. A small file or emery cloth can be used to remove sharp edges or metal burrs. Dirty ropes are subject to accelerated wear, so try to use clean ropes. When a plate is worn by more than one-third the diameter of the original material, it should be discarded.

Several newer linking hardware devices have also become available. One such device is based on the windlass concept. The rope is wrapped around a metal cylinder, thereby causing friction. This device also has a brake mechanism that comes into play if a sudden increase in rope speed occurs, such as in a fall. Such a device is typically used on belay (safety) lines. Another device is a pass-through friction device that is controlled by a dead-man handle. Should the operator accidentally let go, the device will brake itself and stop. Such a device can be used for rappel or belay purposes or as a lowering device.

Mechanical Advantage

Pulleys are constructed of a metal sheave (wheel) mounted on a bearing or metal bushing, and can be aluminum or steel bodied

with a steel axis. They are used for the following applications during rope rescue **Figure 2-58 ▼**:

- Change direction (directional pulley)
- Provide a mechanical advantage
- Reduce friction over an edge
- Provide rope tension

A rescue pulley should have a minimum pulley diameter of four times the rope's diameter. For example, a ½-inch rope requires a 2-inch pulley; a ⅝-inch rope requires a 3-inch pulley. The 2-, 3-, and 4-inch sizes are available as standard pulley models. A 4-inch pulley is much more efficient than a 2-inch pulley because the rope bends less and has less friction.

The minimum breaking strength for pulleys is summarized here:

- Personal-use device: 1200 lbf minimum load test without permanent damage to device or rope; 5000 lbf minimum load test without failure
- General-use device: 5000 lbf minimum load test without permanent damage to device or rope; 8000 lbf minimum load test without failure

Do not drop or throw pulleys. Keep them clean. When inspecting pulleys, ensure that there is proper movement of cheeks and sheave. Any egg-shaped attachment holes indicate the pulley has been overstressed. Check the tightness of nuts or bolts holding the pulley together. Wipe the equipment clean with a cloth.

Remember to use the proper-diameter rope for the size of the pulley (four times the rope diameter). Always back up high-

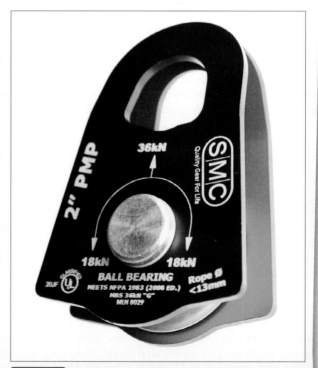

Figure 2-58 Rescue pulleys have a variety of applications in rescue scenarios.

point directional pulleys. Use a steel pulley as a means to change direction of a rope system, because the forces created may be compounded in this situation.

Hand and power winches are also useful in rope rescue and confined-space rescue. These devices provide a built-in mechanical advantage system, so rescuers do not have to construct rope mechanical advantage systems.

Ascent (Grab) Devices

Ascent devices are an auxiliary equipment system component used as a friction or mechanical device to allow for ascending a fixed line Figure 2-59 ▼. They can withstand a minimum test load of at least 1200 lbf without permanent damage to the device or rope.

The type of ascent device most commonly used in rope rescue is the cam (Gibbs) ascender. It is available in cast aluminum, forged aluminum, and forged stainless steel and comes in sizes able to accommodate ⅜-inch to ¾-inch rope. This device can be free running or spring loaded. It consists of a sleeve, cam, pin, and spring (for spring-loaded types). The capacity of a cam ascender depends on its construction. The typical breaking strength for a ½-inch size is as follows:

- 2550 lb: cast aluminum
- 5000 lb: forged aluminum
- 5400 lb: forged stainless

The application of any ascender as a pulling cam or braking cam should be done only in accordance with the manufacturer's recommendations. As a pulling cam, this device pulls the rope into motion. As a braking cam, it stops the rope from moving. When installing a cam ascender on the line, remember that the arrow points toward the load. This positioning should be double-checked by pulling in the intended direction.

Do not drop or throw any ascender. Inspect the cam ascender by looking for worn cam teeth, egg-shaped holes for pin placement, cracks around the holes for pin placement, and a worn cord or chain holding the pin and cam to the sleeve. Keep the cam connected to the sleeve with the pin when storing the unit. Make sure that the pin goes through both sides of the sleeve during use. Keep the cam clean by wiping it with a cloth.

Rope grab devices are another auxiliary equipment system component. They are used to grasp a life safety rope for the purpose of supporting and catching single-person loads. Such a device can also be used for ascending a fixed line. NFPA 1983 specifications state that a rope grab device must withstand a minimum test load of at least 2400 lbf without incurring permanent damage to the device or the rope.

Life Safety Harnesses

Life safety harnesses are used as a quick clip-in point for a belay or emergency rappel, as fall protection, as a work platform, and as a means of transporting victims. Only harnesses certified as NFPA 1983 compliant should be used for rescue work. In NFPA 1983, harnesses are classified as one of three types:

- Class I: Fastens around the waist and thighs or buttocks. Designed for emergency escape with a one-person load.
- Class II: Fastens around the waist and thighs or buttocks. Designed for rescue where a two-person load may be encountered Figure 2-60 ▼.
- Class III: Fastens around the waist and thighs or buttocks and over the shoulders. Designed for rescue where two-person loads and/or inverting may occur Figure 2-61 ▶.

Many rescue organizations use only Class III harnesses (often referred to as *full-body harnesses*) because there is always the possibility of inverting the load, whether intentionally or inadvertently.

Carefully inspect rescue harnesses on a routine basis, as well as before and after use. It is ultimately the responsibility of the user to inspect the harness. Look for worn or broken stitching and rivets torn out of the holes. Check the material for damage from abrasion, cuts, or chemicals. Inspect the hardware

Figure 2-59 Ascent device.

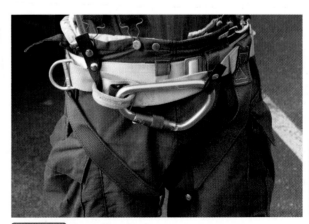

Figure 2-60 Class II life safety harness.

Figure 2-61 Class III life safety harness.

Table 2-1	Effect of Knots on Rope Strength	
Group	**Knot**	**Reduction in Strength**
Loop knots	Figure eight on a bight	20 percent
	Figure eight with a follow-through	19 percent
	Bowline	33 percent

The following knots, bends, and hitches are covered in this chapter:

- Safety knot (overhand knot)
- Clove hitch
- Square knot
- Figure eight
- Figure eight on a bight
- Figure eight with a follow-through
- Double-loop figure eight
- Figure eight bend
- Butterfly knot
- Double fisherman's knot
- Prusik hitch
- Water knot
- Load-release hitch (mariner's hitch)

Terminology

Specific terminology is used to refer to the parts of a rope in describing how to tie knots **Figure 2-62 ▾**. The **working end** is the part of the rope used for forming the knot. The **running end** is the part of the rope used for lifting or hoisting a load. The **standing part** is the rope between the working end and the running end.

A bight is formed by reversing the direction of the rope to form a U bend with two parallel ends. A loop is formed by making a circle in the rope. A round turn is formed by making a loop and then bringing the two ends of the rope parallel to each other **Figure 2-63 ▶**.

components for missing or damaged parts, dents, cracks, or excessive wear. If the harness looks damaged or unsafe, do not use it!

Knots, Bends, and Hitches

Knots, bends, and hitches are prescribed ways of fastening lengths of rope or webbing to objects or to each other. As a rescuer, you must know how to tie and when to use them. It is not unusual for knots, bends, and hitches to be used for one or more particular purposes.

Hitches, such as the clove hitch and prusik hitch, are used to attach a rope around an object or another rope. **Knots**, such as the figure eight and water knot, are used to form loops. **Bends**, such as the figure eight bend, are used to join two ropes together. **Safety knots**, such as the overhand knot, are used to secure the ends of ropes to prevent them from coming untied.

Any knot will reduce the load-carrying capacity of the rope by a certain percentage **Table 2-1 ▶**. You can avoid an unnecessary reduction in rope strength if you know which type of knot to use and how to tie it correctly.

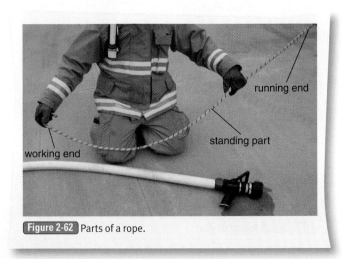

running end

standing part

working end

Figure 2-62 Parts of a rope.

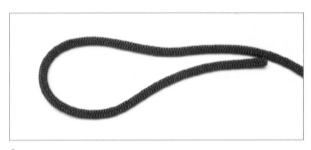

A.

B.

C.

Figure 2-63 **A.** A bight. **B.** A loop. **C.** A round turn.

Safety Knot

A safety knot (also referred to as an overhand knot or a keeper knot) is used to secure the leftover working end of the rope or webbing to the standing part. It provides a degree of safety to ensure that the primary knot will not come undone. A safety knot should always be used to finish the other basic knots.

A safety knot is simply an overhand knot in the loose end of the rope that is made around the standing part of the rope. This secures the loose end and prevents it from slipping back through the primary knot. Follow the steps in **Skill Drill 2-1 ▶** to tie a safety knot:

1. Take the loose end of the rope, beyond the knot, and form a loop around the standing part of the rope or webbing (**Step 1**).
2. Pass the loose end of the rope or webbing through the loop (**Step 2**).

3. Tighten the safety knot by pulling on both ends at the same time (**Step 3**). To test whether you have tied a safety knot correctly, try sliding it on the standing part. A knot that is tied correctly will slide.

Loop Knots

Loop knots are used to form a loop in the end of a rope. These loops may be used for securing a person during a rescue, for securing a rope to a fixed object, or for identifying the end of a rope stored in a rope bag. When tied properly, these knots will not slip and are easy to untie.

Figure Eight

A figure eight is a basic knot used to produce a family of other knots, including the figure eight on a bight and the figure eight with a follow-through. A simple figure eight knot is seldom used alone except when used as a stopper knot at the end of a rope. Follow the steps in **Skill Drill 2-2 ▶** to tie a figure eight:

1. Form a bight in the rope (**Step 1**).
2. Loop the working end of the rope completely around the standing end of the rope (**Step 2**).
3. Thread the working end of the rope through the bight (**Step 1**).
4. Tighten the knot by pulling on both ends simultaneously (**Step 4**). When you pull the knot tight, it will have the shape of a figure eight.

Figure Eight on a Bight

The figure eight on a bight knot creates a secure loop at the working end of a rope. The loop can be used to attach the end of the rope to a fixed object or a piece of equipment, to tie a life safety rope around a person, or to create loops for anchoring systems. The loop may be any size—from an inch to several feet in diameter. Follow the steps in **Skill Drill 2-3 ▶** to tie a figure eight on a bight:

1. Form a bight on the working end of the rope approximately 12 inches long (**Step 1**). The bight will now become the working end of the rope.
2. Loop the working end of the rope completely around the standing end of the rope (**Step 2**).
3. Thread the working end of the rope through the bight created when you doubled the rope in Step 2 (**Step 3**).
4. Tighten the knot by pulling on both ends simultaneously (**Step 4**). When you pull the knot tight, it will have the shape of a figure eight with a loop on the working end.
5. Secure the loose end of the rope with a safety knot (**Step 5**).

Figure Eight with Follow-Through

A figure eight with a follow-through knot creates a secure loop at the end of a rope when the working end must be wrapped around an object or passed through an opening before the loop can be formed. It is very useful for attaching a rope to a fixed ring or a solid object with an "eye." This knot can also be used to tie the ends of two ropes together securely.

Follow the steps in **Skill Drill 2-4 ▶** to tie a figure eight follow-through knot:

1. Tie a figure eight approximately 2 feet (or more if needed) (**Step 1**). Leave this knot loose.

Skill Drill 2-1

Tying a Safety Knot

1. Take the loose end of the rope, beyond the knot, and form a loop around the standing part of the rope or webbing.

2. Pass the loose end of the rope or webbing through the loop.

3. Tighten the safety knot by pulling on both ends at the same time.

2. Thread the working end through or around the object to which you want to attach the rope (**Step 2**).

3. Secure the working end by tracing the rope through the path of the original figure eight from the opposite end (**Step 3**).

4. Once threaded completely through, tighten the knot and place a safety knot on the loose end (**Step 4**).

Double-Loop Figure Eight

The double-loop figure eight is used when there is a desire for greater strength in the loop itself, when constructing a self-equalizing anchor system where loops of two different sizes are needed, or when it is desirable to incorporate a ring directly into the knot (sometimes used for stretcher attachments). Follow the steps in **Skill Drill 2-5 ▶** to tie a double-loop figure eight knot:

1. Form a bight approximately 18 inches long on the working end of the rope (**Step 1**). The bight will now become the working end of the rope.

2. Loop the working end of the rope completely around the standing end of the rope (**Step 2**).

3. Lay the working end of the rope over the bight created when you looped the rope in Step 2 (**Step 3**).

4. Pass your hand through the loop created in Step 1 and grab the working end of the rope (both pieces) through the bight created in Step 2 (**Step 4**).

5. Pass the loop over the loop created in Step 4 and on to the top of the knot (**Step 5**).

6. Tighten the knot by pulling on both ends simultaneously (**Step 6**).

7. When you pull the knot tight, it will have the shape of a figure eight with a double loop on the working end (**Step 7**). If you need loops of different lengths, adjust the loops before tightening.

Figure Eight Bend

This knot is used to join two ropes together. Follow the steps in **Skill Drill 2-6 ▶** to tie a figure eight bend:

1. Tie a figure eight near the end of one rope (**Step 1**).

2. Thread the end of the second rope completely through the knot from the opposite end (**Step 2**).

3. Pull the knot tight (**Step 3**).

4. Tie a safety knot on the loose end of each rope to the standing part of the other (**Step 4**).

Butterfly Knot

The butterfly knot is used to create a loop in the middle of the rope. This can be advantageous when you need a second point of connection for additional safety for the rescuer or victim. Follow the steps in **Skill Drill 2-7 ▶** to tie a butterfly knot:

Skill Drill 2-2

Tying a Figure Eight Knot

1 Form a bight in the rope.

2 Loop the working end of the rope completely around the standing end of the rope.

3 Thread the working end of the rope through the bight.

4 Tighten the knot by pulling on both ends simultaneously.

Skill Drill 2-3

Tying a Figure Eight on a Bight

1 Form a bight approximately 12 inches long on the working end of the rope.

2 Loop the working end of the rope completely around the standing end of the rope.

3 Thread the working end of the rope through the bight created when you doubled the rope in Step 2.

4 Tighten the knot by pulling on both ends simultaneously.

5 Secure the loose end of the rope with a safety knot.

Skill Drill 2-4

Tying a Figure Eight with Follow-Through

1 Tie a figure eight approximately 2 feet (or more if needed). Leave this knot loose.

2 Thread the working end through or around the object to which you want to attach the rope.

3 Secure the working end by tracing the rope through the path of the original figure eight from the opposite end.

4 Once threaded completely through, tighten the knot and place a safety knot on the loose end.

Skill Drill 2-5

Tying a Double-Loop Figure Eight

1 Form a bight approximately 18 inches long on the working end of the rope.

2 Loop the working end of the rope completely around the standing end of the rope.

3 Lay the working end of the rope over the bight created when you looped the rope in Step 2.

4 Pass your hand through the loop created in Step 1 and grab the working end of the rope (both pieces) through the bight created in Step 2.

5 Pass the loop over the loop created in Step 4 and on to the top of the knot.

6 Tighten the knot by pulling on both ends simultaneously.

7 When you pull the knot tight, it will have the shape of a figure eight with a double loop on the working end.

Skill Drill 2-6

Tying a Figure Eight Bend

1 Tie a figure eight near the end of one rope.

2 Thread the end of the second rope completely through the knot from the opposite end.

3 Pull the knot tight.

4 Tie a safety knot on the loose end of each rope to the standing part of the other.

Skill Drill 2-7

Tying a Butterfly Knot

1 Loop the rope around your hand three times.

2 Pass the rope closest to your fingers over the second loop and place it next to the first loop.

3 Grasp the loop now closest to your fingers, pass it over the other loops toward your wrist, and slide this loop beneath the others along your palm until a loop is created on your fingers. Pull the loop and ends in opposite directions to snug the knot up.

4 To complete the knot, grasp the two ends and snap them by quickly pulling in opposite directions.

1. Loop the rope around your hand three times (**Step 1**).
2. Pass the rope closest to your fingers over the second loop and place it next to the first loop (**Step 2**).
3. Grasp the loop now closest to your fingers, pass it over the other loops toward your wrist, and slide this loop beneath the others along your palm until a loop is created on your fingers. Pull the loop and ends in opposite directions to snug the knot up (**Step 3**).
4. To complete the knot, grasp the two ends and snap them by quickly pulling in opposite directions (**Step 4**).

Square Knot

The square knot is used in some victim packaging systems. It must always be backed up with safety knots. Follow the steps in **Skill Drill 2-8** to tie a square knot:

1. Create a bight on one end of the rope (**Step 1**).
2. Pass the other end of the rope through the bight (**Step 2**).
3. Pass the working end around the legs of the bight, and then back through the bight (**Step 3**).
4. A properly tied square knot will have both legs of the same rope passing through the bight from the same side (**Step 4**).

Double Fisherman's Knot

The double fisherman's knot is most commonly used to create a prusik loop, but may also be used to join two ropes. A single fisherman's knot is also used as a safety knot by some organizations. Because it is self-tightening, the double fisherman's knot is one of the few that does not require a safety knot. Follow the steps in **Skill Drill 2-9** to tie a double fisherman's knot:

1. Create a loop with the rope or prusik cord with an approximate 9- to 12-inch overlap (**Step 1**).
2. Place your hand out, palm down, index finger extended, and lay the overlap on your curled fingers (**Step 2**). Grasp the working end by your thumb and wrap it around your index finger from front to back.
3. Again grasp the working end by your thumb and wrap it around your index finger from front to back, but crossing over the first loop in the direction of your hand (**Step 3**).
4. Pass the working end through the loops where your index finger has been while removing your index finger (**Step 4**).
5. Tighten the knot by grasping each end and pulling simultaneously (**Step 5**).
6. Reverse the loop and repeat Steps 1–5.
7. The knot is properly tied and dressed if one side has four parallel lines and the opposite side has two X's. There should be approximately 1 inch of rope exposed at each end of the knot (**Step 6**).

Water Knot

The water knot is used to create a loop in webbing. These loops are used for a variety of purposes, including use as anchors and load-release hitches. Follow the steps in **Skill Drill 2-10** to tie a water knot:

1. In one end of the webbing, approximately 6 inches from the end, tie an overhand knot (**Step 1**).
2. With the other end of the webbing, start retracing from the working end through the knot until approximately 6 inches is left on the other end (**Step 2**).
3. Tie an overhand knot on each tail as a safety (**Step 3**).

Hitches

Hitches are knots that wrap around an object. They are used to secure the working end of a rope or webbing to a solid object.

Clove Hitch

A clove hitch is used in rescue primarily as a stretcher tie-off or in ladder rescue situations. It can be tied anywhere in a rope and will hold equally well if tension is applied to either end of the rope or to both ends simultaneously. There are two different methods of tying this knot. A clove hitch tied in the open is used when the knot can be formed and then slipped over the end of an object. If the object is too large or too long to slip the clove hitch over one end, the same knot can be tied around the object.

Follow the steps in **Skill Drill 2-11** to tie a clove hitch in the open:

1. Make the first loop with the left hand and have the running part of the rope pass over the working part (**Step 1**).
2. Make the second loop with the right hand and have the running part of the rope pass under the working part (**Step 2**).
3. Place the right-hand loop behind the left-hand loop so that the openings are aligned (**Step 3**).
4. Hold the two loops together and slip them over the object (**Step 4**).
5. Tighten the clove hitch by pulling both ends of the rope simultaneously in opposite directions (**Step 5**).

If the object is too large, is too long, or is closed with no end (a stretcher rail for example), follow the steps in **Skill Drill 2-12** to tie a clove hitch around an object:

1. Loop the rope completely around the object, with the working end below the running end (**Step 1**).
2. Loop the working end around the object a second time to form a second loop, slightly above the first (**Step 2**).
3. Pass the working end under the second loop, just above the point where the second loop crosses over the first loop (**Step 3**).
4. Secure the knot by pulling on both ends (**Step 3**).
5. Tie a safety knot on the working end of the rope (**Step 4**).

Prusik Hitch

The prusik hitch is used as a rope grab. It can provide a second point of connection for a rescuer on a rope or be used as a braking system. Dual wraps are used for personal use, whereas triple-wrapped, tandem prusiks are used where multi-person loads might be experienced (systems). A prusik hitch will slide when loose, but grab when loaded. This hitch is also a favorite of many rescue teams because of its low cost and ability to slide a little when shock loaded, thereby reducing the amount of shock placed on the rope system. A properly constructed prusik hitch will be dressed (i.e., tightened and all twists, kinks, and slack removed from the rope) to create maximum surface contact between the prusik cord and the rope.

Skill Drill 2-8

Tying a Square Knot

1   Create a bight on one end of the rope.

2   Pass the other end of the rope through the bight.

3   Pass the working end around the legs of the bight, and then back through the bight.

4   A properly tied square knot will have both legs of the same rope passing through the bight from the same side.

Skill Drill 2-9

Tying a Double Fisherman's Knot

1 Create a loop with the rope or prusik cord with an approximate 9- to 12-inch overlap.

2 Place your hand out, palm down, index finger extended, and lay the overlap on your curled fingers. Grasp the working end by your thumb and wrap it around your index finger from front to back.

3 Again grasp the working end by your thumb and wrap it around your index finger from front to back, but crossing over the first loop in the direction of your hand.

4 Pass the working end through the loops where your index finger has been while removing your index finger.

5 Tighten the knot by grasping each end and pulling simultaneously. Reverse the loop and repeat Steps 1–5.

6 The knot is properly tied and dressed if one side has four parallel lines and the opposite side has two X's. There should be approximately 1 inch of rope exposed at each end of the knot.

Skill Drill 2-10

Tying a Water Knot

1 In one end of the webbing, approximately 6 inches from the end, tie an overhand knot.

2 With the other end of the webbing, start retracing from the working end through the knot until approximately 6 inches is left on the other end.

3 Tie an overhand knot on each tail as a safety.

Follow the steps in **Skill Drill 2-13 ▶** to tie a prusik hitch:

1. Place the knot over the rope to be attached (**Step 1**).
2. Roll the knot around the rope two or three times depending on the intended use (**Step 2**).
3. Grab the prusik cord on one side of the knot and pull so that a loop is created with the knot along the side of the loop (**Step 3**).
4. Dress the knot so that there is maximum surface contact between the prusik cord and the rope. You can tell it is correct when you see parallel lines with no crossover of the prusik cord (**Step 4**).

Load-Release Hitch (Mariner's Hitch)

The load-release hitch (mariner's hitch) is actually a knot system that can be released while under a load. It is used in low-angle belay systems to attach the brake to the anchor. Such a hitch can be used when a brake system gets stuck or is holding a load or when it is necessary to switch the load from one line to another. The load-release hitch is constructed from an anchor strap or webbing loop. It is not the preferred load-release device for use in a high-angle environment unless it is used with a shock-absorbing device with the appropriate rating for the anticipated load.

Follow the steps in **Skill Drill 2-14 ▶** to tie a load-release hitch:

1. Clip one end of the strap or loop into a general-use carabiner and lock the carabiner (**Step 1**).
2. Loop the strap through a second locked carabiner, leaving approximately 12 inches between the carabiners. Bring the strap or webbing up through the first carabiner (**Step 2**).
3. Wrap the strap or webbing in a spiral fashion around the length of the strap or webbing until you reach the opposite end (**Step 3**).
4. Pass the end of the loop through the lengthwise straps or webbing, and secure the loop to the other carabiner (**Step 4**). Lock the carabiner.

Follow the steps in **Skill Drill 2-15 ▶** to release a load-release hitch:

1. Disconnect the carabiner attached to the loop that was passed through the lengthwise strap (**Step 1**).
2. Carefully start to loosen the spiral wraps until you feel movement in the strap or webbing. Continue until the load has been released (**Step 2**).

Skill Drill 2-11

Tying a Clove Hitch (Open Object)

1. Make the first loop with the left hand and have the running part of the rope pass over the working part.

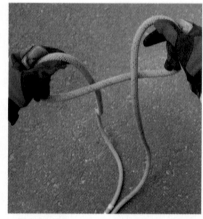

2. Make the second loop with the right hand and have the running part of the rope pass under the working part.

3. Place the right-hand loop behind the left-hand loop so that the openings are aligned.

4. Hold the two loops together and slip them over the object.

5. Tighten the clove hitch by pulling both ends of the rope simultaneously in opposite directions.

■ Dressing Knots

A knot should be properly "dressed" by tightening and removing twists, kinks, and slack from the rope. The finished knot should be firmly fixed in position. The configuration of a properly dressed knot should be evident so that it can be inspected easily. All loose ends should be secured by safety knots to ensure that the primary knot cannot be released accidentally.

With practice, you should be able to tie these knots in the dark, while wearing heavy gloves, and behind your back. Knot-tying skills can be lost quickly without practice.

Skill Drill 2-12

Tying a Clove Hitch (Closed Object)

1 Loop the rope completely around the object, with the working end below the running end.

2 Loop the working end around the object a second time to form a second loop, slightly above the first. Pass the working end under the second loop, just above the point where the second loop crosses over the first loop.

3 Secure the knot by pulling on both ends.

4 Tie a safety knot on the working end of the rope.

Skill Drill 2-13

Tying a Prusik Hitch

1 Place the knot over the rope to be attached.

2 Roll the knot around the rope two or three times depending on the intended use.

3 Grab the prusik cord on one side of the knot and pull so that a loop is created with the knot along the side of the loop.

4 Dress the knot so that there is maximum surface contact between the prusik cord and the rope. You can tell it is correct when you see parallel lines with no crossover of the prusik cord.

Skill Drill 2-14

Tying a Load-Release Hitch

1. Clip one end of the strap or loop into a general-use carabiner and lock the carabiner.

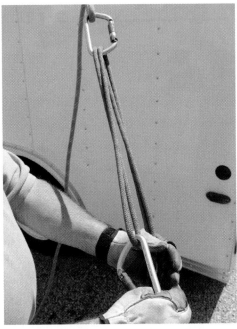

2. Loop the strap through a second locked carabiner, leaving approximately 12 inches between the carabiners. Bring the strap or webbing up through the first carabiner.

3. Wrap the strap or webbing in a spiral fashion around the length of the strap or webbing until you reach the opposite end.

4. Pass the end of the loop through the lengthwise straps or webbing, and secure the loop to the other carabiner. Lock the carabiner.

Skill Drill 2-15

Releasing a Load-Release Hitch

1. Disconnect the carabiner attached to the loop that was passed through the lengthwise strap.

2. Carefully start to loosen the spiral wraps until you feel movement in the strap or webbing. Continue until the load has been released.

Wrap-Up

Chief Concepts

- The type of rescue vehicle is determined less by the size of the vehicle and more by the equipment carried.
- Rescuers should be knowledgeable about the appropriate personal protective equipment to be used in rescue, ensure this equipment is available and in proper working order, and, most importantly, always use the PPE.
- While it is almost impossible to have every piece of rescue equipment ever manufactured available at a rescue scene, maintaining a good selection of the various types will allow for an efficient and safe operation for the vast majority of incidents. Having the equipment is only part of the equation, however: Proper training and maintenance are equally critical.
- Rope rescue equipment consists of hardware and software, which are combined to make a system for raising or lowering rescuers and victims.
- Rescue knots are used not only in rope rescue systems, but also in a variety of other situations where rope is used.

Hot Terms

A-frame A piece of equipment designed to provide vertical lifting capabilities at a point away from a rescue vehicle. It consists of two poles that are connected at the working end, with the base of each pole being connected to the vehicle, and is stabilized by one or more cables attached to the vehicle.

Bend A knot that joins two ropes or webbing pieces together.

Block creel construction A type of rope without knots or splices in the yarns, ply yarns, strands, or braids.

Boom A piece of equipment designed to provide vertical lifting capabilities at a point away from a rescue vehicle. It consists of a telescoping beam that can be rotated 360 degrees.

Cascade system A compressed air system used for refilling SCBA bottles and/or regulating for direct supply to supplied-air respirators.

Dynamic rope A rope that is designed to be elastic and will stretch when it is loaded.

Extrication The removal of trapped victims from an entrapment.

Gin pole A piece of equipment designed to provide vertical lifting capabilities at a point away from a rescue vehicle. It consists of a single pole that attaches to a rescue vehicle, with stabilizing cables being attached to the vehicle on either side.

Hitch A knot that attaches to or wraps around an object such that when the object is removed, the knot will fall apart.

Infection control plan A written plan that provides a comprehensive infection control system to maximize protection against communicable diseases.

Kern The center or core of a rope; it provides approximately 70 percent of the strength of the rope.

Knot A fastening made by tying together lengths of rope or webbing in a prescribed way.

Life safety harness A system component that is an arrangement of materials secured about the body and used to support a person during rescue.

Life safety rope A compact but flexible, torsionally balanced, continuous structure of fibers produced from strands that are twisted, plaited, or braided together. It serves primarily to support a load or transmit a force from the point of origin to the point of application.

Mantle The sheath of a rope; the braided covering that protects the core of the rope from dirt and abrasion. It provides 30 percent of the strength of the rope.

Personal protective equipment (PPE) The equipment provided to shield or isolate personnel from infectious, chemical, physical, and thermal hazards.

Pulley A metal sheave (wheel) mounted on a bearing or metal bushing used during rope rescue to change direction, provide a mechanical advantage, reduce friction over an edge, or provide rope tension.

Respiratory protection A protective device to provide safe breathing air to a user in a hostile or dangerous atmosphere.

Respiratory protection program A written program, with workplace-specific procedures addressing the major elements of the program, established whenever respirators are necessary to protect the health of the employee.

Rope rescue hardware Rigid mechanical auxiliary equipment that can include, but is not limited to, anchor plates, carabiners, and mechanical ascent and descent control devices.

Rope rescue software A flexible fabric component of rope rescue equipment that can include, but is not limited to, anchor straps, pick-off straps, and rigging slings.

Rope rescue system A system consisting of rope rescue equipment and an appropriate anchor system intended for use in the rescue of a subject.

Running end The part of a rope used for lifting or hoisting a load.

Safety knot A type of knot used to secure the ends of ropes to prevent them from coming untied.

Standing part The part of a rope between the working end and the running end when tying a knot.

Static rope A rope that is designed not to stretch when it is loaded.

Utility rope A rope used for securing objects, for hoisting equipment, or for securing a scene to prevent bystanders from being injured. Utility rope is not used for life safety purposes or to support a person's weight in any way.

Victim packaging and removal equipment Equipment used in the process of securing a victim in a transfer device, with regard to existing and potential injuries or illness, so as to prevent further harm to the victim during movement.

Working end The part of the rope used for forming a knot.

Rescue Responder *in Action*

The chief of your department has asked you to be part of a committee that will be making recommendations on the purchase of a new rescue vehicle and equipment. He tells you that selection of equipment needs to be decided before the appropriate vehicle type can be chosen.

1. What would your first consideration be when starting to think about which equipment should be purchased?

   A. Cost

   B. The types of incidents to which your department is called

   C. The locations of other rescue units in your area and the types of equipment and training they have

   D. All of the above

2. Which type of tools are most commonly found on rescue vehicles?

   A. Pneumatic tools

   B. Hand tools

   C. Electrical tools

   D. Hydraulic tools

After the equipment has been chosen, the chief tells the committee that it must be very careful about costs. He is under pressure from the financial office to keep costs as low as possible.

3. What would be your first consideration when choosing the size of the vehicle?
   A. How many people the vehicle can carry
   B. How fast the vehicle will go
   C. Whether the vehicle will carry the necessary tools and equipment safely
   D. Whether the vehicle will provide easy access to the tools

4. You have come up with a specification for the new vehicle, and it costs more than the financial office wants to spend. The chief agrees with the committee's recommendation and has asked for its help in convincing the financial office that it should spend the extra money. What can the committee do to help the chief?
   A. Tell him he should play politics to get the vehicle
   B. Provide him with a list of justifications for the equipment, explaining why this vehicle is needed
   C. Go back to the drawing board to see if anything can be cut out
   D. Have the chief advise the fiscal office that it would be a good idea that financial personnel get some basic rescue training so they will understand the committee's concerns

Rescue Incident Management

NFPA 1670 Standard

4.1 General.

4.1.1* The authority having jurisdiction (AHJ) shall establish levels of operational capability needed to conduct operations at technical search and rescue incidents safely and effectively, based on hazard identification, risk assessment, training level of personnel, and availability of internal and external resources. [p. 70–85]

4.1.4 The AHJ shall establish written standard operating procedures (SOPs) consistent with one of the following operational levels for each of the disciplines defined in this document:

(1)* *Awareness Level.* This level represents the minimum capability of organizations that provide response to technical search and rescue incidents.

(2)* *Operations Level.* This level represents the capability of organizations to respond to technical search and rescue incidents and to identify hazards, use equipment, and apply limited techniques specified in this standard to support and participate in technical search and rescue incidents.

(3) *Technician Level.* This level represents the capability of organizations to respond to technical search and rescue incidents and to identify hazards, use equipment, and apply advanced techniques specified in this standard necessary to coordinate, perform, and supervise technical search and rescue incidents. [p. 84]

4.1.6 The AHJ shall establish operational procedures consistent with the identified level of operational capability to ensure that technical search and rescue operations are performed in a manner that minimizes threats to rescuers and others. [p. 83–84]

4.1.7 The same techniques used in a search and rescue operation shall be considered equally useful for training, body recovery, evidence search, and other operations with a level of urgency commensurate with the risk/benefit analysis. [p. 77–84]

4.1.8 Operational procedures shall not exceed the identified level of capability established in 4.1.4. [p. 84]

4.1.9* At a minimum, medical care at the basic life support (BLS) level shall be provided by the organization at technical search and rescue incidents. [p. 77]

4.1.10 Training.

4.1.10.1 The AHJ shall provide for training in the responsibilities that are commensurate with the operational capability of the organization. [p. 76]

4.1.10.1.1 The minimum training for an organization shall be at the awareness level. [p. 76–77, 84]

4.1.10.1.2 Organizations expected to perform at a higher operational level shall be trained to that level. [p. 76–77]

4.1.10.2* The AHJ shall provide for the continuing education necessary to maintain all requirements of the organization's identified level of capability. [p. 77]

4.1.10.3 An annual performance evaluation of the organization based on requirements of this standard shall be performed. [p. 84]

4.1.10.4* The AHJ shall evaluate its training program to determine whether the current training has prepared the organization to function at the established operational level under abnormal weather conditions, extremely hazardous operational conditions, and other difficult situations. [p. 73–84]

4.1.10.5* Documentation.

4.1.10.5.1 The AHJ shall be responsible for the documentation of all required training. [p. 84]

4.1.10.5.2 This documentation shall be maintained and available for inspection by individual team members and their authorized representatives. [p. 84]

4.1.11 Prior to operating at a technical search and rescue incident, an organization shall meet the requirements specified in Chapter 4 as well as all relevant requirements of Chapters 6 through 15 for the specific technical rescue incident. [p. 77]

4.1.12 Standard Operating Procedure.

4.1.12.1 The AHJ shall ensure that there is a standard operating procedure to evacuate members from an area and to account for their safety when an imminent hazard condition is discovered. [p. 84]

4.1.12.2 This procedure shall include a method to notify all members in the affected area immediately by any effective means, including audible warning devices, visual signals, and radio signals. [p. 84]

4.1.13* The AHJ shall comply with all applicable local, state, and federal laws. [p. 84]

4.1.14* The AHJ shall train responsible personnel in procedures for invoking, accessing, and using relevant components of the *U.S. National Search and Rescue Plan*, the *U.S. National Response Framework*, and other national, state, and local response plans, as applicable. [p. 84]

4.2 Hazard Identification and Risk Assessment.

4.2.1* The AHJ shall conduct a hazard identification and risk assessment of the response area and shall determine the feasibility of conducting technical search and rescue operations. [p. 75–76]

4.2.2 The hazard identification and risk assessment shall include an evaluation of the environmental, physical, social, and cultural factors influencing the scope, frequency, and magnitude of a potential technical search and rescue incident and the impact they might have on the ability of the AHJ to respond to and to operate while minimizing threats to rescuers at those incidents. [p. 74–79]

4.2.3* The AHJ shall identify the type and availability of internal resources needed for technical search and rescue incidents and shall maintain a list of those resources. [p. 76]

4.2.4* The AHJ shall identify the type and availability of external resources needed to augment existing capabilities for technical search and rescue incidents and shall maintain a list of these resources, which shall be updated at least once a year. [p. 76–78]

4.2.5* The AHJ shall establish procedures for the acquisition of those external resources needed for technical search and rescue incidents. [p. 76–78]

4.2.6 The hazard identification and risk assessment shall be documented. [p. 78–79]

4.2.7 The hazard identification and risk assessment shall be reviewed and updated on a scheduled basis and as operational or organizational changes occur. [p. 78]

4.2.8 At intervals determined by the AHJ, the AHJ shall conduct surveys in the organization's response area for the purpose of identifying the types of technical search and rescue incidents that are most likely to occur. [p. 78]

4.3   Incident Response Planning.

4.3.1 The procedures for a technical search and rescue emergency response shall be documented in the special operations incident response plan. [p. 73]

4.3.1.1 The plan shall be a formal, written document. [p. 73]

4.3.1.2 Where external resources are required to achieve a desired level of operational capability, mutual aid agreements shall be developed with other organizations. [p. 76, 83]

4.3.2 Copies of the technical search and rescue incident response plan shall be distributed to agencies, departments, and employees having responsibilities designated in the plan. [p. 83]

4.3.3 A record shall be kept of all holders of the technical search and rescue incident response plan, and a system shall be implemented for issuing all changes or revisions. [p. 73]

4.3.4 The technical search and rescue incident response plan shall be approved by the AHJ through a formal, documented approval process and shall be coordinated with participating agencies and organizations. [p. 83–84]

4.4   4.4 Equipment.

4.4.1   4.4.1 Operational Equipment.

4.4.1.1* The AHJ shall ensure that equipment commensurate with the respective operational capabilities for operations at technical search and rescue incidents and training exercises is provided. [p. 76–78]

4.4.1.2 Training shall be provided to ensure that all equipment is used and maintained in accordance with the manufacturers' instructions. [p. 81]

4.4.1.3 Procedures for the inventory and accountability of all equipment shall be developed and used. [p. 70]

4.5   Safety.

4.5.1   General.

4.5.1.1 All personnel shall receive training related to the hazards and risks associated with technical search and rescue operations. [p. 74–77]

4.5.1.2 All personnel shall receive training for conducting search and rescue operations while minimizing threats to rescuers and using PPE. [p. 76]

4.5.1.3 The AHJ shall ensure that members assigned duties and functions at technical search and rescue incidents and training exercises meet the relevant requirements of the following chapters and sections of NFPA 1500, *Standard on Fire Department Occupational Safety and Health Program*:

(1) Section 5.4, Special Operations

(2) Chapter 7, Protective Clothing and Protective Equipment

(3) Chapter 8, Emergency Operations [p. 83]

4.5.1.4* Where members are operating in positions or performing functions at an incident or training exercise that pose a high potential risk for injury, members qualified in BLS shall be standing by. [p. 77]

4.5.1.5* Rescuers shall not be armed except when it is required to meet the objectives of the incident as determined by the AHJ. [p. 85]

4.5.2   Safety Officer.

4.5.2.1 At technical search and rescue training exercises and in actual operations, the incident commander shall assign a safety officer with the specific knowledge and responsibility for the identification, evaluation, and, where possible, correction of hazardous conditions and unsafe practices. [p. 72–73]

4.5.2.2 The assigned safety officer shall meet the requirements specified in Chapter 6, Functions of the Incident Safety Officer, of NFPA 1521, *Standard for Fire Department Safety Officer*. [p. 72–73]

4.5.3   Incident Management System.

4.5.3.1* The AHJ shall provide for and utilize training on the implementation of an incident management system that meets the requirements of NFPA 1561, *Standard on Emergency Services Incident Management System*, with written SOPs applying to all members involved in emergency operations. All members involved in emergency operations shall be familiar with the system. [p. 70]

4.5.3.2 The AHJ shall provide for training on the implementation of an incident accountability system that meets the requirements of NFPA 1561, *Standard on Emergency Services Incident Management System*. [p. 83]

4.5.3.3 The incident commander shall ensure rotation of personnel to reduce stress and fatigue. [p. 72]

4.5.3.4 The incident commander shall ensure that all personnel are aware of the potential impact of their operations on the safety and welfare of rescuers and others, as well as on other activities at the incident site. [p. 71–74]

4.5.3.5 At all technical search and rescue incidents, the organization shall provide supervisors who possess skills and knowledge commensurate with the operational level identified in 4.1.4. [p. 70]

4.5.4* **Fitness.** The AHJ shall ensure that members are psychologically, physically, and medically capable to perform assigned duties and functions at technical search and rescue incidents and to perform training exercises in accordance with Chapter 10 of NFPA 1500, *Standard on Fire Department Occupational Safety and Health Program*. [p. 82–83]

4.5.5 **Nuclear, Biological, and Chemical Response.**

4.5.5.1* The AHJ, as part of its hazard identification and risk assessment, shall determine the potential to respond to technical search and rescue incidents that might involve nuclear or biological weapons, chemical agents, or weapons of mass destruction, including those with the potential for secondary devices. [p. 76]

4.5.5.2 If the AHJ determines that a valid risk exists for technical search and rescue response into a nuclear, biological, and/or chemical environment, it shall provide training and equipment for response personnel. [p. 76]

Additional NFPA Standards

NFPA 1006, *Standard for Technical Rescuer Professional Qualifications*

NFPA 1250, *Recommended Practice in Emergency Service Organization Risk Management*

NFPA 1500, *Standard on Fire Department Occupational Safety and Health Program*

NFPA 1521, *Standard for Fire Department Safety Officer*

NFPA 1561, *Standard on Emergency Services Incident Management System*

Knowledge Objectives

After studying this chapter, you will be able to:

- Describe the characteristics of the Incident Command System (ICS) and the functions of positions within this system.
- Describe the function of, and list the components of, an incident action plan.
- Describe the administrative and operational aspects of an organization health and safety program.
- List and describe the components of an incident response plan.
- Describe the components of an operational risk/benefit analysis.
- Describe the purpose and benefits of standard operating procedures (SOPs).
- Describe the purpose, components, and benefits of performing a needs analysis.
- List various methods of personnel and equipment accountability, and describe the importance of having accountability systems.

Skill Objectives

After completing this chapter, you will be able to perform the following skills:

- Demonstrate an understanding of the Incident Command System and its functional areas as they relate to rescue.
- Perform a needs assessment for your response area.
- Perform a hazard and risk assessment for a given incident.

ou have been selected to be the first officer of your organization's newly approved rescue company. The first tasks you have been asked to complete are to determine the equipment and training priorities and to begin development of necessary standard operating procedures.

1. What do you believe are important considerations when making these determinations, and why?
2. What SOPs are most important, and why?

Introduction

It is not enough to be proficient in various rescue skills; you must also be able to operate as a member of a team. For a team to function properly, there must be a system in place where roles and responsibilities are identified and assigned. To this purpose, rescuers use an incident command system (ICS). As part of this hierarchy of responsibilities and duties, rescue responders must also be familiar with concepts such as safety, standard operating procedures (SOPs) or guidelines (SOGs), risk-versus-benefit analysis, hazard analysis, personal accountability systems, and equipment inventory, capability, and tracking systems. It is also expected that those involved in the management of the team will be knowledgeable in subjects such as needs identification and response planning.

The ICS used by an organization must meet the requirements of the federal National Incident Management System (NIMS) model. Although most emergency situations are handled by local response organizations, during a major incident those organizations may need help from other local jurisdictions, state responders, and/or the federal government. The NIMS management plan allows responders from different jurisdictions and disciplines to work together more efficiently during the response to natural disasters and emergencies, including acts of terrorism. Utilization of the NIMS plan allows for a standardized approach to incident management; standard command and management structures; and emphasis on preparedness, mutual aid, and **resource management**.

This chapter addresses these subjects, showing how they are interrelated. This chapter is not, however, intended to replace ICS training such as that offered by the U.S. Department of Homeland Security's Federal Emergency Management Agency (FEMA). It is expected that at least basic training in ICS will have already occurred prior to the reader taking this course.

Incident Command System

The **incident command system (ICS)** is a management structure that is based on business management principles and that provides a standard approach and structure to managing operations. Its use ensures that operations are smoothly coordinated, safe, and effective, especially when multiple agencies must work together. By employing an ICS to organize even simple rescue incidents, rescuers gain the necessary skills and confidence needed to manage even the most complex incidents. Although the components of the ICS inevitably reflect the specific needs of the incident and variations do exist, the overall concepts are the same no matter what the situation.

A rescue scene can be chaotic and confusing if a command system is not established early in the incident. The command system must be relatively familiar to the users if it is going to be usable, versatile, and expandable in a logical manner when changing conditions dictate, and it must be readily adaptable to fit any type or size of emergency or incident. Most technical rescue operations will be most efficiently managed without having a separate individual assigned to each functional area within the ICS. Instead, different individuals will take on the responsibilities associated with a number of the positions. This standardized approach is also critical to facilitate and coordinate the use of resources from multiple agencies so that all parties work in a coordinated fashion to achieve their common objectives.

Jurisdictional Authority

An effective ICS clearly defines which agency will be in charge of each incident. Although determination of the jurisdiction in charge is rarely a problem for a structure fire, larger-scale incidents may cross geographic or statutory boundaries. When situations arise where there are overlapping responsibilities, the ICS may employ a **unified command**. Unified command is often used in multijurisdictional or multiagency incident management. It allows agencies with different legal, geographic, and functional responsibilities to coordinate, plan, and interact effectively. In a unified command structure, multiple agency representatives make command decisions, instead of a single incident commander making all decisions on scene.

All-Risk and All-Hazard System

By design, ICS can be used equally well at all types of incidents, including nonemergency events, such as public gatherings

Figure 3-1 ▶. ICS can and should be used for everyday operations as well as major incidents; regular use of the system enhances rescue responders' familiarity with the standard procedures and terminology.

Unity of Command

In a properly run ICS, each person working at an incident has only one direct supervisor. All orders and assignments come directly from that supervisor, and all reports are made to the same supervisor. This supervisor then reports to his or her own supervisor, and so on up the chain of command.

Span of Control

The ICS allows for a manageable span of control of people and resources. In most situations, one person can effectively supervise only three to seven people (with five being the optimal number). In the ICS setting, the **incident commander (IC)** communicates with and receives information from a maximum of five people, rather than assuming responsibility for the assignment of all personnel at the scene. Individual managers of personnel and resources within ICS also work within a manageable span of control. The actual span of control will depend on the complexity of the incident and the nature of the work being performed.

Modular Organization

ICS is designed to be modular. Not all components of an ICS need to be utilized at every incident—only what is appropriate given the incident's nature and size. Additional components can be added or eliminated as needed as the incident unfolds. Some components are used on almost every incident, whereas others apply to only the largest and most complex situations.

Common Terminology

ICS promotes the use of common terminology both within an organization and among all the agencies involved in emergency incidents. This shared language eliminates confusion about what is intended when different things are called by the same name in different jurisdictions, counties, areas, or departments. For example, when identifying the geographical sides of a building, some organizations will identify each side by numbers (e.g., sides 1, 2, 3, or 4), while other organizations may use alphabetical references (e.g., alpha, bravo, charlie, or delta). Using a common terminology increases the level of understanding among the various response agencies working at an incident site.

Integrated Communications

ICS must support communication up and down the chain of command at every level. Messages must move efficiently throughout the system. This consideration is especially important because it ensures that control objectives established by the command staff are effectively implemented by task-level resources. Integrated communications is also necessary so that outcomes produced by these task-level units are reported back up the chain of command, allowing progress toward incident goals to be measured as the incident unfolds.

Figure 3-1 ICS can be used for both emergency and nonemergency events.

Consolidated Incident Action Plans

An ICS ensures that everyone involved in the incident is following the same overall plan. The incident action plan may be developed by the IC alone on smaller incidents or in collaboration with all agencies involved in larger incidents.

Designated Incident Facilities

Development of a standard terminology for commonly needed operational facilities improves operations because everyone knows what goes on at each facility. Examples of such standard terms include the following:

- Base
- Command post
- Staging area

Resource Management

A standard system of assigning and tracking the resources involved in the incident is of critical importance to ensuring an efficient and safe operation. At small-scale incidents, units and personnel usually respond directly to the scene and receive their assignments there. At large-scale incidents, units are often dispatched to a staging area, rather than going directly to the incident scene. Some units are assigned upon arrival, whereas others may be held in reserve, ready to be assigned if needed.

ICS Organization

The ICS structure identifies a full range of duties, responsibilities, and functions that are performed at emergency incidents.

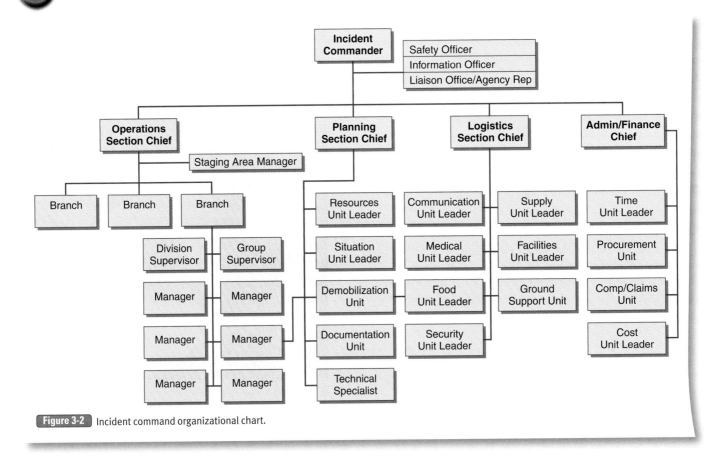

Figure 3-2 Incident command organizational chart.

Its hierarchy is best illustrated by the standard organizational chart shown in **Figure 3-2 ▲**, which clearly defines the positions within the ICS and the chain of command.

The IC is the individual with overall responsibility for the management of all incident operations, including rotating personnel to limit stress. This command position is always filled and is initially established by the first unit on the scene. Ultimately, command is likely to be transferred to the senior arriving officer, unless organization policy or circumstances dictate that someone else would be more appropriate in this position. During transfer of command, a brief current situation status report is given to the new IC and includes the following elements:

- Tactical priorities
- Action plans
- Hazardous or potentially hazardous conditions
- Accomplishments
- Assessment of effectiveness of operations
- Current status of resources

When multiple agencies with overlapping jurisdictions or legal responsibilities are involved in an incident, a unified command structure is sometimes used. The ICS allows for a single IC to manage an emergency incident that has limited complexity. In contrast, a unified command allows for representatives from each agency participating in the response to share command authority at large, complex emergencies; this power-sharing arrangement helps ensure cooperation, avoids confusion, and guarantees agreement on goals and objectives.

The IC does not need to be well versed in technical rescue; however, he or she should be thoroughly familiar with the ICS

management process. The IC should be stationed at a command post that is nearby (relative to the size and scope of the incident), yet set apart from the immediate incident scene. The command post does not need to be directly on scene but should be protected by enforcement of restricted access so the command staff can function without needless distractions or interruptions.

The IC also communicates directly with the other members of the command staff:

- Public information officer
- Safety officer
- Liaison officer

Public Information Officer

The **public information officer (PIO)** interacts with the media and provides a single point of contact for information related to the incident, thereby allowing the IC to focus on managing the incident. The PIO prepares for IC approval any press releases to be issued. Also, prior to any press briefings, the PIO may provide the IC with background information, suggest questions that may be asked, and assist with selection and coordination of photographers.

Safety Officer

The **safety officer** is responsible for enforcing general safety rules and developing measures for ensuring personnel safety. When human resources are limited, the safety officer position may be combined with the technical rescue officer and/or operations positions. The safety officer can bypass the chain of command when necessary to correct unsafe acts immediately and has the authority to stop or suspend unsafe operations, as is clearly stated in national standards such as NFPA 1500, NFPA 1521, and NFPA 1561. In this event, the safety officer must immediately report to the IC any action taken that may affect a strategic goal or tactical objective.

The safety officer, at a minimum, must be knowledgeable in the following areas:

- Strategy and tactics
- Hazardous materials
- Rescue practices
- Building construction and collapse potential
- Departmental safety rules and regulations

More specific responsibilities include the following duties:

- Monitoring environmental hazards
- Conducting an ongoing evaluation of physical and mental state of rescuers
- Ensuring that personnel working near the edge of an elevation are securely tied off
- Performing final safety checks on all knots and rigging
- Developing measures for ensuring personnel safety
- Monitoring sanitary conditions affecting rescuers, including the rescue environment, infectious waste disposal, eating conditions, and rest and sleeping habitat

Liaison Officer

The **liaison officer** is the IC's point of contact for outside agencies and coordinates information and resources between cooperating and assisting agencies. The liaison officer also establishes contacts with agencies that may be capable of providing support or are available to do so.

ICS Sections

Other than those tasks specifically assigned to the command staff, all activities at an incident can be relegated to one of the following four major ICS sections. Each section can be headed by a section chief or by the IC personally, depending on the size and complexity of the incident.

- Operations
- Planning
- Logistics
- Finance/administration

Operations

The **operations section** is responsible for development, direction, and coordination of all tactical operations conducted in accordance with an **incident action plan (IAP)** that outlines strategic objectives and the way in which operations will be conducted. On modest-sized rescue incidents, operations may fulfill the functions of a technical rescue officer. At large-scale rescues, a rescue branch director or rescue group supervisor may be assigned to handle these duties. Operations personnel may also interact with the media and other appropriate agencies as necessary in the absence of the PIO and liaison officer (as directed by the IC). Other operations responsibilities include the following:

- Stabilizing the scene
- Rescuing trapped individuals
- Treating patients

Under the operations chief, certain functional areas and positions are established to allow for a manageable span of control and to organize resources based on incident needs. The first of these areas is staging. All resources in the staging area must be available and ready for assignment. This area should not be used for storing out-of-service resources or for performing logistics functions. Staging areas may be relocated as necessary. After a staging area has been designated and named, a **staging area manager** will be assigned. The staging area manager will report to the operations section chief (or to the IC if an operations section chief has not been designated).

Divisions, groups, and branches are established when the number of resources exceeds the manageable span of control of the IC and the operations section chief. **Divisions** are established to divide an incident into physical or geographical areas of operation. For example, if a flood affected two different streets, then Division 1 might be street A and Division 2 street B. **Groups** are established to divide the incident into functional areas of operation (e.g., medical groups, search and rescue groups).

Branches may serve several purposes and may be either functional or geographic in nature. In general, they are established when the number of divisions or groups exceeds the recommended span of control for the operations section chief or for geographic reasons. Branches are identified by functional name or Roman numerals and are managed by a branch director. They may have deputy positions as required.

Planning

The **planning section** is responsible for the collection, evaluation, dissemination, and use of information and intelligence critical to the incident (unless the IC places this function elsewhere). One of the most important functions of the planning section is to look beyond the current and next operational period and anticipate potential problems, events, and logistical needs to execute the upcoming IAP. Other responsibilities include the following:

- Developing and updating the IAP
- Examining the current situation
- Reviewing available information
- Predicting the probable cause of events
- Preparing recommendations for strategies and tactics
- Maintaining resource status
- Maintaining and displaying situation status
- Providing documentation services

Technical specialists initially report to the planning section and work within that section or are reassigned to another part of the organization. These advisors have the special skills required at the incident. These skills can be in any discipline required—for example aviation, environment, hazardous materials, and engineering.

Logistics

The **logistics section** is responsible for all support requirements needed to facilitate effective and efficient incident management, including providing supplies, services, facilities, and materials during the incident. Key responsibilities include the following duties:

- Communications
- Medical support to incident personnel
- Food for incident personnel
- Supplies
- Facilities
- Ground support

The service branch within the logistics section may include the following units:

- **Communications Unit:** Develops plans for use of incident communications equipment and facilities, and installs, tests, distributes, and maintains communications equipment. This unit also supervises the incident communications center.
- **Medical Unit:** Develops the medical plan, obtains medical aid and transportation for injured and ill incident personnel, and prepares reports and records.
- **Food Unit:** Supplies the food needs for the incident.

The support branch within the logistics section may include the following units:

- **Supply Unit:** Obtains, stores, and maintains an inventory of supplies needed for an incident and services supplies and equipment as necessary. This unit also orders personnel, equipment, and supplies.
- **Facilities Unit:** Is responsible for the layout and use of incident facilities and provides sleeping and sanitation facilities for incident personnel. The facilities unit is also responsible for managing base and camp operations.
- **Ground Support Unit:** Transports personnel, supplies, food, and equipment and implements the traffic plan for the incident. This unit also fuels, services, maintains, and repairs vehicles and other ground support equipment.

Finance/Administration

The **finance/administration section** is responsible for the accounting and financial aspects of an incident, as well as any legal issues that may arise. While not staffed at most incidents, this position accounts for all activities and ensures enough money is made available to keep operations running. The following units are contained within the finance/administration branch:

- **Time Unit:** Is responsible for equipment and personnel time recording.
- **Procurement Unit:** Administers all financial matters pertaining to vendor contracts, leases, and fiscal agreements.
- **Compensation/Claims Unit:** Handles financial concerns resulting from property damage, injuries, or fatalities at the incident.
- **Cost Unit:** Tracks costs, analyzes cost data, creates cost estimates, and recommends cost-saving measures.

■ Additional ICS Terminology

Single Resources and Crews

A **single resource** is an individual vehicle and its assigned personnel. A **crew** is a group of personnel working without apparatus and led by a leader or boss.

Task Forces and Strike Teams

Task forces and strike teams are groups of single resources assigned to work together for a specific purpose or for a specific period of time under a single leader. A **task force** is a group of up to five single resources of any type. A **strike team** is a group of five units of the same type working on a common task or function.

■ Needs Assessment

Why should organizations perform a needs assessment? Often there is a perceived need within an organization that it should have a capability to handle every possible call that could arise. Sometimes this perception occurs because of specific incidents to which the organization has responded. At other times the "need" is based on the availability of grant money to improve response to a specific problem, even though the facts are insufficient to support the need.

While the sentiment of being "all things to all people" is laudable when it comes to rescue capabilities, the reality of the response need can be radically different than what is perceived. The provision of technical rescue services is a rather involved endeavor that should not be taken lightly. Organizations that plan to develop response assets must concentrate on addressing those emergencies that are a priority in their response district.

A number of considerations must be taken into account before an organization commits to providing these services. These considerations include actual need, cost, personnel requirements, and political climate. Additionally, four separate components must be addressed when performing a needs assessment:

- Hazard analysis
- Organizational analysis
- Risk/benefit analysis
- Level of response analysis

■ Hazard Analysis

Hazard analysis is the identification of situations or conditions that may injure people or damage property or the environment. In other words, it determines whether there is a hazard to protect against.

There are a host of possible situations where specialized rescue services could be needed, but the majority fall into the following categories:

- Building collapse
- Rope rescue
- Trench/excavation collapse
- Wilderness search and rescue
- Elevator rescue
- Vehicle and machinery rescue
- Water rescue
- Ice/cold water rescue
- Confined-space rescue

While organizations might argue that most of these types of incidents could occur in their jurisdiction, some incidents might be more likely than others. Building collapse is surely a concern in earthquake-prone areas. A trench/excavation collapse is much more likely to occur in a suburban community experiencing a

Voices of Experience

When we responded to a career department's request for multicounty mutual aid on a fast-developing wildland fire, we did not hesitate to send an engine and crew to help, even though it took us three hours of driving to get there. Conditions upon our arrival should have warned us that the home team did not have control over all the resources coming to their aid. There was chaos and confusion. Our radio check-in on the correct frequency elicited no response, so we proceeded to the area's fire station. There were no line officers present to brief us—only an office staff person who was struggling to keep up. We sat around waiting for an assignment. Some crews who had been there for hours tired of the wait and went back home. Now it was dark, and we were in an area with which we were not at all familiar.

Once we were assigned to an area (with inadequate briefing), we were pretty much on our own. For the next 26 hours, we had next to no contact with any supervisors, no rest, and no food. We patrolled; we fought fire; we advised residents as best we could. While we understood that the size of the incident was extraordinary, this was no way to run an incident. We were completely dependent on ourselves (and the good graces of residents) for all our needs. We never received an update on fire activity. When we finally returned to the incident command post, confusion was no less, just different. The incident was transitioning to a Type II team, and the little organization that had existed previously was being dismantled. Our demobilization was uneventful, in that we just left. After being on the fire for 32 hours, we drove for 3 hours and arrived home at 3 A.M.—certainly a violation of the work/rest rules.

> **"For the next 26 hours, we had next to no contact with any supervisors, no rest, and no food."**

An incident commander's most important job is to account for his people. Don't put anybody in harm's way without making sure that they are competently supervised. If you find yourself part of a poorly organized incident, do everything you can to ensure the safety of your crew. Be proactive and get all the information and support you need to do your job safely. Take care of your responders. Don't assign them to anything until you have the overhead to track them. This process must start before the first unit arrives on scene.

With an emphasis on safety, no scene today should leave responders in harm's way. We lose too many fire fighters to death and injury. It is up to each of us to do everything we can to create safe working environments.

Alan Tresemer
Battalion Chief
Painted Rocks Fire & Rescue Company
Darby, Montana

significant amount of new building construction. It is doubtful that an ice rescue capability would be needed in a tropical environment. The point is that the organization needs to determine not only the possibility, but more importantly the probability of various types of incidents occurring within its jurisdiction. This includes an assessment of the potential for technical search and rescue events that might involve nuclear, biological, and chemical weapons, as well as weapons of mass destruction. In such cases, the authority having jurisdiction (AHJ) must provide equipment and training to responders.

Organizational Analysis

The next phase in the needs assessment process, termed **organizational analysis**, seeks to determine the possibility that an organization could establish and maintain the capability and whether the organization could comply with appropriate rules, regulations, laws, and standards.

The first step is to look at the personnel requirements. Some types and levels of technical rescue are very labor intensive. How many responders are required? What is the availability of the personnel? Can you get the people when you need them? Do the members have the time to develop, train for, and maintain the capability?

Training—both initial and ongoing—is a major concern with technical rescue capabilities. What is the existing versus required level of training? Is the training available? What will it cost to attain and maintain the required level?

Providing technical rescue services can require a significant amount of resources **Figure 3-3 ▶**. You need to determine which resources are available in-house and what is available from outside sources, such as mutual aid, local industry, or private vendors. Does your organization have the financial resources necessary to support the proposed capabilities? Concerns in this area also include equipment, tools and consumable supplies, personal protective equipment (PPE), vehicles, and maintenance.

Risk/Benefit Analysis

Risk/benefit analysis entails an assessment of the risk to the rescuers compared to the benefits that might potentially come from the rescue. All rescue work is a relative risk, but some operations have higher risks than others **Figure 3-4 ▶**. What is the likelihood that the benefits will outweigh the risks? Two important questions should be considered at this phase of the analysis when weighing risk versus benefit:

1. What is the probable (and possible) danger to rescuers?
2. Are the victims likely to be salvageable?

For the hazards identified in your response area, if there is a high level of risk associated with a low probability of a favorable outcome, you should consider not performing that type of rescue.

Level of Response Analysis

After the determination is made that there is a need for a specific technical rescue capability, the next step in the process is to decide which level of service will be provided. Based on an evaluation of the information obtained from the hazard, orga-

Figure 3-3 Organizational analysis examines the available resources and determines which other resources would be required for different rescue incidents.

nizational, and risk/benefit analysis, a three-level system is used to determine the level of response:

- Awareness (or basic)
- Operations (or medium)
- Technician (or heavy)

Figure 3-4 The high risks associated with some incidents, such as those in confined spaces, may outweigh the benefits of an attempted rescue.

These terms are consistent with the levels of capability outlined in NFPA 1670. Because this is an organizational standard, it can be argued that every organization should have a capability at one of these levels unless there is no possibility of an incident of that type occurring (such as ice rescue in the tropics). For example, if there is only a low risk of having a water rescue incident, your organization should be trained to at least the awareness level. If the only excavation occurring in your response area is in trenches less than 8 feet deep, operations level training would be appropriate for trench rescue.

Operations at Technical Rescue Incidents

Operations at technical rescue incidents are no different from operations at other emergency situations, in that a consistent approach in dealing with an incident will produce a more favorable outcome. The following discussion is intended to provide your organization, with the use of the ICS, a step-by-step approach to preparing for, assessing, and responding to technical rescue incidents in a safe, effective, and efficient manner. These steps, in order, are

1. Preparation
2. Response
3. Arrival and size-up
4. Stabilization
5. Access
6. Disentanglement
7. Removal
8. Transport
9. Incident termination
10. Postincident debriefing and analysis

Preparation

In preparing for technical rescue incidents within your response area, it is important to know the different types of rescue situations you could encounter and the terminology used in the field. This knowledge will make communicating with other rescuers easier and more effective. Prior to a technical rescue call, your department must consider these issues:

- Does the organization have the training, personnel, and equipment to handle an incident from start to finish?
- Which levels of training have been completed by personnel within your organization? Where can you acquire higher-trained rescue personnel?
- Which specialized equipment is available from within the organization?
- Where can specialized equipment and resources be acquired from outside organizations?
- Does the department meet NFPA and OSHA standards for technical rescue calls?
- Which resources will the department send on a technical rescue call?
- Do members of the department know the hazard areas of the department's response area? Have they visited hazard areas with local representatives?
- Has any pre-incident planning been done on identified hazards or sites?

Keep in mind that a rescue organization must meet the requirements in Chapter 4 of NFPA 1670 (as well as the requirements for a special rescue incident) before operating at an incident.

Response

If your agency has its own technical rescue team, it will usually respond with a rescue squad, medic unit, engine and/or truck company, and chief. This initial response should satisfy all the basic requirements of any incident. Regarding emergency medical response resources, basic life support (BLS) is the minimum level of medical care required at technical rescue incidents. The ability to supply advanced life support (ALS) assets should be planned as well. Rescuers must be able to manage medical emergencies at technical rescue events, including cervical immobilization, crush syndrome, soft-tissue injuries, orthopedic injuries, and respiratory distress.

Organizations that do not have their own technical rescue team will often respond with a medic unit, engine company, and chief. In this case, the rescue unit will need to come from an outside agency. If the agency will depend on specialized rescue teams from other jurisdictions, it is important to know what the typical response time will be and to understand the team's capabilities.

It is often necessary to notify power and other utility companies that their assistance may be needed during an incident. Many technical rescues involve electricity, sewer pipes, or factors that may otherwise create the need for additional heavy equipment, to which utility companies have ready access. Response agencies should develop joint training programs so that each entity understands the other agencies' needs, capabilities, and limitations. This enhances the efficiency of all parties and leads to better coordination during actual emergency operations.

Arrival, Assessment, Planning, and Size-Up

Immediately upon arrival, the first company officer will assume command. A rapid and accurate size-up is needed to avoid placing rescuers in danger and to determine which additional resources, if any, are needed. The first-arriving unit should follow the steps in **Skill Drill 3-1 ▶**:

1. Establish command.
2. Keep apparatus at an appropriate distance from the scene.
3. Determine utility involvement, if any. (**Step 1**)
4. Establish victim contact.
5. Set up initial safety zones. Often co-workers, family members, and sometimes other rescuers will enter an unsafe scene and become additional victims. (**Step 2**)

As part of the size-up and subsequent planning process, the following issues should be addressed:

- Determine what happened, when, and what was done prior to responders' arrival.
- Identify the number of victims involved and the extent of their injuries. This information will help to determine how many medic units and other resources are needed.

Skill Drill **3-1**

First-Arriving Unit Tasks

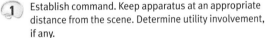

1 Establish command. Keep apparatus at an appropriate distance from the scene. Determine utility involvement, if any.

2 Establish victim contact. Set up initial safety zones. Often co-workers, family members, and sometimes other rescuers will enter an unsafe scene and become additional victims.

- Determine any potential for non-entry or victim self-rescue.
- Identify the victims' locations.
- Identify any hazards.
- Make an initial assessment of the prospects for rescue versus recovery.

When responding to a work site or industrial facility, the officer should make contact with the responsible party. These individuals can often provide valuable information about the work site and the emergency situation.

Once responders are armed with the information from the size-up, a decision can be made to call for additional resources, if necessary, and safe and effective actions can be taken to stabilize the incident. In any event, responders should not rush into the incident scene until an assessment can be made of the situation.

Not only are hazard and risk assessments used in the organizational needs analysis, but they are also an integral part of operations at a rescue incident and should be updated regularly. Four components are taken into account:

1. Preplan information
2. Determination of rescue versus recovery
3. Determination of hazards
4. Determination of risk versus benefit

The first part of the preplan for an event focuses on the information obtained from a site survey. Items that should be included in the site survey are terrain, accessibility, exposures, and other pertinent factors. The second part of the preplan is an analysis of past incidents, including where and how past incidents have occurred and what the probability of survival is. The third part of the preplan is an assessment of available resources. Included in this assessment should be both in-house resources and available outside resources. These resources can take many forms in addition to fire departments, such as specialized rescue teams and individuals with special expertise or equipment. The final part of the preplan includes the lessons you have learned and information obtained from training at the site. This can provide valuable information on conditions and situations you may encounter.

The next component, determining whether the incident is a rescue or a recovery, will make a big difference in how the operation proceeds. **Rescue** involves the moving of live victims to a safer environment. When it is unknown whether a victim is alive, the operation should proceed as a rescue. **Recovery** is the removal of a body from a trapped location to a location where it can be examined and identified. Unfortunately, responders sometimes cannot tell with assurance that there is no possibility of rescue. The following list, which is not intended to be exhaustive, identifies some factors that may influence this decision process.

- **Knowing the victim is alive.** Examples include incidents when you can see or hear the victim or you have a report from a reliable source.
- **High probability the victim is alive.** Examples include incidents when there are no known toxic conditions, air space is available, or the victim fell less than 25 feet onto a moderately soft surface.

- **Low probability the victim is alive.** Examples include incidents where there is a high probability the victim was exposed to toxic or hazardous gases, there is a minimal chance of air space remaining, or a victim fell 50 feet or more onto a moderately hard surface.
- **Certainty that the victim is dead.** Examples include victim exposure to a high concentration of toxic gases, a victim who was trapped with no air voids (e.g., a sand bank or farm grain silo), a victim who fell 100 feet or more onto a hard surface, or a body that is decapitated or dismembered.

The next component is to determine whether any hazards present at the scene may prohibit rescue of live victims. Hazard assessment may dictate whether the incident is a rescue or a recovery. A size-up or scene survey to identify potential or existing hazards can assist the decision-making process and must include the entire area.

The final component is to weigh the risks versus the benefits for a rescue or recovery. All rescue work involves some risk, but some operations are riskier than others. The following issues should be considered when calculating risk versus benefit:

- The number of victims and the determination of whether their rescue is possible
- The danger to rescuers
- The capabilities of the department

Development of the Incident Action Plan

Sound, timely planning provides the foundation for effective incident management. The NIMS planning process represents a template for strategic, operational, and tactical planning that includes all steps an IC and other members of the command and general staffs should take to develop and disseminate an IAP **Figure 3-5 ▶**. The planning process may begin with the scheduling of a planned event, the identification of a credible threat, or the initial response to an actual or impending event. The process continues with the implementation of the formalized steps and staffing required to develop a written IAP.

A clear, concise IAP template is essential in guiding the initial incident management decision process and continuing collective planning activities of incident management teams. The planning process should proceed through five primary phases:

1. Understand the situation.
2. Establish the incident objectives and strategy.
3. Develop the plan.
4. Prepare and disseminate the plan.
5. Evaluate and revise the plan.

The IAP, at a minimum, should include a complete risk/benefit analysis and information on the safety of the scene and general area, hazard mitigation, resource staging, and selection and use of protective systems where appropriate. Additionally, the IAP should establish and confirm the incident strategy and tactics, establish and assign operational and support tasks, and establish a rapid intervention team that must be trained in the type of rescue being performed. Specific components of a complete written IAP and corresponding ICS forms are found in **Table 3-1 ▶**.

City of Centreville
INCIDENT ACTION PLAN

July 10, 2006
Operational Period-0700 to 1900

Tornado Impact
Centreville North Incident

Action Plan Contents

1) Incident Objectives	5) Communications Plan
2) Organizational Chart	6) Transportation Plan
3) Division Assignments	7) Medical Plan
4) Safety Message	

Figure 3-5 Example of an incident action plan. The details of this plan guide the initial incident management decision process and the collective planning activities of incident management teams.

Table 3-1 Components of an Incident Action Plan

Component	ICS Form (if applicable)
Incident objectives	ICS 202
Organizational list or chart	ICS 203
Assignment list	ICS 204
Communications plan	ICS 205
Logistics plan	
Responder medical plan	ICS 206
Incident map	
Health and safety plan	

Other possible components may include the following items:

- Air operations summary
- Traffic plan
- Decontamination plan
- Waste management or disposal plan
- Demobilization plan
- Operational medical plan
- Evacuation plan
- Site security plan
- Investigative plan
- Evidence recovery plan

Stabilization

Once the resources are on the way and the scene is safe to enter, it is time to stabilize the incident. First responders and trained rescue teams will need to establish an outer perimeter to keep the public and media out of the staging area; they must also establish a smaller perimeter directly around the rescue to keep out responders who are not directly involved in the rescue. A **rescue area** is an area that surrounds the incident site and whose size is proportional to the hazards that exist. The zones constituting this area should be established by identifying and evaluating the hazards that are discovered at the scene Figure 3-6 . Specifically, three controlled zones should be established:

- **Hot zone:** This area is for entry teams and rescue teams only. It immediately surrounds the dangers of the incident, and entry into this zone is restricted to protect personnel outside the zone.
- **Warm zone:** This area is for properly trained and equipped personnel only. The warm zone is where personnel and equipment decontamination and hot zone support take place.
- **Cold zone:** This area is for staging vehicles and equipment and contains the command post. The public and the media should be kept clear of the cold zone at all times.

The most common way to establish the boundaries of the controlled zones for an emergency incident site is to use police or fire line tape. Once the controlled zones have been marked off,

responders should ensure that the restrictions associated with the various zones of the emergency scene are enforced.

Lockout/tagout systems should be used at this time to secure a safe environment. Lockout and tagout systems are methods of ensuring that the electricity, gases, solids, or liquids have been shut down and that switches and valves are locked and cannot be turned on at the incident scene. During stabilization, responders should start atmospheric monitoring to identify any environments that are immediately dangerous to life and health (IDLH) for either rescuers or victims. Lockout/tagout control systems are most often used at confined-space incidents and at machinery entrapment situations to eliminate the movement of equipment, the flow of fluids, or the release of stored energy. For example, during a confined-space incident, rescuers would evaluate the presence of any operating machinery such as augers, mixers, or conveyors that may injure a rescuer who must work in close proximity to the machinery. Power for this machinery would be turned off, and if possible locks would be placed on the control switches or valves to eliminate the potential for the machinery to be re-energized.

Gaining Access

With the scene stabilized, responders must gain access to the victim. This access must be appropriate to the type of rescue. For example, when dealing with a trench collapse, considerations would include the placement of pads, ladders, and panels, as well as the presence of voids behind the panels, among other things. In a low-angle rope rescue, considerations would include the establishment of contact with the victim, finding bomb-proof anchors, and setting up a medic line for quick access.

Disentanglement

Once the rescue responders have taken the necessary precautions and identified the reason for entrapment, the victim should be freed as safely as possible. This may involve the lifting or removal of large or heavy objects. Sometimes, however, you may not be able to move or cut the item in which the victim is entangled; in such cases, alternative techniques must be utilized.

Consideration should also be given to how the victim will be removed. Allow EMS personnel access to the victim for medical assessment and stabilization, as disentanglement may sometimes cause life-threatening reactions that, if not treated immediately, could turn the rescue effort into a recovery operation.

A team member should remain with the victim to make sure that the person is not injured further by the rescue operation and to direct the rescuers who are performing the disentanglement. Throughout the course of the rescue operation, which can sometimes span many hours, EMS personnel must continually monitor and ensure the stability of the victim.

Victim Packaging and Removal

Once the victim as been disentangled, redirect efforts toward removing the victim. Preparing the victim for removal involves maintaining continued control of all life-threatening problems, dressing all wounds, and immobilizing all suspected fractures and spinal injuries Figure 3-7 .

Figure 3-6 The rescue area is divided into the hot zone, warm zone, and cold zone.

Figure 3-7 Victim packaging and removal.

Figure 3-8 Incident debriefing.

The overriding objective for each rescue, transfer, and removal operation is to complete the process as safely and efficiently as possible. In some situations, however, a victim may have to be removed quickly (rapid extrication) because the victim's general condition is deteriorating and time will not permit meticulous splinting and dressing procedures, or because safety concerns dictate a speedy egress of both victim and rescuers.

Transport

Once the victim has been removed from the hazard area, EMS transports the victim to an appropriate medical facility. The type of transport used for this purpose will vary depending on the severity of the victim's injuries and the distance to the medical facility.

Incident Termination

Incident termination activities include removing equipment from the scene, ensuring the scene is left in a safe condition, cleaning and replacing equipment on the responding vehicles, completing documentation and reports, incident debriefing, and stress debriefing as appropriate.

Once the rescue is complete, the scene must be stabilized by the rescue crew to ensure that no one else becomes injured. After you have secured the scene and packed your equipment, it is important to return to the station and fully inventory, clean, service, and maintain all the equipment (per manufacturer's instructions) to prepare it for the next call. Some items will inherently need repair, but most will need simple maintenance before they are placed back on the truck and considered in service.

Once back at the station, responders must complete paperwork and document the rescue incident. Record keeping serves several important purposes. Adequate reporting and the keeping of accurate records ensure the continuity of quality care, guarantee proper transfer of responsibility, and fulfill the administrative needs of the department for local, state, and federal reporting requirements. In addition, the reports can be used to evaluate response times, equipment usage, and other areas of administrative interest.

Postincident Debriefing and Analysis

As with any type of call, the best way to prepare for the next rescue call is to review the last one and honestly identify any strengths and weaknesses. What could have been done better? Which equipment would have made the rescue safer or easier? If a death or serious injury occurred during the call, a critical incident stress management (CISM) session may occur to assist rescuers. Reviewing an incident with everyone involved will allow all parties to learn from the call and make the next call even more successful Figure 3-8 ▲ .

Resource Management

Personnel Accountability

The accountability system is the single most important process that any rescuer needs to ensure safety and should be implemented at all emergencies. An accountability system tracks the personnel on the scene, including the following information:

- Responders' identities
- Their assignments
- Their locations

This system ensures that only qualified rescuers who have been given specific assignments are operating within the area where the rescue is taking place. By using an accountability system and working within the ICS, an IC can track the resources at the scene, hand out the proper assignments, and ensure that every person at the scene operates safely.

A number of accountability systems exist, including lists, boards, tags, badges, T cards, bar code systems, and radio frequency identification (RFID) tags Figure 3-9 ▶ . Bar code systems and RFIDs are used with electronic systems, which can give real-time information on personnel, such as their qualifications and assignments. This information can even be transmitted to multiple computers, thereby allowing more than one individual or location to have real-time access to the data (such as the safety officer, IC, or even the emergency operations center).

Figure 3-9 Accountability systems like the accountability tag and passport system are available to keep track of personnel at a rescue.

Agency	ST	Kind	Type	I.D. No.
Order/Request No.		Date/time Check In		
Home Base				
Departure point				
Leader Name				
Crew I.D. No./Name (For Strike Teams)				

No. Personnel	Manifest ☐Yes ☐No	Weight
Method of Travel ☐ Own ☐Bus ☐Air		
Other		
Destination Point		ETA
Transportation Needs ☐ Own ☐Bus ☐Air		
Other		
Ordered Date/Time	Confirmed Date/Time	
Remarks		

Agency	TF	Kind	Type	I.D. No./Name
Incident Location				Time
Status ☐Assigned ☐O/S REST ☐O/S PERS. ☐Available ☐O/S MECH ☐ ETR				
Note				
Incident location				Time
Status ☐Assigned ☐O/S REST ☐O/S PERS. ☐Available ☐O/S MECH ☐ ETR				
Note				
Incident Location				Time
Status ☐Assigned ☐O/S REST ☐O/S PERS. ☐Available ☐O/S MECH ☐ ETR				
Note				
Incident Location				Time
Status ☐Assigned ☐O/S REST ☐O/S PERS. ☐Available ☐O/S MECH ☐ ETR				
Note				

Figure 3-10 Resource tracking systems are used to keep accurate records of rescue equipment.
Courtesy of NIMS/FEMA.

Equipment Inventory and Tracking Systems

In addition to requiring documentation on the purchase, maintenance, and use of equipment, many technical rescue organizations maintain certain records during an incident **Figure 3-10 ▶**. The combination of these records provides a comprehensive resource use and accountability system. Typically used under the supervision of a logistics officer, these systems can be manual or electronic. Manual systems may include lists, sign-out sheets, and T cards. Computerized systems may consist of bar code systems and RFID tags. In some cases, both types of systems may be used simultaneously. Fully integrated systems include all information on all equipment and may use bar codes or RFID tags to identify both the equipment and the personnel signing out that equipment.

Long-Term Operations

Fire fighters are accustomed to incidents that terminate quickly, often within an hour of onset. Many short-term incidents can be effectively directed and controlled using an ICS system using an IC, safety officer, technical rescue officer, and medical officer.

By comparison, technical rescue incidents can extend well beyond these relatively short time periods and be much more complex. The IC needs to consider both immediate and anticipated need for all phases of such an incident. Remember that whichever functions and responsibilities the IC does not assign to others in the ICS system stay with the IC. The IC could become overwhelmed easily in a technical rescue incident if he or she attempts to handle all of these issues for the duration of the incident.

Use of an expanded ICS system will assist the IC with planning and functioning in the long-term incident environment. This should include staffing of the planning, logistics, and finance/administration sections.

A rescue team can typically work up to 24 hours without downtime (depending on the type of rescue), but then must have at least 24 hours for rest and equipment maintenance. If the rescue effort is expected to go beyond 24 hours, plans must be made for long-term operations. This determination can depend on any of the following factors:

- The extent of rescue required
- The training level of the rescue team
- The available resources
- The rescuers' physical condition
- The rescuers' psychological condition
- Needs supported by other ICS functions

Safety

To all organizations and ICs, the health and safety of their personnel are of the highest importance. Members who are injured not only reduce the available personnel by their loss, but can also divert other personnel and resources when they must come to the injured individual's aid. And, of course, no commander wants to be the one to go to the member's family to explain what happened, and most likely why it should not have happened.

NFPA 1500, *Standard on Fire Department Occupational Safety and Health Program*, is a critical document that relates to health and safety programs within the firefighting organization. Contained within this standard are provisions that can be invaluable not only in the development of a health and safety program, but also in the development of response plans, SOPs/SOGs, and risk assessments. This document, in concert with NFPA standards 1250, 1521, and 1561, provides the tools necessary to run a safe and effective organization. The following major chapters are contained within this standard:

- Fire Department Administration
- Training, Education, and Professional Development
- Fire Apparatus, Equipment, and Drivers/Operators
- Protective Clothing and Protective Equipment
- Emergency Operations
- Facility Safety
- Medical and Physical Requirements
- Member Assistance and Wellness Programs
- Critical Incident Stress Program
- Annex A, Explanatory Material
- Annex B, Monitoring Compliance with a Fire Service Occupational Safety and Health Program
- Annex C, Building Hazard Assessment
- Annex D, Risk Management Plan Factors
- Annex E, Fire Fighter Safety at Wildland Fires
- Annex F, Hazardous Materials PPE Information
- Annex G, Sample Facility Inspector Checklists
- Annex H, Informational References

Of particular importance as it relates to on-scene operations, the emergency operations chapter includes provisions for the following issues:

- Incident management
- Communications
- Risk management during emergency operations
- Personnel accountability
- Member staffing and operational requirements
- Control zones
- Traffic incidents
- Rapid intervention teams
- Rehabilitation
- Scenes of violence, civil unrest, and terrorism
- Postincident analysis

Response Planning

Response planning, also known as **pre-incident planning**, is the process of compiling information that will assist the

Figure 3-11 Example of an emergency response plan. The major components of an emergency response plan include the identification of the problem, resource identification and allocation, and procedures.

organization should an incident occur. While there are frequently variables in any incident type, this plan provides a framework for the organization's response. The major components of a response plan include the identification of the problem, resource identification and allocation, and, in some cases, suggested or mandatory procedures **Figure 3-11 ▲**. Additional administrative elements include plan authority and approval, plus who gets a copy of the plan and when the plan is implemented and revised (including those with responsibilities outlined in the plan).

Problem Identification

To write any plan, you must first identify the problem. This is accomplished by conducting the needs assessment discussed earlier in this chapter. This assessment, along with the organization's identified operational capability, forms the basis of what the plan will try to accomplish. Whether a single plan or multiple plans are used will depend on the complexity of your organization's involvement in a given rescue incident type (e.g., structural collapse, swiftwater rescue).

Resource Identification and Allocation

The **pre-incident plan** (also known more informally as a **pre-plan**) will be worthless if there are no resources to implement it. The necessary resources (personnel, equipment, and materials) must be identified, and must be assured of being made available in a timely manner. Frequently, this goal is accomplished through the use of mutual aid agreements, although some resources may require other arrangements. For example, agreements may be established with private vendors such as contractors for heavy equipment, lumber yards, and other private resources (e.g., industrial rescue teams). When at all possible, these arrangements should be made before the incident occurs and should be put in writing to try to avoid any confusion.

Once you have identified these resources, collect the necessary information pertaining to them. This includes procedures for getting the needed resource and contact information such as names and phone numbers. Review this information at least annually to ensure it is still accurate. The pre-incident plan should also include any necessary forms or agreements that might be needed during the incident.

Operational Procedures

The pre-incident plan should identify not only which resources are needed, but how they will be used (deployed). Should any resource not be available, the plan should identify alternative resources. It should identify possible scenarios where needed resources may not be available or may be delayed because of response difficulties or responder capabilities. In this event, alternative plans should address the needs of the incident.

Once the pre-incident plan has been developed, it must be tested to ensure that the plan will work and accomplish the desired goals. Should any deficiencies be found, revise and retest the plan. Periodic testing is also important because of changing personnel or conditions.

Standard Operating Procedures

A **standard operating procedure (SOP)** is an organizational directive that establishes a standard course of action; in other words, it provides written guidelines that explain what is expected and required of emergency services personnel while performing their jobs Figure 3-12 ▸ . SOPs are not intended to tell you how to do the job (technical knowledge and skills), but rather are designed to describe related considerations (the rules for doing the job) such as safety, command structures, and reporting requirements. Technical knowledge and skills, such as how to rappel or operate a specific tool, are obtained through training and technical protocols and are not normally the subject of SOPs.

In today's world, the decisions that personnel make are more complex and can be subjected to intense scrutiny by the public, the government, and the media. SOPs are important because they provide a mechanism to document intentions and strategies and comply with regulatory requirements, thereby helping to improve accountability and operational efficiency as well as to reduce liability. From a legal point of view, the use of the term *SOP* versus *SOG* is not as important as the content of the SOP/SOG and how it is implemented. Typical content covers areas such as development and maintenance, compatibility with regulations and standards, organizational needs, training and competence, and the ability to monitor performance and ensure compliance. The organization (through the AHJ) must ensure that SOPs comply with all applicable local, state, and federal laws, such as occupational health and safety regulations. For example, SOPs directing operations at confined-space rescues must meet the requirements of the federal OSHA standard for permit-required confined-space entry and training.

Organizations performing technical rescue are required to establish SOPs that identify the operational level at which they are expected to perform for each category of rescue to which personnel are required to respond—specifically, at the awareness, operations, or technician level. Awareness is a basic level where responders must be aware of potential hazards and know which resources are required to manage the rescue; personnel at this level are generally not considered rescuers. Operations level requires responders to perform search, rescue, and recovery techniques, while under the supervision of higher-trained and more-experienced responders. Technician level is considered the highest level of performance and indicates that the organization is capable of identifying hazards, using equipment, and applying advanced rescue techniques to coordinate, perform, and supervise rescue operations. The AHJ is responsible for training documentation and review.

The development of SOPs should be done in a coordinated fashion with the development of other plans such as mutual aid agreements, strategic plans, laws, regulations, and standards so that the organization avoids any conflicts that might potentially cause problems. The first phase of the development process is to perform an assessment of which SOPs are needed. Typically, SOPs can be classified into three major categories: management and administration, prevention/special programs, and emergency operations.

Included in the management and administration area are SOPs dealing with general administration, health and employee assistance programs, and planning/preparedness. Organizations that respond to search and rescue missions will need to ensure that written operational procedures are established that define how the organization will implement management processes required by the *U.S. National Search and Rescue Plan*, the *U.S. National Response Framework*, and other state and local response plans. Elements of these plans define how emergency incidents must be managed, how information exchange needs to occur, how resources should be tracked and requested, and how organizations at various levels of government should interact on an emergency scene.

Prevention and special programs might include SOPs on public information and prevention, code enforcement, building inspections, and other special endeavors.

The general section of the emergency operations segment includes items such as emergency vehicle operations, safety, communications, command and control, special operations, and postincident operations. Of particular importance is the need to establish guidelines that define how to evacuate members from an area if hazardous conditions intensify or cannot be

---

FIRE AND RESCUE DEPARTMENT
STANDARD OPERATING PROCEDURE

SUBJECT: RESOURCE DEPLOYMENT	S.O.P. **05.04.01**	
	PAGE **1** OF **4**	
CATEGORY: Suppression	**SUBCATEGORY:** Transfer and Fill-In Procedures	
APPROVED BY: CHIEF, FIRE AND RESCUE DEPARTMENT	**EFFECTIVE DATE:**	

FORMS REQUIRED: None
NOTE: Current forms are located on the department's Intranet.

PURPOSE:
To ensure that apparatus and personnel resources are deployed in a consistent and effective manner.

I.   PREFACE
Operations staff shall be responsible for ensuring that personnel and apparatus are strategically positioned to provide efficient and effective service within response times consistent with department objectives.

II. GUIDELINES FOR APPARATUS DEPLOYMENT
A.   The Assistant Chief of Operations ultimately shall be responsible for managing Operations' resources and informing the Operations' deputy chiefs and the Staff Duty Officer of conditions that will affect operations.

B.   Battalion chiefs shall be responsible for managing resources within their respective battalions and for informing the Operations' deputy chief, or his or her designee, of conditions that may affect countywide Fire and Rescue Department service.

C.   The uniformed fire officer (UFO) at the Department of Public Safety Communications (DPSC) shall be consulted any time units need to be placed out of service or relocated, i.e., units going to the Radio Shop or units being moved for training.

D.   The UFO, in coordination with the field battalion chiefs, shall assume responsibility for relocating units for coverage shortages created by working incidents, vehicles out of service, or other short-term or temporary situations.

E.   When units are out of service for an extended period of time due to prearranged training exercises (OARs, EMSCEP, etc.), staff shortages, or long-term mechanical problems, the Operations deputy chief shall consult with the UFO at DPSC to coordinate unit relocations.

F.   The Apparatus Section's vehicle coordinator (or duty apparatus officer after normal business hours) shall be informed any time reserve apparatus is placed in or out of service.

Figure 3-12   Sample SOP.

---

controlled, and how to account for the safety of those personnel. This standard procedure is especially important because of the high hazard atmosphere and environment in which rescuers must work at technical rescue events. In conjunction with this procedure, the organization must specify methods to notify all members in the affected hazardous area immediately by any effective means, including audible warning devices, visual signals, and radio signals. Many organizations have established evacuation signals such as the sounding of apparatus air horns or the use of whistles or radio-activated signals to order an evacuation of hazardous areas.

Also contained within the emergency operations section is the segment pertaining specifically to technical rescue. Included here are the key concerns of technical rescue risk management, general rescue operations, and special rescue operations. Typically, the risk management section includes SOPs on PPE, air monitoring, and lockout/tagout procedures, among others. The general rescue operations section includes SOPs for more routine operations, such as vehicle and machinery extrication, scene stabilization, and application of rescue equipment. Some law enforcement organizations are involved in search and rescue missions, and their personnel are often required to be armed to carry out their duties. When their presence is required, SOPs must be written that identify the rescue situations when firearms must be secured and disallowed from the operational area for safety reasons. These situations might include a confined-space entry or operations on a structural collapse rubble pile.

The special rescue operations section includes SOPs for specific responses such as structural collapse, trench rescue, water rescue, and ice rescue. This section should detail those areas to which the organization may be called to respond, regardless of whether it has a rescue capability. Even if the organization does not have such a capability, there should be a SOP detailing the procedures to follow regarding the role the organization will play, who will be in charge, methods to obtain the necessary rescue resources, and any other special procedures where appropriate.

Wrap-Up

Chief Concepts

- The Incident Command System (ICS) is a management structure that provides a standard approach and structure to managing operations. The use of the ICS ensures that operations are coordinated, safe, and effective, especially when multiple agencies must work together.
- A number of considerations should be taken into account before committing your organization to providing rescue services, including actual need, cost, personnel requirements, and political climate. The components of a needs assessment include hazard, organizational, risk/benefit, and level of response analysis.
- The 10-step process used to provide a step-by-step approach to responding to rescue incidents is (in order) preparation, response, arrival and size-up, stabilization, access, disentanglement, removal, transport, incident termination, and postincident debriefing and analysis.
- Sound, timely planning provides the foundation for effective incident management. A clear, concise incident action plan (IAP) is essential to guide the incident commander and subordinate staff and to help the continuing collective planning activities of incident management teams.
- Resource management is a critical tool for managing any incident and includes not only personnel but equipment and other assets.
- Rescue incidents often extend into long-term operations (lasting more than a few hours), so consideration must be given to planning for such a possibility.
- The health and safety of members are of the highest importance to all responding organizations. A health and safety program, developed in concert with NFPA standards 1500, 1250, 1521, and 1561, can provide the tools necessary to run a safe and effective organization.
- Response planning is the process of compiling information that will assist the organization should an incident occur. The major components of a response plan include identification of the problem, resource identification and allocation, and, in some cases, suggested or mandatory procedures.

Hot Terms

Branch A segment with the ICS that may be functional or geographic in nature. Branches are established when the number of divisions or groups exceeds the recommended span of control for the operations section chief or for geographic reasons.

Crew A group of personnel working without apparatus and led by a leader or boss.

Division A segment within the ICS established to divide an incident into physical or geographical areas of operation.

Finance/administration section Staff function responsible for the accounting and financial aspects of an incident, as well as any legal issues that may arise.

Group A segment within the ICS established to divide an incident into functional areas of operation.

Hazard analysis The process of identifying situations or conditions that have the potential to cause injury to people, damage to property, or damage to the environment.

Incident action plan (IAP) An oral or written plan containing general objectives reflecting the overall strategy for managing an incident. It may include the identification of operational resources and assignments. It may also include attachments that provide direction and important information for management of the incident during one or more operational periods.

Incident commander (IC) The individual with overall responsibility for the management of all incident operations.

Incident command system (ICS) A management structure that provides a standard approach and structure to managing operations, thereby ensuring that operations are coordinated, safe, and effective, especially when multiple agencies must work together.

Liaison officer The incident commander's point of contact for outside agencies. This officer coordinates information and resources among cooperating and assisting agencies and establishes contacts with agencies that may be capable or available to provide support.

Lockout/tagout systems Method of ensuring that the electricity, gases, solids, or liquids have been shut down and that switches and valves are locked and cannot be turned on at an incident scene.

Logistics section Staff function responsible for all support requirements needed to facilitate effective and efficient incident management, including providing supplies, services, facilities, and materials during the incident.

Operations section Staff function responsible for development, direction, and coordination of all tactical operations conducted in accordance with an incident action plan.

Organizational analysis A process to determine if it is possible for an organization to establish and maintain a given capability.

Planning section Staff function responsible for the collection, evaluation, dissemination, and use of information and intelligence critical to the incident.

Pre-incident plan (preplan) A document containing detailed information about a facility or site that allows staff or first responders (e.g., police, fire fighters, emergency medical staff) to respond to any crisis situation at that location quickly and effectively.

Public information officer (PIO) Staff position that interfaces with the media and provides a single point of contact for information related to an incident.

Recovery The removal of a body from a trapped location to a location where it can be examined and identified.

Rescue Those activities directed at locating endangered persons at an emergency incident, removing those persons from danger, treating the injured, and providing for transport to an appropriate health care facility.

Rescue area An area that surrounds the incident site and whose size is proportional to the hazards that exist there.

Resource management A standard system of assigning and tracking the resources involved in an incident.

Response planning (pre-incident planning) The process of compiling information that will assist the organization should an incident occur.

Risk/benefit analysis An assessment of the risk to the rescuers versus the benefits that can be derived from their intended actions.

Safety officer Staff position responsible for enforcing general safety rules and developing measures for ensuring personnel safety.

Single resource An individual vehicle and its assigned personnel.

Staging area manager Staff position responsible for ensuring that all resources in the staging area are available and ready for assignment.

Standard operating procedure (SOP) A written organizational directive that establishes or prescribes specific operational or administrative methods to be followed routinely for the performance of designated operations or actions.

Strike team A group of five units of the same type working on a common task or function.

Task force A group of up to five single resources of any type.

Technical specialists Advisors who have the special skills required at a rescue incident.

Unified command An incident management tool and process that allows agencies with different legal, geographic, and functional responsibilities to coordinate, plan, and interact effectively to manage emergencies. In a unified command structure, multiple agency representatives make command decisions instead of just a single incident commander.

Rescue Responder *in Action*

You are assigned to a heavy rescue company and during one of your work rotations your unit was dispatched on a mutual aid event to assist another jurisdiction with a worker trapped in a trench that had collapsed. This event was a long-term operation in which a specialized trench rescue team had to be requested from another region of your state because your department did not have the proper level of training or the proper equipment to safely affect the rescue. Your supervisor has assigned you to develop a response plan for trench rescue incidents within your response area.

1. Which of the following elements should be included in the response plan?
   A. Operational procedures
   B. Resource identification
   C. Resource allocation
   D. All of the above

2. The resource identification portion of the response plan should identify:
   A. Personnel requirements
   B. Equipment requirements
   C. Material requirements
   D. All of the above

3. What elements are not required to be part of the trench rescue response plan?
   A. Plan authority
   B. Trench rescue training manuals
   C. List of who is assigned copies of the plan
   D. Operational procedures

Search Operations and Victim Management

NFPA 1670 Standard

4.1.9* At a minimum, medical care at the basic life support (BLS) level shall be provided by the organization at technical search and rescue incidents. [p. 94–95, 98, 104]

4.1.13 The AHJ shall comply with all applicable local, state, and federal laws. [p. 98]

Additional NFPA Standards

NFPA 1001, *Standard for Fire Fighter Professional Qualifications*

NFPA 1006, *Standard for Technical Rescuer Professional Qualifications*

Knowledge Objectives

After studying this chapter, you will be able to:

- Discuss the use of search tactics during search and rescue operations.
- Describe emergency medical response system levels of service.
- List general hazards that may be present during victim management at a technical rescue event.
- Describe victim access considerations and procedures.
- Describe victim stabilization.
- Identify the need for triage at a technical rescue incident.
- List common victim injury patterns related to technical rescue events.
- Describe types of patient packaging devices and their use.
- Explain the safety procedures for working around helicopters.
- Describe response planning and incident management requirements related to victim management at a technical rescue event.

Skill Objectives

There are no skill objectives for this chapter.

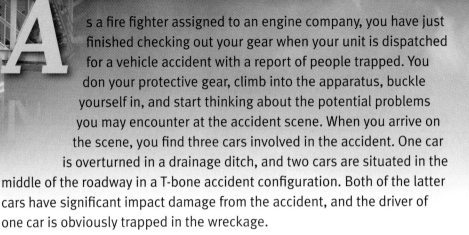

As a fire fighter assigned to an engine company, you have just finished checking out your gear when your unit is dispatched for a vehicle accident with a report of people trapped. You don your protective gear, climb into the apparatus, buckle yourself in, and start thinking about the potential problems you may encounter at the accident scene. When you arrive on the scene, you find three cars involved in the accident. One car is overturned in a drainage ditch, and two cars are situated in the middle of the roadway in a T-bone accident configuration. Both of the latter cars have significant impact damage from the accident, and the driver of one car is obviously trapped in the wreckage.

Your officer tells you to check on the passengers in the overturned car and directs the other fire fighter on the rig to pull a protective hose line; the officer then goes to check on the passengers in the other cars. As you approach the overturned car, you can smell gasoline and the car appears to be unstable, lying on its side. You see two passengers inside the car. One passenger is bleeding from the face and is in obvious pain. The second passenger appears to be unconscious, and you cannot tell from your initial approach whether this person is breathing.

As you start to perform a size-up and gather more information about the event, you realize that a technical rescue incident can result in a variety of injury patterns depending on the type of entrapment. In addition, victim management may be difficult and may require many different types of resources, including specially trained responders and specialized tools and equipment, to mount an effective response.

1. What should you be looking for as you arrive on the scene of a technical rescue event?
2. Which types of hazards might rescuers find at a technical rescue event?
3. Which types of problems might rescuers encounter when trying to gain access to patients?
4. Which types of injuries might patients suffer at technical rescue events?
5. Which patients should be treated first?
6. Which equipment is required to treat, package, and move patients to a transport unit?

Introduction

Victim management refers to all aspects of an incident involving one or more victims: identifying hazards to which a victim may be exposed, accessing a victim, performing triage of multiple victims, assessing patient injuries, stabilizing victim injuries, interacting with victims, and moving and transferring victims requiring technical rescue assistance. **Triage** is the process of sorting victims based on the severity of their injuries and medical needs; it helps establish treatment and transport priorities. Responders at the **awareness level** will from time to time be required to identify the location of rescue victims, such as when first arriving at a vehicle accident. Awareness level is the first level of rescue training provided to all responders, which emphasizes recognizing hazards, securing the scene, and calling for appropriate assistance. There is no actual use of rescue skills at this level.

Awareness level responders may also be required to gain access to the victims' location—for example, when an awareness level responder can safely enter through the window of a crashed vehicle to stabilize a victim's injuries. In most situations, however, awareness level responders assist more qualified rescuers with patient movement and transfer.

This chapter defines the knowledge and skills required of awareness level responders to effectively manage patients in the technical rescue environment.

Search Operations

To determine the status of a patient and the complexities of a rescue event, the patient(s) must first be located—a process known as the **search phase**. The search phase of a technical rescue event in many situations is very simple and straightforward, such as locating victims of a vehicle accident. As soon as responders arrive on the scene of a vehicle **entrapment**, the location of the victims is usually obvious. In contrast, a multi-vehicle entrapment involving large transportation vehicles such as tractor trailers may result in significant damage and heavy entrapment where victims are not easily seen within the wreckage. Entrapment is a condition in which a victim is trapped by debris, soil, or other material and is unable to perform self-extrication. One vehicle accident resulted in such heavy destruction that a passenger who had been riding in the hatchback area of a sedan was completely hidden by the damaged sheet metal of the vehicle. Because it was not anticipated that a victim would be riding in a non-passenger area, the body was not located until a day later, after the vehicle had been towed to a vehicle storage facility.

Search operations involve a combination of detective work, in gathering information about the potential victims; analysis work, in scrutinizing known and unknown victim location information; speculation about potential victim locations within various search environments; and the use of search tools, equipment, and tactics. In some situations, such as a structural collapse, the search phase of a technical rescue event can be significant in terms of its length and the application of resources.

Search Environment

The environment in which a rescue operation will take place determines the difficulty of search operations and the amount of search resources required. Search operation tactics are applied differently depending on the rescue environment encountered.

Water search and rescue situations, for example, may involve an obvious person in distress on the surface of the water, a person being carried by swift water, or a person located clinging to the surface of ice. In these situations, the location of the victim can be seen and the search is somewhat simplistic, although the rescue operations will require specially trained swiftwater or ice rescue assets. If the victim is submerged, however, the search operations become much more complex and require specialized underwater search and recovery teams to analyze the potential location of the victim, to make entry, and to extract the victim.

Search environments such as wilderness rescue require significant resources to gather and analyze known information, to determine potential victim locations, and to implement wilderness search techniques.

Structural collapse search operations require gathering information about the building and last known location of occupants, implementing hailing systems, performing a physical search, and using technical and canine search assets.

Additional information about search operations within specific rescue environments (e.g., trench, structural collapse, wilderness) is discussed within those specific rescue chapters in this textbook.

Search Tactics

The search methods used will depend on the type of rescue operation responders encounter, which in turn drives some decisions regarding which tools or techniques are applied to the problem. Initially, all search operations are driven by the gathering of infor-

mation and intelligence through visualization of an entrapment or an interview and investigation process. Additionally, resources may be used to perform a reconnaissance of an area. **Reconnaissance** (or **recon**, as it is better known) is not in itself a search, but rather an initial survey or rapid visual check of a rescue environment to obtain information that can be used to develop a search plan and direct search and rescue operations.

The size of the recon team is determined by the type, size, and complexity of the rescue situation. Recon teams may be deployed on foot or by ground vehicle, watercraft, or aircraft depending on the size of the rescue environment. Because recon duty is intended to serve as an information-gathering tool to make search and rescue decisions, the recon team members should not normally get bogged down in rescue operations during this task.

For example, in a structural collapse situation, the recon group might be directed to circle the building to gather more details about the collapse situation. If they visualize a trapped person in the rubble pile or hear a victim calling for help, the recon team should report this information to the incident commander (IC) or the person in the command system to which the team is assigned. The recon team personnel may be required to take some limited action to deliver basic life support (BLS) to a trapped victim, but only until additional response personnel arrive to take over patient care duties. The team should then continue with its recon duties.

Rescue Tips

Effective search operations are important to successful structural collapse rescue operations. An adequate and distinct search phase must occur if rescue teams are to limit redundancy in breaching operations at collapse sites. A lot of time can be wasted in blindly penetrating walls, floors, and other obstacles if a thorough search has not been conducted to identify specific areas that could contain victims.

As mentioned earlier, information gathering to determine potential victim locations at a rescue incident relies on a variety of investigative techniques. Typically, responders interview people such as witnesses, friends, relatives, and co-workers to collect important information about the location or potential location of a lost or trapped victim. During the investigation, responders should attempt to determine a victim's last known location or the point last seen within a rescue environment, such as a building or trench collapse. In a building collapse setting, gaining knowledge of the point last seen, or the victim's last known location such as in an office on the second floor, will assist rescuers in planning where to begin the search efforts. During wilderness search and rescue situations, it is important to glean information related to the victim's last known point or last known direction of travel, physical and mental condition, and capability to survive within the rescue environment to which they are exposed. The information about the victim's last known point or location will then be used by the search and rescue team in developing a search plan, especially for deciding on a starting point for the search efforts.

One important outcome of the information-gathering phase is a determination of the urgency of the search mission, which is a critical factor in the development and application of search tactics. Determining the urgency of the search mission helps establish the speed, nature, and level of response to the search problem. Before incident commanders call for significant personnel and specialized search assets, they must have a good understanding of the search problem and the level of urgency of the search operation. The search urgency takes into account factors such as the following:

- Time elapsed
- Victim profile (e.g., age, medical condition, number of missing victims)
- Environmental conditions (short-term versus long-term exposure to the elements)
- Victims' experience and available survival equipment
- Terrain/hazards
- History of incidents in the area/site

The development of the search plan and the application of search tactics are based on various factors as well, including the urgency of the search, the resources available, the existing hazards to responders, and known information about the victim(s).

The development of a search plan may include the step of confining or defining the area to be searched. A search grid may be developed to identify, prioritize, and document areas to be searched. This tactic is used to start a search in a specific area based on the probability that the victim may be found in a certain location—a probability developed during the information-gathering phase. For example, a search for a victim of a structural collapse may be started in a defined area based on that person's routine location in the building at a certain time of day.

Another search tactic involves segmentation of a rescue environment. This allows for a large area to be searched to be divided into smaller, more easily managed areas. Specific search area boundaries are established, search segments or grids are established, and resources are assigned to the various segments based on the priority of the search. At a building collapse, for example, the areas with a higher likelihood for survival may receive more specialized search resources than those areas where the chance for survival is unlikely.

When search assets are assigned to a rescue area, they may be assigned to conduct either a primary search or a secondary search. A **primary search** (often referred to as a hasty search) involves a deployment of search resources that will initiate a rapid search of an area that is thought likely to contain survivors. A **secondary search** is a slower, more methodical, and detailed search of a rescue environment to ensure that no victims were overlooked during the initial search operation. For example, at a building collapse incident, a primary (hasty) search might be initiated to scan the perimeter of a building, the debris pile, and selected void spaces rapidly to identify visible victims or those victims who can be heard calling for help. This search may be accomplished by personnel performing physical searches, possibly in conjunction with canine search teams, and with the use of technical search equipment.

Search Resources

The type and complexity of the rescue problem determine the search resources needed to manage a technical rescue incident. Search resources may include personnel trained in various search techniques, canine search teams, and technical search equipment such as search cameras and acoustic search tools. Personnel resources might include incident management personnel and support teams, trained searchers, specialized searchers, and emergency medical services (EMS) personnel. Search dogs and handlers should be trained and certified to nationally accepted standards, such as those developed by the FEMA urban search and rescue (USAR) response system.

Specialized vehicles may also be needed to assist with large-area searches, such as those involving a person lost in the wilderness or a large-scale natural disaster situation. These vehicles could include helicopters, off-road vehicles, horses, and watercraft. Response organizations should identify where search assets can be located and should have a procedure in place so responders can request assistance of these special resources.

Levels of Emergency Medical Service

Emergency medical services are provided at the BLS level and at the advanced life support (ALS) level Figure 4-1 ▼.

Basic Life Support

Basic life support (BLS) is the level of medical care that can be provided by persons trained as medical first responders or as EMT-Basics; it requires a limited set of emergency medical skills Table 4-1 ▶. In addition to the skills listed in Table 4-1, BLS providers can perform cardiac defibrillation to return an irregular heartbeat to a normal rhythm by using an automated external defibrillator (AED). BLS services do not, however, include the administration of medications beyond assisting the patient with his or her prescribed medications or the administration of intravenous fluids.

Figure 4-1 EMS care can be provided by BLS and ALS providers.

Table 4-1   BLS and ALS Skills

BLS	ALS
▪ Scene control	▪ Advanced airway management techniques, including endotracheal intubation to keep the airway open and help the patient breathe
▪ Evaluating conditions for responder safety	
▪ Patient assessment	
▪ Basic airway management techniques	▪ Administering intravenous fluids to treat shock
▪ Cardiopulmonary resuscitation (CPR)	▪ Administering medications for various reasons
▪ Administering oxygen	▪ Monitoring and interpreting heart rhythms
▪ Splinting	
▪ Controlling external bleeding and bandaging	▪ Electrically pacing the heart using medical intervention
▪ Treating patients for shock	▪ Defibrillating the heart
▪ Lifting and moving patients	▪ Removing trapped air from the chest
▪ Transporting patients to an appropriate medical facility	

■ Advanced Life Support

Advanced life support (ALS) providers have extensive training in advanced life-saving procedures. ALS systems, which require a greater investment in equipment and personnel training than BLS systems do, operate as the extension of a physician, using standing orders, protocols, and/or radio direction from the physician to ensure uniform and correct treatment. At times, providing ALS care requires more personnel at an emergency scene because the care the patient receives is more complex. ALS providers are trained as EMT-Paramedics, although EMT-Intermediate levels of training include limited ALS skills.

▌Hazards and Hazard Assessment

Because of the variety of technical rescue situations that fire fighters, EMS personnel, and rescue teams may face, all responding personnel must be cognizant of potential approach hazards while responding to an event. Approach hazards vary depending on the type of emergency. In fact, often the specific type of technical rescue is a secondary factor or cascade event created by a larger event. For example, a swiftwater rescue may be caused by a tropical storm, or a building collapse rescue may be caused by an earthquake or tornado event.

In such cases, the primary cause of the technical rescue situation can create other hazards the rescuer may face while responding to the call **Figure 4-2 ▶**. For example, rescuers may face the challenge of flooded roadways, washed-out bridges, downed electrical wires, or debris in the roadway due to the tropical storm. A tornado or other natural disaster may cause broken gas lines, fires, hazardous materials releases, or instability of nearby buildings. Additionally, rescuers must be aware of their potential to *create* hazards at a rescue site.

Upon arrival at the scene, rescuers should perform an adequate size-up that includes a comprehensive hazard and

Figure 4-2   Often the need for technical rescue arises because of a natural disaster.

risk assessment. This assessment helps identify the hazards that rescuers will be exposed to, the hazards that victims must be protected from, the level of personal protection required, and the safety procedures that must be implemented.

Vehicle accidents, for example, are often associated with hazards such as fuel spills, jagged metal, broken glass, uncontrolled traffic flow, unstable vehicles, air bag systems, hydraulic systems, and possibly hazardous cargo **Figure 4-3 ▾**. Trench collapse incident hazards include the potential for a secondary collapse that may entrap rescuers, atmospheric hazards, and exposure to broken utility lines. Confined-space hazards may include exposure to oxygen-deficient, explosive, or toxic atmospheres; engulfment potential from flowing products such as grain or gravel; uncontrolled energy sources; reduced visibility; and slippery surfaces.

Safety Tips

Responders face a variety of victim management challenges at technical rescue emergencies. For this reason, all responders must take the time to evaluate hazards properly prior to initiating patient care, then have a crew member stay disconnected from patient care to evaluate the environment continuously for any changing conditions.

Figure 4-3   An adequate size-up reveals the potential hazards at a rescue incident.

The hazard and risk assessment may identify a variety of needs—for example, the need for rescuers to don special technical rescue protective gear, the need to place and staff protective hose lines if a fire hazard is identified, the need to control broken utilities or uncontrolled energy sources, or the need to get a flotation device to a water rescue victim. Likewise, the hazard and risk assessment may reveal the need for additional or specialized resources. These resources could include representatives from the local electrical power company or gas company to control utilities, police assistance for traffic and crowd control, or additional fire department assets to control fire spread.

It is extremely important that responders take enough time to evaluate hazards properly prior to initiating patient care. Once patient care begins, rescuers become focused on managing the victim and it becomes very difficult to maintain good situational awareness of hazards in the environment. Structural instability at a collapse site may worsen, traffic flow around a vehicle accident may increase, or the speed of water movement may accelerate at a water rescue event. Given a potentially dynamic situation, at least one member of the crew should stay disconnected from patient care and evaluate the environment continuously for any changing conditions.

Victim Management

Victim Access

To assess patient injuries and initiate patient care, rescuers must first access the victims. *Gaining access* refers to the process in which a rescuer is able to approach a victim who is trapped or situated within a technical rescue environment safely and effectively. It is not always a simple matter; indeed, the complexity will vary depending on the type of entrapment. Gaining access may be as straightforward as crawling through the window of a damaged vehicle to reach the victim of a vehicle entrapment, or it may be as complicated as crawling through debris inside a dangerous void space where a patient is trapped under concrete at a building collapse site. The type of technical rescue will determine the difficulty of gaining access to the victim. Of course, this statement assumes that the victim has been located during the search phase. Once the victim's location is determined, the hazards must first be controlled or reduced. At the very least, the rescuers' exposure to extreme hazards must be limited.

Victim access procedures, access equipment, and access techniques vary widely, from lowering rope rescue equipment over the edge of a rock face at a high-angle event to the use of watercraft, personal flotation devices, and rope systems to gain access to swiftwater rescue victims. For responding personnel, the proper level of personal protective equipment (PPE) must be worn and protection from bloodborne pathogens must always be assured prior to gaining access and delivering patient care.

Gaining access to a victim trapped in a trench collapse requires that rescuers first control the hazards of a potential secondary collapse by installing shoring to stabilize the trench walls. Only then may rescuers use ladders to access the patient to begin the delivery of medical care. Similarly, gaining access to

Figure 4-4 Specialized equipment may be needed to gain access to victims—for example, a tripod in a confined-space rescue.

a victim trapped inside a confined space may require the use of breathing apparatus, tripods, mechanical advantage equipment, and other rope rescue equipment **Figure 4-4 ▲**.

While gaining access to a patient, rescuers should continuously look for hazards that may not have been visible during the initial hazard assessment. Secondary collapse hazards may become evident as rescuers are navigating through the interior of a partially collapsed building. As rescuers are negotiating a rock face at a high-angle rescue event, they may come across loose rocks that could be a falling hazard either for the victim or for other rescuers located below the victim.

One goal of victim access is to secure and stabilize the victim and to protect the victim from further harm. The rescue environment dictates how this goal can be accomplished effectively. Some environments require significant hazard control before victim access can occur. An in-depth hazard and risk assessment is the key to effectively managing the risks to rescuers and the patient(s).

In addition, because rescuers must enter a hazard zone during victim access, they must always identify an escape route prior to making entry. This step is especially important at building collapse and trench collapse incidents. Rescuers must know that they can get out of the hazard zone from a different direction than they entered in case a secondary collapse eliminates the possibility of exiting from their initial entry point.

Each technical rescue situation has its own unique environmental problems that limit how rescuers can access victims. Preplanning access methods, having the right equipment, and ensuring that rescuers are properly trained are necessary to be effective at gaining access to patients.

Victim Assessment

One of the first priorities for rescuers, once they have accessed victims of a technical rescue event, is to assess patient injuries and identify any adverse medical conditions patients may be suffering from. This assessment is necessary to determine whether each victim is a viable patient and, if so, how treatment priorities

Voices of Experience

On a quiet fall evening, the driver of a vehicle containing five teenagers who were out joy-riding lost control of the vehicle as the car crested the top of a hill. The vehicle ran off the roadway, went through a fence, hit a tree hard enough to knock the tree over, and bent the car into a U shape so that the headlights and tail lights were pointing in the same direction. The impact was so severe that one occupant was ejected from the car, the driver's head was trapped between the roof, the tree, and the dashboard, and another front seat occupant was crushed between the front seat and the dash. In addition, two rear-seat occupants had their legs trapped in the floor system sheet metal due to the crushing impact of the accident.

First responders had their hands full when they arrived on the scene. The location of the accident was a dark, rural area with limited access. The impact of the crash had ruptured the gasoline tank, releasing fuel onto the ground and creating a fire hazard. Fire fighters had to position and staff protective hose lines throughout the event. In addition, first responders had to triage the five patients and determine who needed immediate life-saving medical care.

Awareness level responders realized that the accident would require additional resources. ALS units and a specialized rescue company were requested. Command officers were also dispatched to assist with incident command and coordination of the rescue efforts.

> *The incident required the use of just about every hand and power tool carried on the rescue vehicle.*

The rescue company worked for more than five hours to extricate the victims. Additional trained rescuers were requested to help with the rescue operation. The incident required the use of just about every hand and power tool carried on the rescue vehicle, including hydraulic cutters and spreaders, reciprocating saws, pneumatic chisels, hand tools, and stabilization devices.

This event was one of the most daunting rescue operations I have ever been involved in. The personnel assigned to the rescue company, however, were well versed in the use of the tools and equipment and luckily had adhered to continuous training exercises in the past to properly prepare them for this type of event.

Robert Rhea
Battalion Chief
Fairfax County Fire and Rescue Department
Fairfax, Virginia

should be established. During the assessment phase, rescuers also determine which resources will be required to extricate patients from any entrapment and which equipment and staffing will be required to package and move patients out of the hazard zone. For this reason, rescuers must have a good knowledge of the internal resources available from their response organization as well as the response times when requesting specialized resources from outside the organization, such as cranes, specialized rescue teams, or utility company assistance.

Patient assessment methods and patient treatment must always follow the medical protocols established by the authority having jurisdiction (AHJ), which in turn must comply with all applicable local, state, and federal laws. Protection from bloodborne pathogens must be ensured through the use of proper PPE and adherence to written protocols related to **body substance isolation (BSI)**. Body substance isolation is an infection control concept that treats all bodily fluids as potentially infectious.

Victim Stabilization

Initial medical care to a patient (BLS) includes establishing and maintaining an adequate airway, providing respiratory ventilation to maintain air flow, controlling severe bleeding, and maintaining the circulatory system. In addition, spinal immobilization must be assured, and the victim should be treated for shock if this situation is present or suspected. In these situations, local medical protocols must always be followed. If the rescuers are not trained in emergency medical care, they must know where to get these resources and be prepared to assist the emergency medical staff with patient packaging and movement. **Patient packaging** refers to the process of preparing the victim for movement as a unit, often accomplished with a long spine board or similar device.

Triage

Some technical rescue events, such as a building collapse or vehicle accident involving a bus, may result in multiple victims who need medical care Figure 4-5 ▼ . Whenever rescuers find multiple patients at a technical rescue incident, the victim

assessment must include determining which patients are most severely injured, then prioritizing which patients receive treatment first and which patients can wait to receive delayed care. In general, this process (referred to as *triage*) sorts patients based on the severity of their injuries or medical conditions.

Triage is essential at all **mass-casualty incidents** because an increased number of patients places an increased demand on resources. At small-scale incidents, the first provider on scene with the highest level of training usually begins the triage process. Victims are ranked in order of severity of their injuries, and the victim with the most severe injury is given priority attention. After determining the total number of victims and requesting additional assistance, initial assessment of all victims should begin. As more resources arrive, the triage personnel should assign crews to give assistance to those patients in most need first.

At a large-scale mass-casualty incident, triage should occur in several steps. The following triage steps are accepted by the majority of larger-scale mass-casualty operations:

1. Life-saving care is rapidly administered to those in need.
2. Color coding is used to indicate priority for treatment and transportation at the scene. Red-tagged victims are the first priority, yellow-tagged victims are the second priority, and green- or black-tagged patients are the lowest priority Figure 4-6 ▼ .
3. Rapid removal of red-tagged victims for field treatment and transportation occurs as ambulances become available.
4. A separate treatment area is set aside to care for red-tagged victims if transport is not immediately available. Yellow-tagged victims can also be monitored and cared for in a treatment area while waiting for transportation.

Figure 4-5 A rescue event involving multiple victims will require triage.

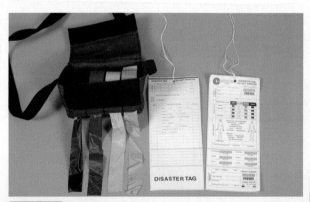

Figure 4-6 Triage tags are color coded to indicate treatment priority.

5. When there are more victims waiting for transport than there are ambulances, the transportation sector officer decides which victim is next to be loaded.

6. Specialized transportation resources (e.g., air ambulances, paramedic ambulances) require separate decisions when these resources are available but limited.

Follow local medical protocols during triage to determine which patients have the highest priority and require immediate ALS treatment.

■ Common Injury Patterns

To anticipate the traumatic injuries or serious medical conditions from which victims of technical rescue events may be suffering, rescuers should have a good understanding of some of the more common injury patterns and medical conditions associated with technical rescue situations.

Injury patterns common during vehicle accident entrapment include severe trauma, which may present itself in almost all areas of the body, but most commonly takes the form of soft-tissue injuries of the head, face, and upper torso. Lower leg fractures are also common. Trauma-related blood loss, both internally and externally, may be a concern, and the victim may be in respiratory distress as a result of traumatic injuries to the chest.

Confined-space emergencies typically result in patients being exposed to an oxygen-deficient or toxic atmosphere within the confined space. Respiratory protection and respiratory support will be required in such cases. Notably, a lack of oxygen and exposure to higher levels of toxic gases may lead to cardiac arrest in confined-space victims. In some situations, a worker inside a confined space may suffer a traumatic injury after falling off a ladder because of a lack of oxygen within the space. Confined-space emergencies can also result in traumatic injuries to extremities, which may come into contact with mechanical devices (e.g., augers or rotating vanes) inside the confined space. Thermal variations due to the type of operations carried out within a confined space (process vessels, industrial smoke stacks, or boilers), in combination with elevated air temperatures, may expose victims to heat-related emergencies such as heat exhaustion or heat stroke.

Trench and structural collapse injuries are likely to be of the traumatic type due to the mechanism of a heavy weight and significant force collapsing onto a victim. Rescuers should be prepared to deal with significant soft-tissue injuries, fractures, severe blood loss, respiratory distress due to pressure on the patient's chest, and potentially the serious medical situations known as **crush syndrome** (a condition wherein a large body part is released after a prolonged period of compression) and **compartment syndrome** (compression of nerves and blood vessels within an enclosed space).

A victim who is trapped or lost in the wilderness is at risk for a wide variety of medical conditions, including hypothermia, overexposure to heat and sun, snake bites, traumatic injuries from falls, dehydration, or drowning.

Water rescue situations vary, but may include ice rescue, surf rescue, still water conditions, and swiftwater rescue scenarios. Each of these events has the potential for a victim to suffer from exposure to the elements. Hypothermia and drowning will be the primary medical considerations at these events, although hypothermia should be a concern in all rescues with extended extraction times.

■ Victim Packaging, Movement, and Transfer

After rescuers gain access to victims and assess and treat their injuries, they must properly package the victims to limit further harm to them during their movement and transfer to a safe environment. Packaging is often accomplished using a long spine board or similar device. Packaging will often include such steps as implementing spinal immobilization, stabilizing or splinting fractures, placing the victim onto a stretcher device, and securing the patient to the device to limit movement that might otherwise exacerbate injuries.

Rescuers must choose the appropriate packaging device for the rescue environment where the patient is found Figure 4-7 ▶ . Rescue environments and their appropriate packaging devices are listed in Table 4-2 ▶ .

The primary objective for each rescue, transfer, and patient removal is to complete the process as safely and efficiently as possible while limiting further injury to the patient. The key to performing this task safely is to ensure that an adequate number of rescuers are assigned to help move the packaged patient through debris, lift the victim over barriers, navigate through water, or carry the victim across unstable surfaces without overexertion that could lead to rescuer injury. For example, patient movement and transfer through rough terrain, such as that often encountered during wilderness rescue, may require passing a stretcher up and over rocks and other obstacles, fording streams, or moving through narrow enclosures.

In certain situations, rescuers may need to use ladders, rope rescue equipment, tripods, and/or other mechanical hoisting equipment to help move patients. This is often the case when victims must be removed from a trench collapse, building collapse, wilderness environment, or confined-space environment.

Rescuers transfer patients to EMS responders for transport to a medical facility. This hand-off, known as **patient transfer**, occurs after the patients have been removed from the hazardous environment. It is important for rescuers to follow local medical protocols and effectively document which patient treatment was delivered while the patient was in their care. This treatment information must be clearly communicated to the EMS personnel so they are fully aware of the patient's condition when the rescuers first made contact and any treatment that was given to the patient by the rescue team. Local medical protocols will identify which type of documentation or patient care reporting must be completed by responders who deliver patient care.

Safety Tips

Responders need to limit their potential for experiencing overexertion during patient movement and transfer. To do so, ensure adequate staffing for patient movement and apply proper body mechanics during lifting and movement of patients.

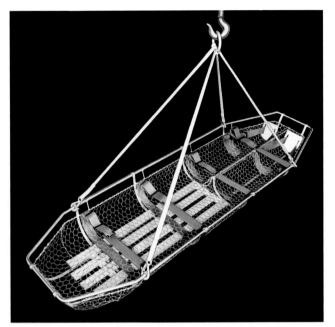

A.

B.

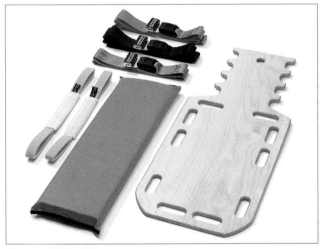

C.

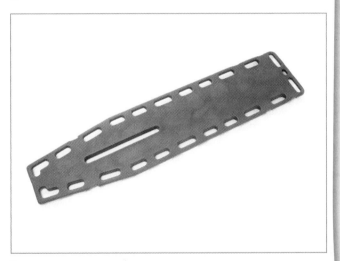

D.

Figure 4-7 Different rescue environments require different packaging devices. **A.** Basket stretcher. **B.** Flexible encapsulating stretcher. **C.** Half spine board. **D.** Long spine board.

Table 4-2   Packaging Devices for Various Rescue Environments

Packaging Device and Design Characteristics	Rescue Environment
Basket stretcher: rigid basket design for use where access to patient is not confined.	Rope rescue, wilderness rescue, trench rescue
Flexible encapsulating stretcher: flexible stretcher that wraps around the patient and creates a narrow profile for use in confined areas. Popular commercial brands are SKED, Half SKED, and Reeves Sleeve.	Confined-space rescue, structural collapse, rope rescue
Half spine board device: designed for use where initial spinal immobilization must occur in a confined area but the patient will be secured to a full board splint when moved to a less restrictive space. Popular commercial brands are LSP Half-Back and Oregon Spine Splint. The LSP Half-Back model can also be used with rope rescue equipment to raise a patient out of a confined space.	Confined-space rescue, vehicle rescue, trench rescue, structural collapse rescue, wilderness rescue
Long spine board device: used to stabilize suspected spinal injuries and/or other fractures. It is used in conjunction with basket stretchers and flexible encapsulating stretchers for moving patients.	Wilderness rescue, vehicle rescue, trench rescue, structural collapse rescue, water rescue

Resource Requirements

Emergency Medical Services Resources

EMS requirements can be significant at a technical rescue event. Both BLS- and ALS-trained personnel may be required to treat victims of vehicle accidents, trench collapse, confined-space emergencies, or building collapse incidents. Depending on the type of technical rescue, these events may result in a mass-casualty situation that will require the need for bulk EMS supplies such as backboards, cervical collars, trauma dressings, and burn care supplies. An enhanced response from EMS providers will be needed as well. For instance, patients trapped within a collapsed structure or pinned in a trench collapse will require ALS intervention while they are still trapped to increase their chances for a successful outcome. A mass-casualty incident can place such great demand on equipment or personnel that the system is stretched to its limit—or beyond.

Specialized Victim Packaging and Transfer Equipment

Some technical rescue events, such as confined-space and structural collapses, require the use of specialized patient packing devices designed to fit in small, space-limited areas. For example, a small-diameter packaging device such as a SKED stretcher might be required for use inside a sewer, narrow tunnel, or building collapse void. The packaging device must allow for proper management of the victim, proper packaging for stabilizing injuries, and proper patient movement through narrow openings and over rough terrain. A technical rescue event with multiple patients requires multiple patient packaging devices on the incident scene. If an adequate number of devices is not available, patient movement and transfer to a medical facility will be delayed.

Some technical rescue situations, such as those involving a climber trapped on a rock face, a worker trapped in a utility vault, or a victim of a building collapse, may require a patient packaging device that can be attached to a rope rescue system for lowering or hauling the victim to a safe location. The chosen device must allow for proper stabilization of the patient and be specifically designed and rated for use with a rope rescue system.

Transport Vehicles

In most technical rescue situations, rescuers must deal with only one or two patients in need of treatment and transportation to a medical facility. In these situations, a BLS or ALS transport vehicle such as an ambulance or medic unit is required Figure 4-8 ▸ . If more than two victims are encountered, however, multiple transport units may be required to deliver the patients to a medical facility.

In some situations, such as a bus accident that produces multiple patients, a mass-care vehicle may be required. Many jurisdictions have specially designed medical buses that have been outfitted for this purpose Figure 4-9 ▸ . In other situations, commercial buses or school buses can be used to transport multiple patients who may be suffering minor injuries.

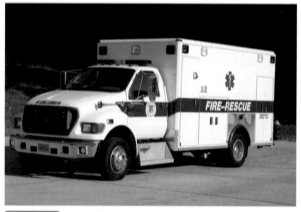
Figure 4-8 An ambulance can transport one or two patients.

For those patients with serious injuries, burns, or other serious medical conditions, the use of aircraft such as helicopters may expedite the transport time to distant hospitals or to specialized medical facilities such as trauma centers or burn centers.

Helicopter Operations
Technical Rescue Helicopter Usage

It is important for rescue workers to know when the use of helicopters will prove advantageous in the management of an incident. Recognizing the need for helicopter operations and knowing how to request a helicopter can go a long away toward ensuring that the victim of a technical rescue incident has the best chances for survival.

Figure 4-9 Many jurisdictions use medical buses to transport several patients at once.

Helicopters are an especially valuable resource when used as part of large-area search operations. Their benefits have been proven time and again at incidents such as earthquakes, tornados, and hurricanes. Helicopters can be used to perform initial reconnaissance of an impacted area after a disaster situation, or to help locate a person lost in a wilderness setting. The advantages of using a helicopter to perform recon missions and search a much larger area in a shorter time than would be possible by personnel on the ground cannot be overstated.

Helicopters are also an excellent tool for transporting rescue teams over long distances and inserting the team members into isolated areas that might be time-consuming and difficult to access by ground. For example, an island may become inaccessible by ground if the bridges and roadways are damaged by a hurricane. An USAR mission to the island may, therefore, require a team to be airlifted by helicopter.

Helicopters are also a valuable tool during flooding and swiftwater rescue scenarios **Figure 4-10 ▼**. For example, a person stranded in a vehicle by flood waters or isolated on a rock outcropping in the middle of a raging river may require the assistance of a helicopter staffed by a trained rescue crew to access the patient, secure the individual in a lifting device, and lift the victim to safety.

Helicopters are also a common mode of transportation when serious medical emergencies arise. Local medical protocols will specify when rescuers may use helicopters for transporting patients. Typically, helicopters are used when the transport of trauma patients to a trauma center is justified, when the travel distance to the appropriate medical center is significant, or when there is a need to transport a patient to a distant specialty treatment center such as a burn center.

Helicopter Resources

Not all helicopters are created equal. Specialized resources are needed to conduct helicopter search and rescue operations, for instance. Although it is not realistic to expect every rescuer to have an in-depth understanding of the capabilities of the various helicopters on the market, it is important to recognize that when requesting the services of a helicopter, you must be very specific about what you expect the helicopter to accomplish. The agency (law enforcement, military, private) that supplies the helicopter to assist with search and rescue will be in the best position to determine which type of helicopter is appropriate for the proposed mission.

There are distinct differences between a helicopter that can transport one or two medical evacuees and a helicopter that can transport several members of a rescue team and all of their associated equipment to a disaster site. In addition, helicopters that can perform water rescues via winch capabilities or that can lift and transport heavy equipment are not always readily available from the local or regional response area. Given these constraints, it is important to identify the locations and response times of such resources during the response planning phase for technical rescue events. Likewise, every rescue response agency should have written standard operating procedures for requesting the assistance of a helicopter and identifying how responders operate in conjunction with a helicopter crew.

Helicopter Safety and Landing Zone Management

Although awareness level responders typically will not work as a member of a helicopter flight crew, they may be exposed to helicopter hazards and must interact with aircraft crews while operating in the vicinity of helicopter landing zones. There are significant hazards involved in the operation of any helicopter, including some hazards directly related to approach, landing, and takeoff from landing zones. Weather conditions such as temperature extremes, rain, wind, snow, and fog all add to the difficulty of flight operations. The altitude at which the helicopter must function may also affect the aircraft's ability to operate efficiently—an important consideration for helicopter usage during search and rescue operations in high-altitude mountain environments.

Ground-level obstructions such as trees, forest canopies, towers, utility poles, and wires are other hazards that may be encountered during landing zone operations. The downwash of air created by the helicopter's rotating blades can cause trash and other lightweight materials to fly up into the air and possibly strike responders, bystanders, or the aircraft.

Properly trained awareness level responders may be involved in identifying, procuring, and securing landing zones during helicopter operations. Many agencies have written procedures for establishing landing zones and often have these zones pre-identified for their response areas. When choosing a landing zone for a helicopter you should take the following points into consideration:

- The proximity of the landing zone to the rescue area
- The potential of exposing the aircraft to any hazardous materials release that may have occurred
- Ground-level obstructions

Figure 4-10 Helicopters are valuable tools for use during water rescue operations.

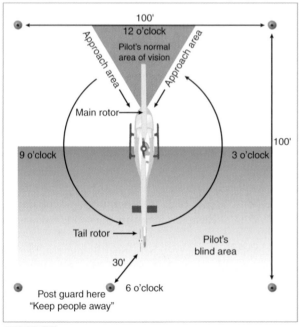

100'
12 o'clock
Pilot's normal area of vision
Approach area
Approach area
Main rotor
100'
9 o'clock
3 o'clock
Tail rotor
Pilot's blind area
30'
6 o'clock
Post guard here "Keep people away"

Figure 4-11 A helicopter landing zone with cones/warning devices in place.

A good landing zone will also be large enough that it includes an approach and departure zone, so that the helicopter does not have to take off and land straight up and down **Figure 4-11 ▲**. As a rule of thumb, the touchdown area for the helicopter should be a minimum of 100 ft × 100 ft. The ground surface should be firm, with vegetation less than 2 feet high and a slope of no more than 5 degrees.

Rescue workers may help load patients into helicopters or board and exit helicopters during rescue missions. As a consequence, all rescuers must have a good understanding of the safety procedures for operating around helicopters. Rescuers who work around helicopters must always wear proper PPE—at a minimum, a helmet, eye protection, hearing protection, safety footwear, gloves, and clothing appropriate for the mission.

Safety Tips

Always follow the direction of pilots and flight crews when operating near helicopters.

Rescuers should also have a thorough understanding of the proper procedures for working around a helicopter, for establishing communications with the helicopter crew, and for working with the pilot and helicopter crews. These procedures should include at least the following guidelines:

- Always follow the pilot's and flight crew's instructions.
- Stay a safe distance away from the aircraft.
- Never approach the aircraft from the rear.
- Never duck under the body of the aircraft.
- Walk in a crouched position when approaching the aircraft.

- Beware of uneven terrain where main rotor blades could dip close to the ground.
- Control all loose clothing, gear, and equipment around the helicopter and in the area of the prop downwash.
- Identify and control loose debris in the landing zone area.
- Cordon off the landing zone to limit incursion by unauthorized personnel or vehicles.
- Keep bystanders at least 200 feet away from the landing zone and emergency responders at least 100 feet away.
- Identify the locations of any obstructions, and communicate this information to the pilot.
- Limit radio traffic and maintain communications with the helicopter crew during takeoffs and landings.

Although awareness level responders will have limited opportunity to fly in a helicopter, helicopters may sometimes be used to transport responders to a disaster site or to a remote rescue site. In these situations, the helicopter pilot or crew will usually perform a safety briefing if a responder assists a helicopter as a passenger. This safety briefing will normally cover the following points:

- Rotor hazard zones
- The way to walk to and from a helicopter safely
- Use of safety belts
- Operation of doors and emergency exits
- Location of survival equipment
- Location and use of required flotation equipment
- Location and operation of fire extinguishers
- Location and operation of emergency locator transmitter
- Location and operation of emergency fuel shut-off
- Crash and emergency procedures

Incident Management Requirements

The strategic objective when operating at a technical rescue incident is to control and manage the event effectively by carrying out the following responsibilities: evaluating the scene and identifying potential victims and their location; initiating operations to minimize hazards to operating personnel and trapped victims, to perform a search, to treat patient injuries, and to perform rescue and removal of trapped victims; and minimizing further injury to victims during search, rescue, and removal operations. These objectives are achieved through the Incident Command System (ICS).

Staffing of ICS positions is determined by the scope of the event. Large-scale technical rescue events such as a building collapse may require response personnel to fill the positions of incident commander, operations section chief, planning section chief, logistics section chief, and safety officer. In situations where multiple patients are encountered, an EMS branch director or group supervisor may also be required. In a large-scale technical rescue event, the operations section chief will most likely have various branch directors and/or division/group supervisors assigned to manage specific problems or geographical areas. The planning and logistics sections will assign knowledgeable skilled personnel to fill positions such as those focusing on the situation status, resource status, service branch, and/or support branch as required by the event.

Incident management personnel are responsible for strategic and tactical management of the technical rescue emergency response. One of these responsibilities is the development of an incident action plan (IAP), which identifies the overall control objectives for the emergency. Large-scale events such as a building collapse may last multiple days. To deal with this situation, long-term staffing considerations are developed and included in the plan. The IAP covers issues such as the assignment of personnel to various management positions responsible for plan implementation. Tactical assignments and resource assignments, such as a medical team to perform patient care within an assigned geographical area, are listed in the IAP as well. Operational work periods are identified and safety messages are included. A medical plan that identifies the location of aid stations, transport resources assigned to the event, and available treatment facilities is also included in the IAP.

The successful management of a multiple-patient event requires not only adequate numbers of EMS personnel to treat, package, and transfer victims, but also mass-casualty management processes. These processes include triage to determine which patients are most severely injured, assignment of patients to various treatment areas depending on the severity of their injuries, determination of the treatment centers to which the injured are to be transported, and management of transportation vehicles.

Response Planning

Responders must know how to initiate the emergency response system to ensure that appropriate resources are deployed to a technical rescue event and that operational guidelines are followed properly. Many response organizations perform needs assessments within their response areas on a regular basis to determine their vulnerability to various technical rescue events. As part of this assessment, an organization should consider the types of technical rescue exposures that are present within its jurisdiction, the potential for mass-casualty events, and the emergency response resource needs. The emergency response system should include written procedures to request agency resources and mutual aid resources; contracts with private-sector companies, if necessary; and memoranda of agreement with regional, state, or federal assets. Private-sector agreements may be established to acquire resources that cannot be supplied by the response agency (e.g., EMS resources, ground transport units for victims, helicopters for victim transport, and mass-care supplies).

The capability of agency personnel should also be identified in advance of an actual emergency. This capability assessment should include the level of performance at which agency personnel are trained, such as the operations or technician level for technical rescue personnel, and the BLS or ALS level for emergency medical services personnel.

As part of the emergency response system, rescuers must also know the location, capabilities, and response times for regional, state, and federal search and rescue teams and medical teams. Written standard operating procedures should be established to identify how on-scene response personnel can request these assets. Dispatch procedures should be in place to ensure that adequate resources are deployed when a technical rescue event occurs, allowing effective search, rescue, and victim treatment operations to begin.

Wrap-Up

Chief Concepts

- Conducting interviews and gathering information about the location of potential victims are important search tactics.
- The specific rescue environment, such as a building collapse or a confined-space emergency, will determine the search resource requirements.
- Victim management requirements at technical rescue incidents can vary from a single patient at a vehicle accident, to multiple persons trapped in a sewer, to hundreds of people injured during the collapse of a high-rise building.
- Victim management resource requirements needed to effectively manage a technical rescue event vary with the type of entrapment problem and number of victims facing rescuers. All technical rescue incidents require resources to perform a search; deliver patient care; perform a rescue; package, transfer, and transport victims; and eliminate hazards.
- Effective planning for a technical rescue emergency incorporates risk assessment, resource assessment, and development of written procedures. It is also necessary to implement an Incident Command System to be able to control this type of emergency.
- Rescuers must be knowledgeable about injury patterns commonly encountered in the various types of technical rescue events.

- Proper patient treatment, packaging, and transfer are critical to the positive outcome of victims at technical rescue emergencies. Patient transfer is the process of removal of a patient from a hazard zone and the subsequent hand-off of the patient to a transport vehicle.
- Effective management of landing zone operations and safe work practices around helicopters is extremely important to responder safety.

Hot Terms

Advanced life support (ALS) Advanced life-saving procedures, such as cardiac monitoring, administration of IV fluids and medications, and use of advanced airway adjuncts.

Awareness level The first level of rescue training provided to all responders, which emphasizes recognizing hazards, securing the scene, and calling for appropriate assistance. There is no actual use of rescue skills at the awareness level.

Basic life support (BLS) Noninvasive emergency life-saving care that is used to treat airway obstruction, respiratory arrest, or cardiac arrest.

Body substance isolation (BSI) An infection control concept that treats all bodily fluids as potentially infectious.

Compartment syndrome A medical condition in which nerves and blood vessels are compressed within an enclosed space.

Crush syndrome A medical condition in which a large body part is released after a prolonged period of compression.

Entrapment A condition in which a victim is trapped by debris, soil, or other material and is unable to perform self-extrication.

Mass-casualty incident An emergency situation that involves more than one victim and that can place such great demand on equipment or personnel that the system is stretched to its limit (or beyond).

Patient packaging The process of preparing a victim for movement as a unit, often accomplished with a long spine board or similar device.

Patient transfer The process of removing a patient from a hazard zone and the subsequent hand-off of the patient to a transport vehicle.

Primary search The deployment of search resources to initiate a rapid search of an area that is thought likely to contain survivors.

Reconnaissance (recon) An initial survey or rapid visual check of a rescue environment to obtain information that can be used to develop a search plan and direct search and rescue operations.

Search phase The phase of a rescue incident in which the victims are located.

Secondary search A slower, more methodical, and detailed search of a rescue environment to ensure that no victims were overlooked during the initial (primary) search operation.

Triage The process of sorting victims based on the severity of their injuries and medical needs, with the goal of establishing treatment and transport priorities.

Victim management All aspects of identifying hazards to which a victim may be exposed, gaining access to a victim, performing triage of multiple victims, assessing patient injuries, stabilizing victim injuries, communicating with victims, and moving and transferring those victims who require technical rescue assistance.

Rescue Responder *in Action*

Your engine company is dispatched to a medical call for two workers injured at a construction site. Two emergency medical response units with BLS capability are also dispatched to assist you and your crew. During the response, the dispatcher tells you that the location of the incident is a construction site where workers were building a concrete cinder block wall, that the wall fell during a gust of wind, and that at least two workers are suffering from head injuries. Your officer requests the assistance of an ALS unit.

When your engine company arrives at the construction site, you find a very chaotic scene where many people are yelling and several workers are digging through the rubble of the collapsed wall. Your officer steps out of the engine, and the construction foreman grabs his coat; he tells the officer that there are 8 injured employees, including 2 who are unconscious, and that at least 10 other workers are buried under the wall debris. The officer gets on the radio and asks for additional resources.

1. Which initial actions should be taken to manage the eight victims who are not trapped?
- **A.** Do not attempt to assess the victims who are not trapped. Instead, concentrate on digging out the trapped victims.
- **B.** Triage the injured and accessible victims.
- **C.** Perform a thorough hazard assessment.
- **D.** Both B and C.

2. Which additional resources should be requested to assist with this event?
- **A.** Additional emergency medical resources, both ALS and BLS
- **B.** Command staff
- **C.** Structural collapse rescue resources
- **D.** All of the above

3. Which potential hazards does the rescue environment pose for personnel who perform victim management?
- **A.** Additional collapse hazards
- **B.** Bloodborne pathogens
- **C.** Loose debris creating tripping hazards
- **D.** All of the above

Introduction to Technical Rope Rescue

NFPA 1670 Standard

5.1 General Requirements.

5.1.1 Organizations operating at rope rescue incidents shall meet the requirements specified in Chapter 4. [p. 112]

5.1.2* The AHJ shall evaluate the need for missing person search where rope rescues might occur within its response area and shall provide a search capability commensurate with the identified needs. [p. 114–115, 118]

5.2 Awareness Level.

5.2.1 Organizations operating at the awareness level for rope rescue incidents shall meet the requirements specified in Section 5.2. [p. 110–112]

5.2.2 Organizations operating at the awareness level for rope rescue incidents shall develop and implement procedures for the following:

(1) Recognizing the need for a rope rescue

(2)* Identifying resources necessary to conduct rope rescue operations

(3)* Carrying out the emergency response system where rope rescue is required

(4)* Carrying out site control and scene management

(5)* Recognizing general hazards associated with rope rescue and the procedures necessary to mitigate these hazards

(6)* Identifying and utilizing PPE assigned for use at a rope rescue incident [p. 110–112, 115]

Additional NFPA Standards

NFPA 1001, *Standard for Fire Fighter Professional Qualifications*

NFPA 1006, *Standard for Technical Rescuer Professional Qualifications*

Knowledge Objectives

After studying this chapter, you will be able to:

- Identify the need for rope rescue equipment and the application of rope rescue skills at a rescue operation.
- Recognize and identify fall hazards and other hazards commonly found at rope rescue incidents.
- Identify various types of rescue incidents that require rope rescue equipment and rope rescue skills.
- Describe the rope rescue resources needed to conduct various search and rescue operations.
- Describe response planning and incident management requirements related to a search and rescue incident that requires rope rescue teams.
- Describe site control operations at a high-angle rescue incident.
- Describe non-entry rescue considerations at a high-angle rescue incident.

Skill Objectives

There are no skill objectives for this chapter.

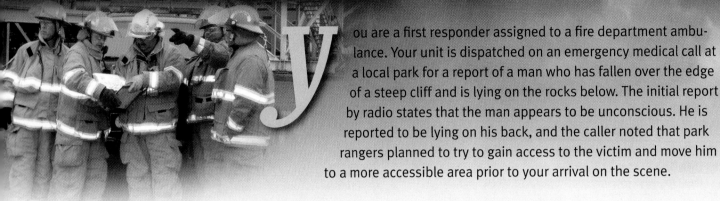

You Are the Rescue Responder

You are a first responder assigned to a fire department ambulance. Your unit is dispatched on an emergency medical call at a local park for a report of a man who has fallen over the edge of a steep cliff and is lying on the rocks below. The initial report by radio states that the man appears to be unconscious. He is reported to be lying on his back, and the caller noted that park rangers planned to try to gain access to the victim and move him to a more accessible area prior to your arrival on the scene.

As you approach the scene with your vehicle, you notice a park maintenance worker waving for you to follow him down a pathway that leads to where the patient is located. You and your partner park your vehicle, grab your medical equipment, and follow the park worker along a dirt path to the edge of a steep drop-off. This area is a sheer rock face that is often used by recreational climbers. The rock face is too steep to descend without specialized equipment. You see the victim located about 75 feet below the top of the cliff, lying face down on a rock outcropping that is still elevated more than 50 feet from the bottom of the cliff. The park rangers who planned to try to access the patient are at the bottom of the cliff and

cannot access him. As you perform a size-up of the event, you realize that this type of incident requires specially trained responders and specialized tools and equipment to operate effectively.

1. What should you be looking for during a high-angle rope rescue size-up?
2. Which hazards might rescuers find at this type of event?
3. Which actions can first responders take at this incident to more effectively manage the event?
4. Which non-entry actions can you take at an event that requires rope rescue equipment and skills?
5. Which types of resources are needed to effectively manage a high-angle rope rescue event?

Introduction

Emergency responders may encounter a variety of situations where rescuing injured or trapped patients requires rope rescue equipment and techniques. These types of emergency operations are commonly referred to as **rope rescue incidents**. Most rope rescue incidents involve victims who are trapped or injured and are in normally inaccessible locations, such as on a mountainside or on the outside of a high-rise building. At the awareness level, rescuers must be aware of the many different elements involved in technical rope rescue. Their responsibilities at these scenes, however, are limited to isolating the incident, protecting the victim from further harm, and recognizing the need for and summoning specialized rescue resources.

Situations Requiring Rope Rescue

Rope rescue incidents are categorized as *low-angle* or *high-angle rescue events*. **Low-angle rope rescue** operations are situations where the slope of the ground over which the rescuers are working is less than 45 degrees. In this instance, rescuers depend

on the ground for their primary support, and the rope system becomes the secondary means of support **Figure 5-1 ▶**. A good example of such a situation is a vehicle accident that results in a car going over an embankment and where rescuers subsequently use a rope to assist in carrying a victim up a slightly sloped hillside. Keep in mind, however, that even low-angle rope rescues can be complicated by the ground surface. Operations are much more difficult on surfaces that are wet, icy, or covered in loose material than they are on surfaces that are dry and hard.

High-angle rope rescue operations include situations in which rope is used to support the rescuers' load and the slope of the surface where rescuers are operating exceeds 45 degrees. In these situations, the rope system is the primary load-carrying support system; indeed, in some cases, it is the entire load-carrying support system. An example of such a situation would be an incident in which a rescuer is lowered over the edge of a building wall to access and retrieve a window washer who is hanging from a malfunctioning fall protection system. **Fall protection** is a system comprising rope hardware and software that is used to protect workers from falling from an elevated position. It also refers to the equipment to which workers are

Figure 5-1   A low-angle rescue.

Industrial settings may also necessitate that rescue workers use rope rescue **mechanical advantage systems** and descent control techniques to move victims from dangerous environments. Mechanical advantage systems create a leverage force through the use of levers, pulleys, or gearing **Figure 5-3 ▾**. The mechanical advantage of a system is usually expressed as a ratio of output force to input force.

attached while working at elevations or near a fall hazard, where the equipment is meant to capture them should they fall.

Responders may, for example, encounter situations requiring high-angle rescue when they respond to injured or ill workers at a construction site. The victims may be found below grade in a basement of a building under construction, in a trench or excavation, or inside a shaft. Victims may also be found at an elevated height, such as on a rooftop or in the operator's compartment of a tower crane **Figure 5-2 ▾**. The victim may be suffering from a life-threatening illness such as a heart attack, or the individual may be injured through a fall or by having an object fall on him or her from a higher floor. In any of these situations, rope rescue equipment such as life safety rope, rope hardware, and victim packaging devices may be required. In such circumstances, rescuers often have to lower themselves to the location of the victim using a system of ropes, anchors, webbing, carabiners, and other devices to render treatment to the victim.

A successful rescue requires rescue teams or individuals who have been trained to use this equipment and to construct and operate **rope rescue systems**. A rope rescue system is a constructed system consisting of rope rescue equipment and an appropriate anchor system intended for use in the rescue of a subject.

Figure 5-2   Specialized training and equipment are required to perform a high-angle rope rescue.

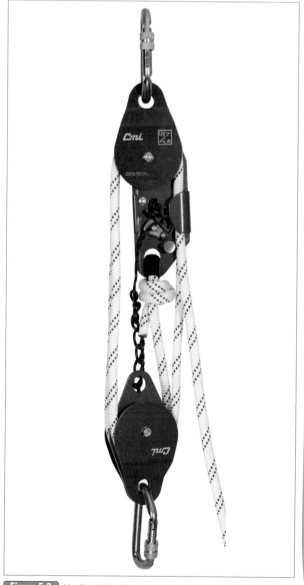

Figure 5-3   Mechanical advantage system.

Descending is a means of safely traveling down a fixed rope with the use of a descent control device. This rescue technique may be required in confined-space enclosures such as storage tanks, tunnels, or process vessels. Industrial site rope rescue may also occur where workers become injured while making repairs on electrical, water storage, or communications towers. In these situations, special rope rescue teams will be required to access the victims safely, properly package them in specialized patient packaging devices, and attach rope descent systems for lowering the victims to ground level. **Packaging** refers to the process of preparing a victim for movement as a unit, often accomplished with a long spine board or similar device.

> ## Safety Tips
>
> Industrial complexes are rife with hazardous situations for rescue workers. These hazards may include hazardous materials exposure, confined spaces, machinery entrapment, falls, and extreme noise levels. Rescuers must be prepared to protect themselves from these hazards through use of the appropriate personal protective equipment.

Emergency responders in many rural settings may encounter situations where recreational climbers have fallen down cliffs, hills, or rock faces and require rescue. In addition, rural area responders may encounter situations where entry into caves and mines is required to retrieve injured patients. Each of these situations requires specialized rescue teams who have been trained to operate in these unique environments and to build and operate rope rescue systems **Figure 5-4 ▶**.

Urban area responders may encounter situations where maintenance workers, such as window washers, become stranded or injured while working at heights on the outside of a high-rise building. Again, specialized rope rescue equipment and specially trained personnel will be required to remove the trapped workers successfully.

Other situations where rescuers may need to apply rope rescue tools and rope rescue skills include transportation accidents where a vehicle goes over an embankment and rope systems are required to access and remove patients; building collapse events, where ropes may be needed for lowering victims down from the top of a rubble pile; and water rescue environments, where personnel may need to use ropes to help victims stranded in swiftwater situations.

Applicable Standards

Emergency responders can find guidance on how to manage technical rescue incidents by referring to the National Fire Protection Association (NFPA) Standards 1006, *Standard for Technical Rescuer Professional Qualifications*, and 1670, *Standard on Operations and Training for Technical Search and Rescue Incidents*.

NFPA 1006 establishes the job performance requirements (JPRs) needed for responders to perform technician level rescue skills. These skills include preplanning a high-angle rope rescue;

Figure 5-4 Rural rope rescue.

performing a site assessment; hazard recognition and control; application of appropriate rope rescue skills; and patient assessment, packaging, and transfer techniques.

NFPA 1670 establishes organizational requirements to operate safely and effectively at rope rescue emergencies. This standard identifies the requirements for risk and hazard assessment, written procedures, safety protocols, personal protective equipment (PPE), establishment of operational response level, and resource considerations. The general requirements in Chapter 4 of NFPA 1670 must be met by any organization operating at a rope rescue incident.

Resource Requirements

Equipment Resources

Rope rescue tools and equipment that may be required for high-angle rescue scenarios are classified into two categories known commonly as hardware and software. **Rope hardware** is rigid, mechanical auxiliary equipment that can include, but is not limited to, anchor plates, carabiners, and mechanical ascent control devices **Figure 5-5 ▶**. A **carabiner** is a piece of metal hardware used extensively in rope rescue operations. It is generally an oval-shaped device with a spring-loaded gate that can be used to connect pieces of rope, webbing, and other rope hardware.

Other hardware devices include descent control devices such as a figure eight plate, brake rack, or brake tube. These devices are used to control the descent speed of a load, such as a rescuer or victim attached to a rope. The rope is attached to the descent

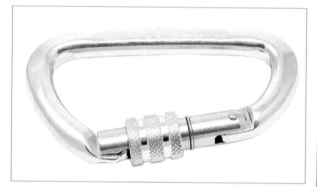

A.

Figure 5-6 One example of rope software—an adjustable pickoff strap.

B.

Figure 5-5 Rope hardware. **A.** Carabiner. **B.** Rope clamp.

Rescue Tips

All rope rescue equipment must be protected from damage during storage, training, and actual rescue operations. Any equipment that is damaged or potentially damaged should be removed from service immediately. All rope rescue hardware and software should be routinely inspected for wear, tear, and damage. Written records, such as rope history logs that track the use and history of rope software, are important adjuncts in assessing the safety of rope equipment.

control device; as the rope then feeds through the descent control device, a varying amount of friction is created between the device and the rope, which controls the descent speed.

Pulleys are rope hardware that is used to change the direction of a rope system or to impart mechanical advantage within a rope system. A variety of specialty devices may also be used to attach to ropes or to help distribute loads within a rope system.

Rope software comprises flexible fabric components of rope rescue equipment; it includes rope, accessory cordage, webbing, anchor straps, and harnesses **Figure 5-6 ▶**. Of course, rope itself is the primary work tool in the rope rescue system. A **harness** is a piece of safety equipment that can be worn by a rescuer and attached to a rope rescue system, thereby allowing the rescuer to work in a high-angle environment safely during rope rescue operations.

All rescue-related rope equipment—especially rope—must be manufactured to specific manufacturing standards to be used for rescue purposes. These manufacturing standards can be found in NFPA 1983, *Standard on Life Safety Rope and Equipment for Emergency Services.*

Specialized equipment that may also be required during rope rescue situations includes portable anchor devices such as tripods, bipods, Larkin Frames, and A-frame devices such as the Arizona Vortex. These devices are used to create an overhead anchor for the attachment of rope retrieval systems, in areas where a stationary anchor may not exist. Tripods are often used in confined-space rescue for entry into manholes and utility vaults.

The Arizona Vortex and Larkin Frame are commonly used in wilderness settings, such as at the top of a rock face, where the portable anchor point created by the device will be projected out and above the edge where rescuers will be raised or lowered **Figure 5-7 ▶**. Other specialized rope equipment includes **edge protection**, a device for protecting software such as webbing and rope from abrasion and friction damage as the rope or webbing passes over sharp or jagged edges such as rock or concrete.

Patient packaging equipment will also be required for use at rope rescue events. Rescuers must choose the appropriate packaging device for the rescue environment. Packaging devices and the rescue environments for which they are designed are discussed in Chapter 4.

■ Rescue Team Resources

Rescue situations in which responders must gain access to a patient located above or below grade, then raise or lower the victim to a safe environment using rope systems, require specially trained personnel or teams. Unique training and skills are necessary to use rope rescue tools and equipment correctly and to build

Figure 5-7 The Larkin Frame may be used in wilderness rescues as substitutes for tripods.

and operate rope rescue systems safely. Rope rescue techniques include establishing anchor systems, establishing fall protection, building and operating lowering systems, building and operating mechanical advantage systems, building and using high-line systems, and **ascending** rope. Ascending is a means of safely traveling up a fixed rope with the use of ascent devices. Any of these skills may be required at a high-angle rope rescue incident.

Awareness level responders are not properly trained to operate rope rescue equipment or to use any of the previously listed rope rescue techniques. Instead, awareness level is the first level of rescue training provided to all responders; it emphasizes the ability to recognize hazards at an incident, secure the scene, and call for appropriate assistance and resources.

Rescue Tips

Rope rescue tools and techniques may be required to move patients effectively within a variety of rescue environments, including swift water, trench collapse, and confined spaces. Awareness level responders must recognize the type of rescue environment involved and ensure that rescue teams crossed-trained for both rope rescue and the rescue environment are requested and made available.

Emergency Medical Services Resources

Emergency medical services (EMS) resources are required so rescue teams can hand off patients for treatment and transport to medical facilities once the victims are retrieved utilizing rope

rescue skills and equipment. High-angle rope rescue victims may be suffering from traumatic injuries and/or other medical emergencies. Advanced life support (ALS) capabilities are required to treat fall victims effectively once they are moved to a safe location.

Incident Management Requirements

The strategic objective when operating at an event that requires rope rescue equipment and personnel trained in rope rescue is to control and manage the event in such a way as to evaluate the scene and identify potential victims and their location; to initiate operations to minimize hazards to operating personnel and trapped victims, to search the area, and to rescue and remove trapped victims; and to minimize further injury to victims during search, rescue, and removal operations.

The Incident Command System (ICS) positions are staffed as determined by the scope of the event. In most cases, at a minimum, management personnel will be required to fill the positions of incident commander (IC), technical rescue group supervisor, and safety officer.

The technical rescue group supervisor is responsible for implementing tactical rescue decisions to support the strategy established by the IC Figure 5-8 ▾ . These tactical benchmarks include hazard mitigation, entry team readiness, rapid intervention capabilities, and emergency medical care for the patient.

Incident management personnel are responsible for strategic and tactical management of the emergency response. One of their responsibilities is the development of an incident action plan (IAP), which identifies the overall control objectives for the emergency. Because an emergency requiring specialized rope rescue skills may include unique operations and hazards related to another causative factor (e.g., confined space, building collapse, water rescue), the event may require additional specialized resources. These specialized resources could include swiftwater rescue teams, building collapse assets, and/or trench rescue teams. The incident management personnel may need to gain input from other knowledgeable people, including personnel who have been trained in specific areas of technical rescue, such as technical search personnel, rescue team managers, structural engineers, or emergency medical professionals.

Figure 5-8 Team briefing at a technical rescue event.

Response Planning

Responders must know how to initiate the emergency response system to ensure that the appropriate resources are deployed to the event and that operational guidelines are initiated correctly. On a routine basis, response organizations should perform needs assessments within their response area to determine their vulnerability to events that might require rope rescue equipment and skills. As part of this assessment, the organization should consider the types of emergencies they might encounter that would require the use of rope rescue skills and equipment, the type of hazards that might be encountered, and the emergency response resources needed to handle such an incident. The emergency response system should include written procedures to request agency resources and mutual aid resources; contracts with private-sector companies; and/or memoranda of agreement with regional, state, or federal assets. Private-sector agreements may be established to acquire resources that cannot be supplied by the response agency, such as specially trained rescuers and specialized rope rescue tools and equipment. In wilderness settings where wilderness search capabilities may be a factor in achieving a successful outcome, specialized search assets should be factored into the plan.

The capability of agency personnel should be known in advance of an actual incident. A capability assessment should determine the level of performance at which agency personnel are trained, such as the operations or technician level for high-angle rope rescue. As part of the emergency response system, it is also important to know the location, capabilities, and response times for mutual aid, regional, state, and private-sector rope rescue teams. Procedures should be established to request these assets when necessary. Dispatch procedures should be in place to ensure that adequate and proper resources are deployed as requested so effective search and rescue operations can begin.

Hazards and Hazard Assessment

Each high-angle rope rescue incident has its own specific problems and hazards. To protect personnel from being injured or killed, personnel must be well trained to recognize and understand the hazards they may encounter.

Scene Assessment

When emergency responders encounter a high-angle rescue event, one of the primary hazards they face is a fall hazard. In many situations, rescuers must work close to the edge of a building or the edge of a rock outcropping to search for and locate a person who may have fallen. This situation places responders in danger of slipping and falling over the edge themselves. A significant fall is likely to result in a responder suffering a significant traumatic injury or death. All rescuers who are required to work within 10 feet of an edge where they could suffer a fall should be connected to some type of fall protection device for their own protection Figure 5-9 ▸.

High-angle rescue events can also involve victims who are trapped hanging from a rope system or fall protection system or who are lying in a dangerous position along a ledge or rock outcropping. Rescuers may locate the victim while operating

Figure 5-9 At a rope rescue scene, rescuers at risk of a significant fall should be equipped with fall protection equipment.

from below the point where the victim is trapped. In these situations, responders must be aware of the potential for falling debris such as rocks, tools, or other items that may fall unexpectedly and strike rescuers positioned below. Responders should always wear head protection and position themselves so that they are not located directly below where the victim is located in case something does fall.

Bystanders, onlookers, and friends or co-workers may also create hazards for rescuers by interfering with rescue operations or getting too close to an unprotected edge. Crowd control is important to alleviate this problem and to ensure that bystanders do not become secondary victims or cause responders to get injured.

High-angle rope rescue situations in the industrial setting may also expose rescuers to energy hazards such as exposed electrical wires, hydraulic fluids, steam lines, or operating machinery Figure 5-10 ▸ . In such cases, utilities and energy sources must be identified and controlled if present. Rescue environment hazards must also be assessed depending on the primary cause of the emergency. At building collapse, trench collapse, water rescue, and confined-space rescues, the hazards specific to those opera-

Rescue Tips

When operating at a scene where a victim is stranded on the side of a building, attempt to locate the victim from the exterior of the building at ground level first, then access the roof and proceed cautiously to the edge if necessary to communicate with the victim. If you must operate on the roof of a building and you make entry to the roof by way of an internal stairwell, do not run to the edge to determine the location of the victim. Lie down and/or shift your body weight to your rear foot as you go near an edge; this stance limits the potential for a fall. You should make such an approach only if necessary to locate the victim initially and to communicate with the victim. If equipment is available, always use fall protection when operating near a drop-off or where a fall hazard exists.

Voices of Experience

On January 24, 2008, at 0744, the Salt Lake City Fire Department Dispatch Center received a 911 call indicating that a crane operator had fallen and injured his back atop a tower crane operating at a construction site at the University of Utah campus. Our heavy rescue team (HRT), along with other apparatus from the SLCFD, were dispatched to the scene.

The first arriving fire fighters learned that a 46-year old male had been climbing within the ladder cage of the crane when he slipped on the icy rungs. He fell approximately six to ten feet before his safety lanyard arrested his fall. Injured and unable to climb up or down, he was lifted onto a small platform below the crane's control booth by a coworker.

"Injured and unable to climb up or down, he was lifted onto a small platform below the crane's control booth by a coworker."

At 0803, we arrived and the HRT was assigned Rescue Group. Two HRT members, one officer and one paramedic, began the 110-foot climb up the crane with some basic rope rescue equipment. They found the patient supine, awake, alert and oriented, and complaining of pain in his upper back, neck, and shoulders. He had no obvious signs of trauma and moved all extremities well. He was lying on the steel grating within an enclosed portion of the crane housing that contained the large rotary gear and motors. Serving as the Rescue Officer on the ground, I ordered the crane's power supply shut off and tagged out. The jib section was locked to prevent any turning or rotation.

As the HRT medic continued to assess the patient, the topside HRT officer climbed up and out of the gear housing and onto the next platform section immediately below the control booth. He lowered a rope to the ground and pulled up a Spec-Pak patient packaging device. Due to the close confines of the tower and the nature of the latticework, full spinal precautions and packaging in a stokes litter was impossible. Meanwhile, on the ground, an aerial ladder was raised and a second two-man HRT rescue team began to climb the crane with more rescue equipment. Due to the patient's location, the ladder was unable to reach high enough and was then rotated out of the way. As the second rescue team reached the patient location, one member assisted the HRT medic with patient packaging while the other continued up past the next platform and control booth to the top walkway. There, he and the topside HRT officer established a belay line with tandem prusiks and a working line with a Petzl I'D as a fixed-brake descent control device. A third line was raised to the walkway and fixed to be used as a tracking/guiding line for the patient. The belay line was lowered down the outside of the tower and brought back inside at the platform just above the patient's location inside the gear housing. The patient was placed on belay and raised vertically through the top opening of the gear housing and onto the platform just below the control booth using a mini 4:1 pulley system attached to the Spec-Pak. The working line and a bight of the guiding line were pulled between the latticework onto the platform. A pulley was reeved onto the guiding line and connected to the Spec-Pak along with the working line.

At 0905, the patient was slid feet first between the latticework to the outside of the tower. The mini 4:1 was used to lower him out in a controlled manner until his weight could be transferred completely onto the working line. Once outside the tower structure, he was lowered using the Petzl I'D. The guiding line was manned and used to pull the patient away from the tower. As he neared the bottom, the guiding line team was able to vector him far enough to clear the construction debris at the base of the crane and an eight-foot chain link fence. He was guided directly onto a waiting backboard and ambulance gurney where he was reassessed and disconnected from the rope system. At 0908, his care was transferred from the Rescue Group (HRT) to the Medical Group. He was transported to the hospital and was released two days later.

Steve Crandall
Heavy Rescue Captain
Salt Lake City Fire Department
Owner, HeavyRescueTraining.com
Salt Lake City, Utah

Figure 5-10 Rope rescue in an industrial setting may need to take into account additional energy hazards.

Figure 5-11 Use barrier tape to establish initial edge control at rescue incidents.

tional environments must be assessed and controlled before any rope rescue activities can take place. Review the hazards sections of the related chapters in this book for further information.

Basic Hazard Mitigation

At the awareness level, responders can take some basic hazard mitigation actions to control the site prior to the arrival of trained rope rescue teams. The general area around a rope rescue scene is considered the entire area within 300 feet of the event. The size of this general area may be enlarged for incident management purposes as required by the IC. Scene control to remove unnecessary bystanders from the actual operational area and to establish a physical barrier (such as banner tape) is important in limiting the potential for further victims. This scene control barrier should be placed at least 100 feet from the operational area where responders initiate rescue actions. Within this operational area, awareness level responders should control traffic, eliminate vibration sources, identify and monitor existing hazards, and remove bystanders and unnecessary personnel.

Awareness level rescuers can also establish initial edge control by using barrier tape to identify positions beyond which

initial responders should not venture without proper fall protection **Figure 5-11**. If initial responders must go near the edge of a dangerous drop-off to locate and/or communicate with the victim, they must be attached to a competent fall protection system. They should then position themselves near the edge of the drop-off by lying down and creeping forward or staying in a crouched position with their body weight shifted to the rear foot to limit the potential of falling over the edge.

Basic Rescue Procedures

Upon their arrival, trained rope rescue teams should perform a scene assessment to further identify hazards, establish more effective fall protection for the personnel who need to operate near the edge, and ensure the use of proper personal protective gear.

After the size-up, the rescue team gathers information from initial responders, initiates a search for victims if required, and locates the victim(s). Once a victim is located, it is important to communicate with that person if possible to reassure him or her that rescuers are on the way to deliver care. Awareness level rescuers could be asked to assist trained rope rescue teams with movement of tools and equipment, securing of the perimeter, helping operate haul systems, or patient transfer. For their part, the members of the rescue team gain access to the patient by utilizing the most appropriate rope system based on the specifics of the situation. It is extremely important that rescuers gain access to the patient as early as possible to deliver patient care.

Securing the patient to ensure that he or she will not suffer any additional fall injuries is critical. Once this step is completed, the rescue team must determine the most appropriate patient packaging device based on the environment, package the victim, and attach the victim to a rope rescue system for removal to a safe location. The rope rescue system used depends on the types of difficulties encountered and could entail a combination of lowering systems, mechanical advantage systems, and high-line systems.

Patient Care Considerations

Initial medical care for a patient consists of basic life support (BLS). Basic life support includes establishing and maintaining an adequate airway, providing respiratory ventilation to maintain air flow, controlling severe bleeding, and maintaining the circulatory system. In addition, BLS providers must ensure spinal immobilization, and treat the victim for shock if present or suspected. In these situations, local medical protocols must be followed. If rescuers are not trained to provide emergency medical care, they must know where to obtain these resources and be prepared to assist the emergency medical staff with patient packaging and movement.

Patient movement and transfer through rough terrain—an operation that is often required at wilderness high-angle rescue incidents—may necessitate passing a stretcher up and over rocks and other obstacles, fording streams, or moving through narrow enclosures. The primary objective for each rescue, transfer, and patient removal process is to complete these steps as safely and efficiently as possible while limiting further injury to the patient. The key to performing this task safely is ensuring that adequate numbers of rescuers are assigned to help move the packaged patient through debris, lift the victim over barriers, navigate through water, or carry the victim across unstable surfaces without overexertion that could lead to injuries to the rescuers.

Wrap-Up

Chief Concepts

- Rescuers may encounter rope rescue situations where patients are located hanging on the side of a building from a fall protection harness, dangling from a ledge on the side of a rock face, located on the top of a building collapse rubble pile, situated at the bottom of an excavation, or trapped on an upper floor of a building under construction. Rescuers must have a clear understanding of how to recognize which type of rescue environment they are working in and which hazards may be present.
- High-angle rope rescue emergencies require the use of specialized resources, including rope rescue hardware and software and specially trained rope rescue teams. Responders should have a good understanding of where to acquire the proper resources to perform a successful rope rescue.
- Rope rescue hazards can include fall hazards, falling debris, energy hazards, environmental hazards, and crowd control issues. Hazard mitigation efforts are required before responders can initiate actions to access, treat, package, and remove a patient. Awareness level responders must have a thorough understanding of which high-angle hazards may be present and know how to identify them.
- Awareness level responders may be able to initiate some initial scene control activities such as crowd control, edge control, and fall protection. However, at no time should they attempt to gain access to a patient or to remove a patient in a situation where fall protection is required.

Hot Terms

Ascending A means of traveling up a fixed rope safely with the use of an ascent device.

Awareness level The first level of rescue training provided to all responders, which emphasizes recognizing hazards, securing the scene, and calling for appropriate assistance. There is no actual use of rescue skills at the awareness level.

Carabiner A piece of metal hardware used in rope rescue operations. It is generally an oval-shaped device with a spring-loaded gate that can be used to connect pieces of rope, webbing, and other rope hardware.

Descending A means of traveling down a fixed rope safely with the use of a descent control device.

Edge protection A device that prevents damage to rope or other rope software from sharp or jagged edges or from friction.

Fall protection A system of associated rope hardware and software used to protect workers from falling from an elevated position. Fall protection also refers to equipment to which workers may be attached while working at elevations or near a fall hazard, which is meant to capture the individuals if they fall.

Harness A piece of safety equipment that can be worn by a rescuer and attached to a rope rescue system, thereby allowing the rescuer to work safely in a high-angle environment during rope rescue operations.

High-angle rope rescue A rope rescue operation where the angle of the slope is greater than 45 degrees. Rescuers depend on the rope for the primary support mechanism rather than a fixed support surface such as the ground.

Low-angle rope rescue A rope rescue operation where the slope of the ground is less than 45 degrees. Rescuers depend on the ground for their primary support, and the rope system becomes the secondary means of support.

Mechanical advantage system A system that creates a leverage force using levers, pulleys, or gearing and is described in terms of a ratio of output force to input force.

Packaging The process of preparing a victim for movement as a unit, often accomplished with a long spine board or similar device.

Rope hardware Rigid mechanical auxiliary equipment that can include, but is not limited to, anchor plates, carabiners, and mechanical ascent and descent control devices.

Rope rescue incident A situation where rescuing an injured or trapped patient requires rope rescue equipment and techniques.

Rope rescue system A system consisting of rope rescue equipment and an appropriate anchor system intended for use in the rescue of a subject.

Rope software A flexible fabric component of rope rescue equipment that can include, but is not limited to, anchor straps, pick-off straps, and rigging slings.

Rescue Responder *in Action*

Your engine company is dispatched to a vehicle accident on a rural highway on the outskirts of your response area. Your unit has a significant response time to this location, and your officer requests the assistance of a rescue company in case any of the potential victims are trapped in the wreckage. As you approach the scene, you notice several cars pulled off on the side of the roadway, but none of the vehicles appears to be damaged. As you and your engine company officer get off of the rig and walk up to the scene, you notice that a crowd has formed near the edge of the roadway and that the guardrail has been significantly damaged. You and your officer walk up to the crowd near the edge of the roadway, and you realize that a large tractor-trailer truck has gone off the roadway and has crashed through the guardrail. It is now lying on its side about 100 feet down a very steep embankment. Bystanders tell you they could hear someone screaming for help from the crash site initially, but that they have not heard anything for a few minutes.

1. Which additional resources should be requested to help manage this emergency?
   A. A rescue company trained in vehicle extrication
   B. An incident command officer
   C. Rope rescue resources
   D. All of the above

2. Which initial actions should awareness level responders take at this event?
   A. Secure the scene, move bystanders away from the danger zone, and attempt to communicate with the victim
   B. Assess the scene for hazardous materials and other hazards
   C. Attempt to remove the victim from the wreckage and haul him or her up the embankment
   D. Both A and B

3. Which knowledge and/or skills will the rope rescue team need to operate effectively?
   A. Rope rescue techniques
   B. Vehicle rescue hazard awareness
   C. Water rescue skills
   D. Both A and B

Structural Collapse Search and Rescue

NFPA 1670 Standard

6.1 **General Requirements**

Organizations operating at structural collapse incidents shall meet the requirements specified in Chapter 4. [p. 126]

6.2 **Awareness Level**

6.2.1 Organizations operating at the awareness level for structural collapse incidents shall meet the requirements specified in Sections 6.2 and 7.2 (awareness level for confined-space search and rescue). [p. 136]

6.2.2 Organizations operating at the awareness level for structural collapse incidents shall implement procedures for the following: [p. 125–141]

(1)   Recognizing the need for structural collapse search and rescue

(2)* Identifying the resources necessary to conduct structural collapse search and rescue operations

(3)* Initiating the emergency response system for structural collapse incidents

(4)* Initiating site control and scene management

(5)* Recognizing the general hazards associated with structural collapse incidents, including the recognition of applicable construction types and categories and the expected behaviors of components and materials in a structural collapse

(6)* Identifying the five types of collapse patterns and potential victim locations

(7)* Recognizing the potential for secondary collapse

(8)* Conducting visual and verbal searches at structural collapse incidents, while using approved methods for the specific type of collapse

(9)* Recognizing and implementing a search and rescue/ search assessment marking system, building marking system (structure/hazard evaluation), victim location marking system, and structure marking system (structure identification within a geographic area), such as the ones used by FEMA Urban Search and Rescue System

(10) Removing readily accessible victims from structural collapse incidents

(11)* Identifying and establishing a collapse safety zone

(12)* Conducting reconnaissance (recon) of the structure(s) and surrounding area

Additional NFPA Standards

NFPA 1001, *Standard for Fire Fighter Professional Qualifications*
NFPA 1006, *Standard for Technical Rescuer Professional Qualifications*

Knowledge Objectives

After studying this chapter, you will be able to:

- Identify the need for structural collapse search and rescue operations.
- Describe the various types of structural collapse events.
- Describe the resources needed to conduct a structural collapse search and rescue operation.
- Describe response planning and incident management requirements related to a structural collapse search and rescue incident.
- Describe site control operations at a structural collapse incident.
- List general hazards associated with a structural collapse incident.
- Identify building construction types and their associated collapse characteristics.
- List the five types of collapse void patterns.
- Describe the FEMA USAR search, building, victim location, and structure marking systems.

Skills Objectives

After studying this chapter, you will be able to perform the following task:

- Remove readily accessible victims from a structural collapse incident.

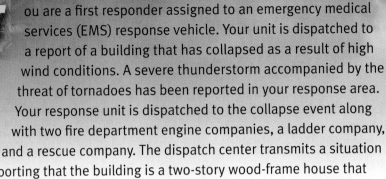

ou are a first responder assigned to an emergency medical services (EMS) response vehicle. Your unit is dispatched to a report of a building that has collapsed as a result of high wind conditions. A severe thunderstorm accompanied by the threat of tornadoes has been reported in your response area. Your response unit is dispatched to the collapse event along with two fire department engine companies, a ladder company, and a rescue company. The dispatch center transmits a situation update reporting that the building is a two-story wood-frame house that was under construction. Witnesses report that high winds from the thunderstorm have blown the house over and that two construction workers are trapped under the debris.

As you are responding to the event, you realize that your unit will be the first responders to arrive on the scene. You know that a structural collapse incident can be very dangerous for responders. You also know that specialized resources are required to conduct search and rescue operations at a structural collapse event.

You arrive to find a wood-frame structure, under construction, that has suffered a lateral rack-over collapse of the first floor. Two construction workers approach your unit and are bleeding from the head. Another construction worker is lying on the ground and appears to be in pain. One of the walking wounded tells you that two of his buddies are still trapped under the collapsed house, but he is not sure where they are located. As you start to perform a size-up of the event, you remember that a structural collapse incident is resource intensive and that specially trained responders and specialized tools and equipment are needed to operate effectively.

1. What should you be looking for during a structural collapse size-up?
2. Which hazards might rescuers find at a structural collapse event?
3. Which actions can first responders take at this incident to more effectively manage the event?
4. How should you search for victims at a structural collapse event?
5. Which victims should be treated first?

Introduction

Structural collapse operations cover a wide range of incident scenarios. These scenarios vary in difficulty based on the significance of the collapse and the size, type, and occupancy of the building involved. Given the inconsistency of their presentation, it is necessary to implement a standard but flexible plan of action that can be used for rescue operations at all structural collapse situations.

Difficult collapse rescue operations require a combination of specialized tools and equipment, as well as specially trained emergency responders who apply unique techniques, for their effective mitigation. Acquiring the appropriate resources as early as possible is an important step in the successful outcome of structural collapse rescue. For this reason, responders must recognize the need for search and rescue operations at a building collapse event quickly and to take the appropriate actions immediately.

Structural collapse emergencies have a variety of causes. Some collapses are a result of natural occurrences such as hurricanes, tornadoes, and other high wind conditions. In addition, earthquakes, mudslides or landslides, floods, and snow or ice loads on roofs are natural causes of building collapses **Figure 6-1 ▶**.

Buildings may also collapse due to fire damage when a fire weakens structural support components such as floors, roofs, walls, or columns. Interior or exterior blast loading on a structure can also result in a collapse situation if building support

Figure 6-1 Structures can collapse for a variety of reasons.

members are severely damaged. This situation may occur because of a gas leak explosion, following a rupture of hazardous building contents (as in a boiler explosion), or as a result of terrorist activity.

Human error can be a contributing factor in a building collapse as well. For example, total or partial building collapse may occur as a result of an accidental impact of a building by a transportation vehicle such as a car, truck, train, or plane. Most buildings are not designed to resist the impact load of a vehicle that runs off of a roadway and into the wall of a building. Engineering errors have also resulted in building collapse: Poor design of a building, poor construction methods, and/or improper building materials can lead to collapse, particularly where building codes are lacking or are not enforced.

Applicable Standards

Emergency responders can find guidance on managing these types of incidents by referring to the National Fire Protection Association (NFPA) Standards 1006, *Standard for Technical Rescuer Professional Qualifications*, and 1670, *Standard on Operations and Training for Technical Search and Rescue Incidents*.

NFPA 1006 establishes the job performance requirements for responders who perform technician-level rescue skills. These skills include hazard recognition and mitigation, building

stabilization techniques, breaching building materials, performing lifting operations to effect rescues, and victim packaging and transfer techniques.

NFPA 1670 establishes organizational requirements to operate safely and effectively at structural collapse emergencies. Specifically, this standard identifies the requirements for risk and hazard assessment, written procedures, safety protocols, personal protective equipment (PPE), establishing operational response level, and resource considerations.

Resource Requirements

As a responder to a structural collapse incident, you must know which resources are required to manage the incident. Large building collapse incidents will likely have a long duration. In addition, specialized resources are required to perform search and rescue operations safely and effectively. The need for significant oversight of responders and long-term planning will dictate whether a large management staff is required to command the various responders to the emergency. Command vehicles such as mobile command units will enhance responders' capabilities to develop management plans for this type of event. Although most search, rescue, and hazard mitigation duties will be assigned to specially trained personnel, awareness level responders can assist in this effort by serving as equipment runners, helping to move and care for casualties, maintaining perimeter control, providing command post assistance, directing traffic, and performing other important tasks within their training level.

At a collapse event, resources will be required to perform hazard control as early as possible. These resources may include firefighting assets to control actual or potential fire problems. Utility company assistance should be requested as early as possible to control any broken, disrupted, or endangered utilities (e.g., gas and electric lines). **Secondary collapse** hazards—those resulting from a partial or total failure of structural members from a building that had been damaged from an earlier collapse event—may be present as well. Such scenarios require shoring supplies, plus specially trained responders to install these supplies. **Shoring** refers to the tools and equipment used to support damaged buildings and damaged structural members, thereby stabilizing the area involved and helping to avoid further collapse. Hazardous materials response teams will be required if a hazardous materials release has occurred, containers are damaged, or hazardous building contents are present at the site.

An uncontrolled incident scene will be one of the initial hazards to responders that must be controlled. To impose order on the scene, police, sheriffs, or other law enforcement resources should be requested to control crowds and traffic and to begin establishing a controlled incident perimeter.

At some incidents, search and rescue resources specially trained in structural collapse rescue will be required. It is important to know which resources are available and what the search and rescue team's capabilities are. For this reason, responders should have knowledge of the standardized resource typing used in their response area and should be familiar with the various structural collapse rescue assets they can request. **Resource typing** refers to the process of categorizing, by capability, various resources that may be used at an incident scene. Measurable standards identifying resource capabilities and performance levels serve as the basis for these categories. For example, some resources may be trained and equipped to operate at only light-frame and unreinforced masonry (URM) structures; such teams are referred to as **operations level** teams in NFPA 1670. In contrast, structural collapse rescue teams that are equipped and trained to perform rescues in steel and reinforced concrete structures are considered **technician level** teams in NFPA 1670. It is important for responders to be aware of other resource typing guidelines that are used in their jurisdiction. Regardless of level, responders must meet the requirements in NFPA 1670 Chapter 4.

Search assets that may be required at a building collapse emergency include **physical search teams**, **canine search teams**, and search teams using technical equipment. Physical search can be performed by responders who have been trained to operate in a collapse environment. Such a search does not require specialized equipment. Physical search techniques will be discussed later in this chapter. Canine search assets include specially trained dogs, dog handlers, and, in some cases, personnel who direct the canine search teams. These resources can be single-dog teams or can consist of multiple dogs and handlers working in concert with one another. In either case, it is important for responders to have knowledge of where to acquire this important asset.

Technical Search Resources

Technical search includes the use of mechanical search equipment—that is, visual, thermal, seismic, and acoustic devices used to locate victims who are trapped within a collapsed structure. Technical search resources also include specially trained personnel who can operate this equipment effectively.

Acoustic and seismic devices utilizing sensors are attached by cables to a control panel; the operator uses such a device in an attempt to hear noises from within a collapsed building (acoustic) or to sense vibrations or movement from around the building (seismic) **Figure 6-2 ▾**. A highly trained operator will likely place the sensors at various locations to triangulate the sounds that may be heard. An acoustic device may also include noise filters that allow the user to differentiate between a victim making noise inside a building and other extraneous background noise. By moving the sensors around the site, various quadrants of the building can be checked for noises or movement. The source of a noise that is heard over a wide area can be difficult

Figure 6-3 Thermal imagers indicate temperature differences.

to pinpoint, however, so triangulation of sounds is often used to focus in on the location of a victim trapped in a collapse.

Visual search devices include tools such as search cameras, fiber optics, and thermal imagers. The capabilities of these devices are also enhanced when they are paired with a highly trained operator. Search cameras and fiber-optic visual devices allow the user to insert the camera end of the device through existing crevices or holes created by rescuers to see below a collapsed floor or into a collapse void. Thermal imagers can be used to assess void spaces and determine the presence of a temperature differentiation that may indicate a victim location **Figure 6-3 ▲**.

Rescue Resources

Structural collapse incidents can create very difficult extrication problems for rescuers. **Extrication** refers to actions taken by responders to remove a victim from an entrapment at a building collapse. **Entrapment** is a condition in which a victim is trapped by debris, soil, or other material and is unable to extricate himself or herself. Extrication activities may include lifting heavy objects, cutting through building materials, and breaching structural building components.

Collapse victims can be pinned under collapsed walls, floors, or other building components, as well as by building contents that have shifted. To deal with these conditions, rescuers need to be trained to use hand and power tools to lift and stabilize heavy objects such as concrete floor slabs. They also must have a good understanding of mechanical advantage, classes of levers, pneumatic lifting bags, and cribbing stabilization systems. Finally, rescuers need to be skilled in the application of these tools to lift, stabilize, and possibly move heavy loads safely so as to access victims.

Structural collapse rescue teams need to be proficient in breaching skills when it is necessary to cut through wood, steel, and concrete construction materials to access victims **Figure 6-4 ▶**. In addition, rescuers must be prepared to breach and cut apart common building contents such as furniture, appliances, mattresses, and carpeting that represent obstacles for rescuers who seek to gain access to trapped victims. To accomplish this task, rescuers operate a variety of power tools,

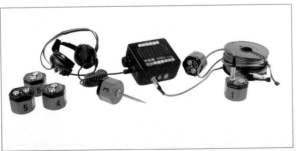

Figure 6-2 Acoustic devices can be used to triangulate a victim's position.

Figure 6-4   Structural collapse rescue teams may be required to cut through wood, steel, and concrete construction materials to access victims.

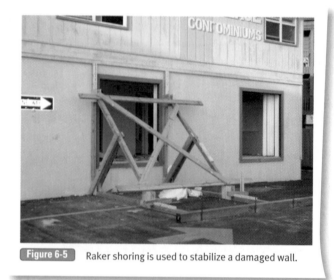

Figure 6-5   Raker shoring is used to stabilize a damaged wall.

such as electric, pneumatic, and/or hydraulic breakers and drills. Reciprocating saws, chain saws, and rotary disc saws are used by rescuers for breaching purposes as well. Torch operations may be required by some rescue teams to cut and remove steel plating, steel I-beams, and other steel building components. Various types of torches may be used by the rescue teams for this purpose, including oxy-acetylene, exothermic, oxy-gasoline, or plasma cutters. Structural collapse rescue teams are also trained to use a variety of special-purpose tools, such as rebar cutters, core drills, concrete anchors, and rigging equipment, to assist in search and rescue efforts at collapsed structures.

Given the large number of specialized tools used by rescuers and the extensive training required for qualification as a member of a structural collapse rescue team, such teams may not be readily available in every jurisdiction. Instead, it may be necessary to request these resources from a regional, state, or federal level.

Another rescue resource required at a collapse incident is the capability of building shoring. In most situations, the structural collapse rescue team is trained and equipped to deliver this service. In other scenarios, it may be necessary to take advantage of local community resources, such as construction companies, to assist first responders with this skill. The types of shoring systems commonly required at a collapse event include vertical shores, horizontal shores, raker shores, sloped floor shores, laced post shores, window/door shores, and cribbing systems. Each of these shoring systems has a specific application. For example, vertical shores are used to support a damaged roof or floor, and a raker shore is built to stabilize a damaged wall **Figure 6-5 ▶**. Rescue teams are trained to construct shoring systems out of

lumber as well as to use mechanical shoring devices such as pneumatic shores.

Once rescuers are able to gain access and extricate entrapped victims, they use a variety of victim packaging and transfer devices to remove those victims, such as SKED™ stretchers, Reeves Sleeve™, and Stokes baskets. **Victim packaging** refers to the process of preparing the victim for movement as a unit, a feat that is often accomplished with a long spine board or similar device. If the victim is located above or below grade, the situation may require rope rescue skills, such as the ability to establish rope anchors, operate lowering systems, use mechanical advantage raising systems, and implement safety belay lines.

Emergency Medical Services Resources

Emergency medical services (EMS) requirements are significant considerations at a collapse event. In particular, personnel trained in basic life support (BLS) and advanced life support (ALS) care are often required to treat victims of building collapse emergencies. EMS command staff personnel will be required to assist in managing these assets and to assist with command decision making at these events.

Common injuries and medical complications observed at building collapse events include fractures, respiratory distress, dehydration, compartment syndrome, and crush syndrome. Depending on the type of building, the size of the collapse area, and the occupancy load, a structural collapse can result in a mass-casualty event that requires bulk EMS supplies such as backboards, cervical collars, trauma dressings, and burn care supplies. An enhanced response from EMS providers may be needed as well. Victims trapped in a collapsed structure typically require the delivery of ALS interventions while they are still trapped to increase their chances of a successful outcome **Figure 6-6 ▶**.

Hazardous Materials Resources

Failure of hazardous materials containers such as propane tanks, insecticide bottles, or gasoline containers requires the intervention of personnel trained to perform hazardous

Figure 6-6 Victims trapped in a collapsed structure may require ALS interventions prior to their removal.

materials assessment and spill containment. Likewise, broken fuel lines, natural gas lines, sewage lines, or other chemical pipelines can necessitate the services of hazardous materials response personnel. These hazardous materials response assets should include personnel who are trained and equipped to detect chemical, biological, nuclear, and other **weapons of mass destruction (WMD)** hazards, and to take control actions as required. WMD include, but are not limited to, four main categories of threats: chemical, biological, radiological/nuclear, and explosive. The release of a chemical, nuclear, or biological hazard could be the result of a natural or human-made structural collapse event.

To effectively perform a hazards assessment of a collapsed building, trained hazardous materials technicians will use atmospheric monitors to determine the presence of dangerous atmospheres and, if necessary, establish control zones. These control zones are used to identify unsafe areas, areas where operations can commence when specific hazard control measures are implemented, and safe zones. Hazardous materials responders also research the hazardous materials suspected to be present and then specify the need for any specialized protective gear, equipment, or procedures.

Law Enforcement Resources

Police, sheriffs, or other law enforcement agencies are needed at building collapse incidents to help establish perimeter control, perform site security, control traffic flow, and maintain force protection for emergency responders. Trained law enforcement investigators may also be helpful in interviewing bystanders, co-workers, or family members as part of the victim location and search planning function.

Logistical Support Resources

Large-scale structural collapse events often have a long duration. In these situations, logistical support resources are required for sustaining long-term rescue operations. Night-time lighting, bathroom facilities, and food and shelter items for rescue workers must be properly supplied and managed. Apparatus maintenance and refueling, field repair of tools and equipment, and the procurement of rescue tools, supplies, and equipment are also part of long-term rescue missions. Additionally, victims in multiple-casualty incidents may benefit from the presence of an emergency room physician and nurses on site. County or state participation may dictate the use of extra resources as well.

> **Rescue Tips**
>
> Structural collapse events are resource-intensive emergencies that take a long time to organize and will generally be long-duration incidents. Responders should recognize that they will not be able to manage these emergencies quickly and that significant hazard mitigation will be required before trained rescuers can make entry into the more dangerous areas of the building to effect rescues.

Incident Management Requirements

The strategic objective when operating at a structural collapse incident is to control and manage the event in such a way as to evaluate the scene and identify potential victims and their location and to initiate operations to minimize hazards to operating personnel and trapped victims, search the structure, rescue and remove trapped victims, and minimize further injury to victims during search, rescue, and removal operations.

The Incident Command System (ICS) positions are staffed based on the scope of the event. In most cases, at a minimum, management personnel are required to fill the positions of incident commander (IC), operations section chief, planning section chief, logistics section chief, safety, and public information officer.

The operations section usually has various branch directors and division/group supervisors assigned to manage specific problems or geographical areas. The planning and logistics sections assign knowledgeable, skilled personnel to fill the necessary section positions, such as those dealing with the situation status, resource status, service branch, and/or support branch, as required by the event.

Incident management personnel are responsible for strategic and tactical management of the structural collapse emergency response. Some of these responsibilities include the development of an **incident action plan (IAP)**. The IAP identifies the overall control objectives for the collapse emergency. Because a collapse emergency may last multiple days, long-term staffing considerations are developed and included within this plan. The IAP includes provisions dealing with the assignment of personnel to the various management positions responsible for implementing the plan. Tactical assignments and resource assignments, such as use of a canine search team to perform a search within an assigned geographical area, are listed in the IAP as well. Operational work periods and safety messages are also identified in the IAP.

The successful management of a large-scale collapse event requires not only incident management personnel, but also incident facilities support. These facilities may include a fixed site for the incident command post, an area in which to stage personnel and equipment that are awaiting an assignment, feeding and sleeping areas, and toilet facilities.

Response Planning

Responders must know how to initiate the emergency response system to ensure that appropriate resources are deployed to the event and that operational guidelines are identified and followed. Many response organizations perform needs assessments within their response areas to determine their vulnerability to structural collapse events. As part of this assessment, the organization should consider the types of buildings that may fail and the types of construction materials that will be encountered, and then determine the emergency response resource needs based on those considerations. The emergency response system should include written procedures to request agency resources as well as resources that will respond through mutual aid agreements, contracts with private-sector companies, and any memoranda of agreement with regional, state, or federal assets. Private-sector agreements may be established so that the organization can acquire resources that cannot be supplied by the response agency, such as heavy equipment, lumber and construction tools, structural engineers, food, sanitation supplies, and housing.

The capability of agency personnel should be identified as well. This capability assessment should include the level of performance at which agency personnel are trained, such as the operations or technician level for structural collapse rescue. As part of the emergency response system, the location, capabilities, and response times for regional, state, and federal search and rescue teams should be determined, and procedures should be established to request these assets. Similarly, dispatch procedures should be in place to ensure that adequate and proper resources are deployed when a structural collapse event occurs, allowing effective search and rescue operations to begin as quickly as possible.

Building Construction Types and Expected Behaviors

Emergency responders should have a good understanding of the different types of building construction and the way in which each type of building will react during collapse events. Buildings can be categorized into four distinct groups: light frame, heavy wall, heavy floor, and pre-cast reinforced concrete. Each of these categories may include buildings used for residential, commercial, and industrial purposes.

Light-Frame Buildings

Light-frame buildings are primarily constructed of wood frame. They are typically one to four stories in height. Light-frame buildings used for residential occupancy can be built for single-family occupancy, such as a one-story house, or can be designed to contain more than 100 residential units, such as an apartment building **Figure 6-7 ▶**.

Figure 6-7   A light-frame building.

These types of building have an inherent lateral stability weakness. This lateral weakness is especially notable at the foundation, at connections between building components such as wall-to-floor connectors, and at sites where walls are weakened by window and door openings. When light-frame buildings suffer lateral movement from earthquakes or other ground movement, the weak walls may rack over and cause one floor to become offset from another floor in multistory buildings. Significant rack-over will cause a building to collapse. As light-frame buildings experience lateral movement and an upper floor becomes offset from a lower floor, the lower floor may experience failure if it becomes overloaded from more stress than the floor is designed to accommodate.

A light-frame house built on a crawl space may suffer lateral movement and slide or shift off its foundation. Other types of damage commonly encountered with light-frame buildings include brick veneer that separates from wood-frame walls due to connector failure and masonry chimneys that collapse. Light-frame buildings with add-on attachments such as garages, porches, or room additions may suffer separation where floor and roof offsets between the main building and the additions are connected. In other words, the add-on attachment may project away from the main building and collapse out into the yard or street.

When responders arrive at a collapse incident involving a light-frame building, they should assess the building stability by looking for the following characteristics:

- Cracked or leaning walls
- Offset of the building from its foundation
- Leaning of the first story in a multistory building
- Cracked or leaning chimney or loose masonry in the chimney
- Cracked or loose brick veneer
- Separation between main building and additions such as porches

Heavy Wall Buildings

Heavy wall buildings are primarily characterized by URM walls, tilt-up concrete walls, steel or concrete frames with masonry

Figure 6-8   A heavy wall building.

Figure 6-9   Tilt-up heavy wall collapse.

infill walls, and other low-rise designs that incorporate concrete and/or masonry walls Figure 6-8 ▲. These buildings may be used for residential, commercial, or industrial purposes and are typically one to six stories in height. The principal weaknesses with heavy wall buildings are the lack of lateral strength of the walls and the weakness of the connections between the walls and the floors and roofs.

URM buildings may have load-bearing walls made of brick, hollow concrete block, or stone. Collapse problems noted to occur frequently with URM buildings include walls collapsing due to inadequate anchors, inadequate reinforcement, and poor condition of masonry and/or mortar, as well as roof and floor failure due to loss of support or connector failure.

Falling hazards are very common when URM buildings suffer damage. Broken bricks and collapsed masonry walls are often found in the street, where people may be trapped in cars covered by the debris. Many times victims are found under this rubble on sidewalks next to the building. In steel-frame buildings with infill walls, the URM infill may shake loose and fall off the steel frames and out into the street. Loose parapet walls, building ornamentation, or hanging signs may break loose and create falling hazards as well.

As rescuers approach the scene of a URM building collapse, they should look for cracked walls at the building corners and near wall openings such as windows and doors. In addition, responders should look for unsupported and partially collapsed floors or roofs and for damaged connections between walls or columns and the floors or roof structures.

Another type of heavy wall building that responders may encounter is a tilt-up (TU) concrete wall building. These buildings are usually only one story, but the height of the TU walls may be as high as a three-story structure. These structures may include wood floors, concrete floors, steel framing with concrete-filled metal deck floors, or lightweight concrete fill on wooden floors.

The most common problem encountered with a TU wall building is separation of the roof/floor structure and the concrete TU wall. This breach may result in a localized partial collapse or total building collapse. The major weight of these buildings is normally in the walls, so most failures are limited to exterior bays of the buildings, which are supported by the walls. This situation occurred to almost 400 TU buildings during the Northridge earthquake in California, for example Figure 6-9 ▲.

Heavy Floor Buildings

Heavy floor buildings are made of a concrete frame and can range from 1 to 13 stories in height Figure 6-10 ▼. Some of these buildings are built with URM infill walls. The heavy floor buildings with the weakest design typically are built with high, open ground-level floors (such as open-front retail buildings) and buildings with high open areas on the corners of a multistory structure. These high open fronts or corners on the lowest level of a heavy floor building, combined with the inherent weight of the upper floors, create a situation referred to as a *soft first-story collapse*.

The principal weakness in these types of buildings derives from the inadequate concrete column reinforcement and inadequate connection between floors and columns. With these types of structure, the columns typically fail where they intersect with a floor beam. In addition, shear failure can occur at

Figure 6-10   A heavy floor building.

floor and column intersections as a consequence of punch-through failure.

Rescuers should assess stability by observing for the confinement of concrete within the column reinforcement. If reinforcing steel is exposed and the concrete has fallen off the column it should be considered dangerous. Look for concrete cracking, in the columns at the floor line both above and below the floor, diagonal shear cracking in beams where they are supported by columns or walls, cracking in flat concrete slabs adjacent to supporting columns, loose infill masonry walls, and cracks in concrete shear walls and/or stairs.

Pre-cast Buildings

Pre-cast concrete buildings may be one or more stories and consist of individual building components such as walls, floors, columns, and beams that are premade at a factory and then assembled on site **Figure 6-11 ▼**. The individual components are shipped to the construction site, set in place with a crane, and connected together with various connection devices. Parking garages are commonly encountered pre-cast structures.

The primary weakness in a pre-cast building occurs at the interconnection of the various components, such as where beams connect to columns or floor slabs connect to walls. As rescuers approach a pre-cast building during a structural collapse emergency, they should assess its stability by observing the connections for broken welds, cracked corbels, or broken connector bolts. Columns should be assessed for any cracking of the concrete, wall panel connections should be assessed for adequate support, and badly cracked walls should be identified.

Figure 6-11   A pre-cast building.

Collapse Patterns and Potential Victim Locations

When buildings fail and collapse, void spaces are created, and victims may be located in them. These void spaces are classified into one of five categories: lean-to, V-shape, pancake, cantilever, or A-frame. Some of these void spaces that are created during the collapse are more survivable than others. Victims may be located in certain areas of the voids depending on the type of void.

For instance, lean-to void spaces are typically created when one side of a floor system has lost its support point, subsequently dropping that side down to the floor below, while the upper side of the floor remains attached at the original upper position **Figure 6-12 ▶**. When this type of void is created, victims will most likely be located under the floor within the void space, or on top of the collapsed floor at the lower end where the debris slides down to the bottom.

V-shape collapse voids are created when a floor or roof assembly is overloaded in the middle or when a middle support such as a wall or column is damaged. As a result, the floor or roof assembly fails in the middle, and this portion of the floor or roof drops down to a lower level while both outside connections to the walls are maintained **Figure 6-13 ▶**. When this type of collapse occurs, victims can be found under each side of the created V-shape near the exterior walls. Victims may also be found on top of the collapsed floor in the middle of the created V-shape along with all of the floor contents, which will shift to the lowest position when the floor fails.

Pancake collapse voids are very common in heavy reinforced concrete floor collapses. These voids often result from floor slab-to-column connector failure. Concrete columns will sometimes punch through concrete floor slabs, resulting in pancake collapse voids. This type of void can also occur in a wood-frame structure when the building fails when it leans over or racks over due to lateral movement of the building **Figure 6-14 ▶**. Many wood-frame buildings are not designed to withstand this kind of lateral stress.

In most cases, pancake collapse voids do not result in very many survivable spaces. Many times the only survivable space within the void exists where a floor assembly falls and comes to rest on furniture or appliances.

Some collapse scenarios result in floor or roof assemblies coming to rest in a cantilever position. A cantilever void is very unstable and will present itself as a floor, leaning wall, or roof that is not supported on one end **Figure 6-15 ▶**. The other end will still be supported by the original building connections or by the weight of debris within the building that has come to rest on one end of the cantilevered assembly. Cantilever voids are inherently unstable because only one end is supported. Building

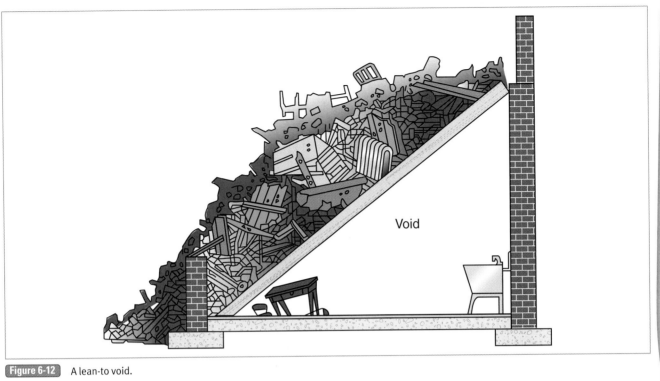

Figure 6-12   A lean-to void.

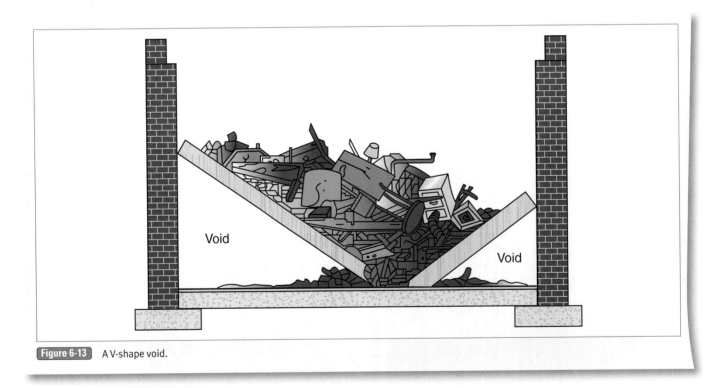
Figure 6-13   A V-shape void.

stabilization techniques such as shoring are needed to create a safer void space in which rescue responders can work. Victims are often found under the unsupported cantilever in very dangerous locations.

A-frame collapse voids are found when a floor or roof section loses its side wall connections on both sides but continues to be supported by a middle support. In this situation, both sides of the system fail and fall to a lower level, even as the middle of the floor remains at an elevated location **Figure 6-16 ▶**. This very unstable void space that results will be difficult to support effectively. Victims are typically found under the void spaces created on either side of the middle support area. In addition, victims may be found outside the void spaces near the side walls where the floors failed at the connection.

Figure 6-14 A pancake collapse void.

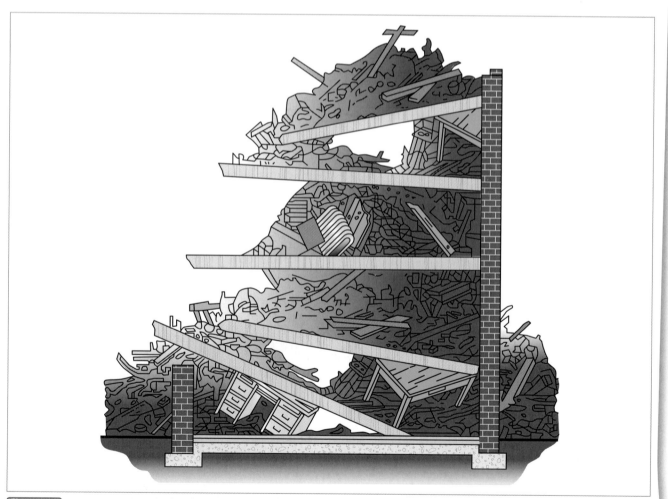

Figure 6-15 A cantilever collapse void.

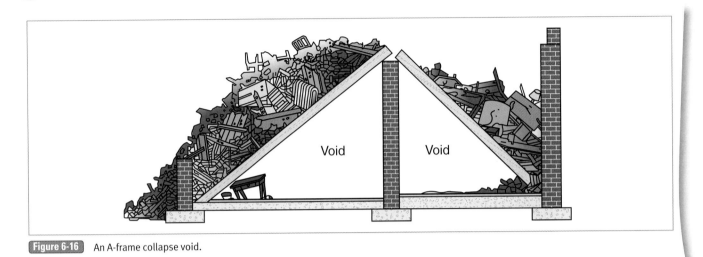

Figure 6-16   An A-frame collapse void.

Hazards and Hazard Assessment

Approach Hazards

First responders should anticipate the possibility of encountering various hazards during their approach to a structural collapse event. The likelihood of encountering **approach hazards** changes depending on the primary cause of the collapse. Approach hazards are those hazards that rescuers may encounter as they travel from their current location, such as a fire station, to the actual collapse site.

Safety Tips

Rescue workers should always assess the response route and the weather conditions when responding to a building collapse that results from a weather-related event. Downed power lines, flooded roadways, or broken utility lines can all endanger a responder prior to arrival at the collapse site.

Approach hazards can be created by natural or human-made causative factors such as ground movement from earthquakes, high wind damage from tornadoes, or terrorist bombings that damage multiple building. These hazards may include damage to local infrastructure such as damaged roadways that limit vehicle approach to the incident area, destroyed bridges as a result of flooding, downed electrical wires blocking roadways, or collapsed-building debris blocking roadways **Figure 6-17 ▶**. All of these hazardous conditions can limit responders' ability to reach the incident location quickly and safely.

If the cause of a collapse event is weather related, such as a hurricane or tornado, responders must also consider the possibility that dangerous weather conditions might still be present in the area. This situation may expose responders to high winds and flying debris, which could result in injury. The most important thing for responders to do during response to a collapse scenario is to be aware of the potential for approach hazards, maintain a cautious approach, and make sure that appropriate safety procedures are followed consistently.

Safety Tips

Responders should never rush into a structural collapse event. Instead, they should approach the incident scene calmly and methodically, in such a way that they can view the entire area and get a better understanding of what the situation is and which hazards can be expected.

Scene Assessment

When responders initially arrive on the scene of a collapse event, they will often be confronted with a chaotic and uncontrolled incident scene. The infrastructure damage mentioned earlier and on-site downed power lines may limit effective approach and positioning of responding emergency apparatus. For this reason, responders should slow down as they arrive at the actual collapse

Figure 6-17   Approach hazards may include downed power lines blocking roadways or access to a rescue site.

scene, thereby ensuring that they can assess the situation more effectively. This action is often referred to as reconnaissance. Performing a reconnaissance of the collapse site will help rescuers gain a better understanding of the overall problem that they are confronted with, including the potential number of victims, the hazards faced by responders, and the resources required to effectively manage the event. Awareness level responders must perform reconnaissance activities from a safe location and away from any potential secondary collapse hazards.

Because there is a high potential for other building utilities to be disrupted and a chance for a secondary collapse, responders should initially position their response vehicles at least 100 feet away from the collapse area. This practice will help avoid the chance of locating a vehicle and a response crew in an area where a gas leak or water main break may be encountered.

Do not feel compelled to rush into a collapsed or partially collapsed building to immediately initiate rescue operations. Instead, take the time to assess the area for hazards and determine which hazard mitigation actions will be required to make the operation safer for responders to operate effectively.

The initial scene assessment time should also be used to identify access problems to the site for other responding apparatus and to determine the most appropriate location for staging incoming emergency apparatus until a full scene assessment can be performed. It is best to maintain priority positioning for special rescue equipment and/or hazard control resources and to determine the best route of travel for incoming resources as early as possible.

During the initial arrival and scene assessment, responders should recognize that they may be quickly overwhelmed by people requesting their assistance. Walking wounded may begin to approach the responders looking for patient care. Witnesses, co-workers, and bystanders will be reporting where victims are trapped or where more wounded occupants are located. Co-workers or bystanders who rush in, attempting to help the trapped and injured, may become secondary victims and add to the complexity of the event. For all these reasons, it is very important for response personnel to keep calm and to gather sufficient information and intelligence about this type of emergency so that they can make sound decisions. Early initiation of a strong ICS will help to organize responders and lay the groundwork for gaining control over a chaotic incident such as a building collapse.

Rescue Tips

Although the scene of a building collapse will generally be very chaotic, responders should always attempt to complete an adequate reconnaissance of the site to get a full understanding of the hazards involved, and to determine the potential number and locations of victims.

Building Utility Hazards and Other Hazardous Materials

When a structure collapses, it is probable that the building utilities—such as electrical conduits, natural gas lines, other fuel lines, chemical lines, steam lines, water lines, and sewage lines—will be broken or damaged. This situation can result in

a release of a utility product, which in turn leads to fires, building settlement, or exposure to hazardous atmospheres. Broken natural gas lines provide a fuel source that may subsequently be ignited by broken electrical lines, which serve as an ignition source for the gas.

Any hazardous materials that are stored in the collapsed structure can also be damaged when containers are broken or destroyed, with the subsequent release of the containers' products. The potential for a hazardous materials release depends on the type of occupancy and the associated hazardous materials storage. A variety of hazardous materials may be stored at commercial businesses and industrial complexes, such as oil, gasoline, fertilizers, insecticides, or other chemicals **Figure 6-18 ▼**. Even collapses that occur at residential structures can result in a hazardous materials spill when household chemicals such as bleach, lawn care products, paints, and similar product containers are broken.

As part of their hazard identification and risk assessment, responders should also determine whether structural collapse incidents might involve nuclear or biological weapons, chemical agents, or WMD, including the potential for use of secondary devices. If a valid risk exists that a technical search and rescue response might be required in a nuclear, biological, and/or chemical hazard environment, responders must ensure that properly trained and equipped response personnel work within this contaminated environment and that hazardous materials mitigation efforts are initiated and maintained throughout the event.

Given the potential for a hazardous materials spill and/or broken utility lines at a structural collapse site, first responders should anticipate the potential for a fire to occur. Many fuel and ignition sources may be present after the collapse. Broken and splintered wood and combustible debris may promote ignition of the collapse site when natural gas leaks and uncontrolled electrical lines are combined in the debris pile. This fire problem will be compounded by the fact that any built-in fire protection

Figure 6-18 Commercial and industrial buildings may contain a variety of hazardous substances that could be released in a collapse.

systems, such as sprinkler and standpipe systems, will likely not be functional. Collapsed sections of the building and dense debris piles will make it difficult to extinguish fires due to responders' inability to access the seat of the fire.

Prior to responders working inside or along the outside perimeter of a collapsed building, it is important to assess the overall stability of the damaged structure. As mentioned earlier, secondary collapses of an unstable structure or falling debris from elevated areas of the damaged building can kill or injure emergency responders. For example, overhead utility wires such as electrical cables or telecommunications lines may become detached from the building, fall, and injure responders. Loose debris that is precariously perched near the edge of upper floors or the roof may fall. Debris may be blown from the building by high wind conditions. All of these situations are considered overhead hazards, necessitating that responders assess the scene for the potential of additional collapse or falling debris.

Several collapse indicators can be used to determine the secondary collapse potential Figure 6-19 ▾ :

- Leaning walls
- Smoke or water seeping through breaks or joints in a wall
- Sounds such as creaking, groaning, snapping, or falling debris
- Sagging floors or roof
- Missing, strained, or damaged building components or connections
- Excessive loading of structural elements due to debris shift
- Racked (leaning) or twisted structure
- Significant vibration sources in the area, such as railroad tracks
- Aftershock conditions or potential
- Active fire in the building

Surface Hazards

Responders must also consider the potential for risk of injury at ground level. **Surface hazards** result from the presence of unstable debris piles, loose and uneven rubble, and sharp materials at a collapse event. When responders must walk through collapse debris and operate on top of debris piles, they are exposed to uneven walking surfaces, debris that may slide and shift, and/or spilled liquids. All of these situations can result in personnel slipping and falling and suffering twists, strains, sprains, or other injuries.

Collapse debris piles may also hide void spaces located beneath the debris surface. Rescuers could potentially fall through loose debris into the void space below and become injured. The collapse material that rescuers must work around at ground level will likely contain broken glass, splintered wood, sharp metal edges, and rough masonry. All these materials can create puncture wounds, laceration injuries, or other soft-tissue injuries.

Void Space Hazards

During a structural collapse incident, rescue workers may be required to enter collapse voids to perform search and rescue operations. When responders must enter a collapse void, the risk involved increases significantly. To enter void spaces such as a lean-to or pancake voids, rescue responders must cut through building materials and building contents to gain access to trapped victims. Secondary collapse during this process is a real concern. Rescuers must assess the stability of the void space and determine whether any of their rescue actions might cause a collapse within the void space. Using an electric saw to cut apart a piece of furniture to free a trapped victim, for example, could cause a secondary collapse if the furniture is supporting a partially collapsed wall or floor section.

While operating inside a void space, rescuers often need to crawl through the collapse debris. Sharp objects in this debris—for example, broken glass, exposed nails, splintered wood, and jagged sheet metal—can cause injury to rescuers. Pooled liquids from broken containers or broken water and sewage lines can create hazards in the collapse voids as well. Slippery conditions can create difficulty for rescuers as they try to navigate the space, and electrical shock can occur if rescuers drag power cords for their electric tools through pooled liquids.

Rescuers should also consider the potential for hazardous atmospheres when entering void spaces. Because many void spaces will have limited access, egress, and air flow, the potential for a hazardous atmosphere to develop as a result of gas leaks, oxygen deficiency, and hazardous materials is increased. For this reason, all awareness-level rescuers must be trained to at least the awareness level for confined-space rescue, so that they understand the potential hazards of entering voids at a structural collapse event. Atmospheric monitoring equipment must often be used prior to entry into below-grade areas such as basements, and collapse voids must be checked for the presence of explosive and toxic gases and oxygen deficiency.

Because collapse void spaces are inherently confined work areas, hazards to rescuers increase due to the diminished lighting, limited work space, and inability to exit the space quickly if an evacuation is required. In particular, the confined work areas commonly encountered as part of collapse voids increase the hazard of operating power tools. Because of the limited work area, rescuers may not be able to position themselves for proper power tool operation. They may be forced to lie on their backs or

Leaning walls are at risk for secondary collapse.

Voices of Experience

As fire fighters, are we really prepared for every incident that we might encounter?

It happened one July night. At approximately 0200 hours, the fire department received a call for a natural gas leak inside a building. This is a common, everyday occurrence for most every fire department in the country. The following few minutes would change the response dramatically, however. While the duty shift was preparing for a response, there was a sudden and loud explosion. A few windows in the fire station were blown out from the force. Those of us who were off duty were awakened by the explosion as it resonated throughout the downtown area. The force activated the master box in the bank across the street from the building. The engine pulled out of headquarters and turned right to find the street blocked from the debris of the building lying across the street.

What happens now? Do we know how many victims might be home? Which hazards might be encountered upon our arrival? What was the actual cause of the explosion? As the first-arriving companies started gathering information, we soon found that there were approximately 12 people who lived in the residence. With the help of the local police, we identified and found all but four of these occupants. As the companies were completing a size-up and surveying the building for surface victims, we had two adult occupants crawl out of a small void space on the B side of the building. They looked up at us and stated that their two young children were still in there, and they were alive and talking to them just before they left to crawl out. We made a decision to make an entry and attempt to locate the two children.

A few department members were part of a trained district technical rescue team, and they were immediately activated for a response. We started setting up a staging area for the equipment to be parked near the scene. The ICS was in use, and early positions established were the IC, operations, rescue team leader, and safety officer. We found a need for early shoring operations, but also found a lack of available lumber to meet these needs.

Upon arrival of the technical rescue team, we set up our command staff positions. We went about shoring operations with what we had available and made entry into the collapse area. During entry operations, the team was evacuated from the building due to an active natural gas leak on the D side of the structure. What we found was that the collapse had landed the building on top of the shut-off valves; there would be a significant time lapse waiting for the leak to be shut off from a remote location.

> **"Their two young children were still in there, and they were alive and talking to them just before they left to crawl out."**

We learned that most departments don't have a trained team available or know how to locate a team should one be needed. The names and contact numbers for after-hours acquisition of materials, such as lumber, were not active or up-to-date. We fell back on the five basic stages for structural collapse incidents. We had the initial response on scene; we could easily do a search for and safe removal of surface victims. Once we had a technical rescue team on scene, we could start an interior search of void spaces. We did not have to get to selected or general debris removal. The technical rescue team did make a recovery of both children as soon as the natural gas leak was shut down.

The final stage of our response was to have all the members operating on scene go through critical incident stress debriefing (CISD) prior to being demobilized. By establishing the command presence early, we kept our members safe and injury free, which made it a successful response and recovery for the team.

Francis M. Clark
Fire Fighter/EMT
Hopkinton Fire Department
Massachusetts Fire District 14 Technical Rescue Team
Hopkinton, Massachusetts

sides and operate a saw or breaker over their heads. In addition, rescuers may not be able to keep a safe distance from the power tool. This situation may lead to injury if a tool kicks back or tool rotation occurs, because the power tool may strike the rescuer.

Basic Hazard Mitigation

Awareness level responders can take some basic actions to minimize their risk at structural collapse incidents. One of the most important things responders can do is to recognize and avoid collapse hazards. By having a clear understanding of the types of hazards found at a collapse site, rescuers can maintain a safe distance and avoid these common hazards. Upon recognizing the presence of collapse site hazards, responders should notify the other responders about the danger and mark the area with barrier tape or other marking devices to identify the unsafe condition or site.

Responders can also limit exposure to hazardous areas by denying entry and establishing collapse safety zones. Some organizations have written standard procedures that outline how these collapse zones should be established. A general guideline to use is that the safety zone should be the height of the building or unstable building component, plus one-third of the height. Thus, if the building is 60 feet tall, the collapse safety zone would be 80 feet. An exclusion zone would be established at this distance from the building. Establishing operational zones—such as hot, cold, and warm zones—will also help to limit exposure to other responders.

If a collapse is suspected to be the result of a WMD event, this dynamic and continually evolving situation will involve responses by multiple agencies. When responding to incidents involving a WMD, a unified command system should be established early and multiple events or secondary devices should be anticipated. First responders may be targets for terrorists and terrorism events. As such, they must maintain situational awareness and assure operational security throughout the duration of the incident **Figure 6-20 ▶**. Structural collapse events involving a WMD have the potential for injuring or killing a significant number of people. Incidents of this type are similar to hazardous material incidents and have the potential to become mass-casualty incidents.

Depending on the resources and capabilities of responders, additional hazard control methods may include basic utility control such as shutting off a water leak at a street valve or the deployment of firefighting hose lines to extinguish or control small fires.

Basic Search Methods

Physical Search

Physical search at a collapse site is similar to physical search in a fire-ground situation. Responders will enter an area around or within a structure and perform a methodical search of each area of the building that is accessible. They will also enter rooms or other areas and perform a right-hand or left-hand search, visually checking all areas and listening for sounds that may indicate the location of a trapped victim. Physical search around the exterior

Figure 6-20   Responders at WMD incidents must reevaluate scene safety constantly.

perimeter of a collapsed building will also occur during the initial reconnaissance of the site.

Physical search is typically carried out early on at a collapse incident in the safer areas of a damaged building, where secondary collapse or significant overhead hazards are not present or where the hazards do not pose a high likelihood of danger to the search crews. Physical search in the less-damaged areas of a partially collapsed building can be completed using response personnel with only limited structural collapse rescue training. In these situations, it is important that the personnel who perform the physical search operations are supervised by personnel who have significant structural collapse rescue knowledge or that the search teams have a very good understanding of collapse hazards.

As each area of the building that can be safely entered is searched, this information must be documented so that the overall search results are known and so that rescue responders can confirm that a thorough, methodical search has been completed. Responders who perform the physical search should use recognized search marking systems to indicate to other responders that a search has been completed in rooms, areas, or the entire structure.

Hailing System

When responders believe that victims may be trapped within a rubble pile or in areas of a collapsed building that are too dangerous to enter, they can use a hailing system search method in hopes of locating victims. The hailing system search method is a basic search technique that involves positioning responders around a collapse site to serve as listeners. One or more members of the search group call out into the collapse rubble and direct anyone who is trapped and still able to speak to call out for help. During this hailing system search, outside noise sources need to be reduced. This effort may involve shutting down operating tools and machinery to limit the background noise as search teams attempt to make verbal contact with trapped victims.

After the background noise is reduced, the search personnel positioned around the collapse site listen for cries for help or victims tapping or making other noises to attract attention. By positioning personnel around the collapse site, noises that are heard by the various personnel from different locations can be used to locate a victim more precisely by triangulation of the sounds. Using these multiple approximations, responders can focus on the location where they believe a victim is located under the collapsed rubble.

Structural Collapse Marking Systems

The Federal Emergency Management Agency (FEMA) has established an Urban Search and Rescue (USAR) response system to be activated for a variety of large-scale disaster events, including collapsed structure response. As part of the FEMA USAR response procedures, several marking systems have been developed to identify hazards and search results at building collapse events. These marking systems include building and structure hazard marking systems, structure identification systems, search assessment marking systems, and victim location marking systems Figure 6-21 ▶ .

Building and Structure Marking System

The FEMA USAR building and structure marking system has several purposes. In the case of widespread damage to a neighborhood, such as may occur after an earthquake or hurricane, the marking system identifies buildings and locations within a geographical area. This information is important when original addresses cannot be identified because of the wide extent of the damage. In these situations, street names and building addresses may be created for the purpose of identification and assignment of resources to a location for search and rescue operations. This information can then be consolidated at the command post, areas assigned search resources can be prioritized, and search outcomes over a large geographic area can be tracked and documented. When street names and addresses are created for identification purposes in heavily damaged areas, international orange spray paint can be used to mark buildings and streets in the devastated area.

Another very important use for the building and structure marking system is to display the results of a structure and hazard assessment by structural engineers, and the presence of hazardous materials by trained hazardous materials response personnel. Orange spray paint is used to mark the buildings

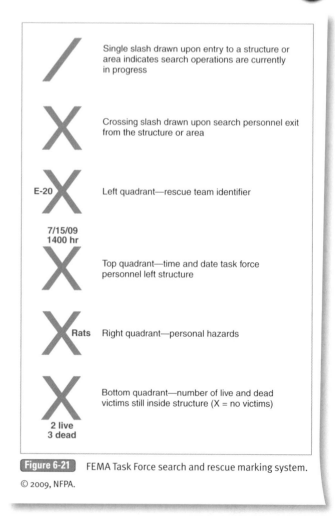

Figure 6-21   FEMA Task Force search and rescue marking system.
© 2009, NFPA.

with general information related to the results of the structure/ hazard assessment. This marking system provides information about the stability of the building and indicates whether rescuers can safely enter. It also identifies whether hazardous materials are present and lists other hazards that were identified during the assessment.

This marking system is a very important communications tool: Search and rescue personnel who arrive on site after the assessment has been completed will know whether an assessment has been completed and which problems they may encounter at the collapse site. To identify that a structure has been assessed, rescue responders spray-paint (international orange) a 2 foot × 2 foot box on or near the building or debris pile. If the box is empty, it means that the building is relatively safe. A single diagonal slash inside the box means that building is significantly damaged and some parts will require shoring to stabilize it, or that other hazards must be removed or controlled. An X inside the box indicates that the building is not safe for rescue operations. Additional information about the structure assessment is written next to the box, including the date and time of the assessment, hazards identified, and the assessment team identifier.

Search Assessment Marking System

The search marking system is intended to identify either that a search is in progress or that a search of an area has been completed. In addition, it identifies who performed the search, when the search was completed, which hazards were identified, and what the result of the search was. This system is an important tool during large-scale collapse incidents where multiple response agencies are operating. When implemented correctly, it helps alleviate unnecessary redundancy in search operations—an important consideration when resources are scarce, hazards to search teams are high, and there are a large number of buildings that require search.

To utilize the search marking system, search teams use orange spray paint and make a large single slash on an exterior surface upon their entry into the building. This mark indicates to others that a search is in progress. A crossing slash resulting in an X will be made when the search teams exit the building. Information may also be added in the quadrants of the X as messages to other teams whose members may have to enter the building at a later time; this information includes the team identifier, date, and time that the search team left the building, hazards, and information about victims.

Victim Location Marking System

The victim location marking system is a method used and applied in closer proximity to where a victim is actually trapped. Whereas the search marking system will be placed on the outside of a structure and will identify the total number of live or dead victims within the building, the victim location marking system is placed next to the actual internal or external location of each victim. Because a victim may not be immediately visible, the victim location marking system can be used to identify where a victim is trapped, whether it is believed the victim is alive or dead, and whether the victim has been removed **Figure 6-22 ▶**. Initially rescuers who locate a victim will spray-paint a large V near the location of the victim. Inside the V, the name of the rescue team is added; a circle is placed around the V when the presence of a live victim is confirmed, or a line is drawn through the markings when a deceased victim is confirmed. A large X over the initial information indicates that the victim has been removed. This system also helps eliminate redundancy in assignment of search and rescue assets.

Patient Care Considerations

Awareness-level responders should anticipate that multiple injured people will need patient care at a structural collapse emergency. First responders may be inundated upon their arrival at the scene with a large number of walking wounded. In addition, responders may need to take some type of action to remove surface victims to a safe location for treatment.

The walking wounded whom responders could encounter may be suffering from typical collapse injuries, including fractures, soft-tissue injuries, respiratory distress, burns, or other traumatic injuries. Initial responders need to ensure that adequate EMS resources are responding. If responders are trained to deliver some level of emergency medical care, then local treatment protocols should be followed to care for patients. Triage and treatment protocols and procedures should be established, and walking wounded should be directed away from the collapse hazard zone to a safe treatment area.

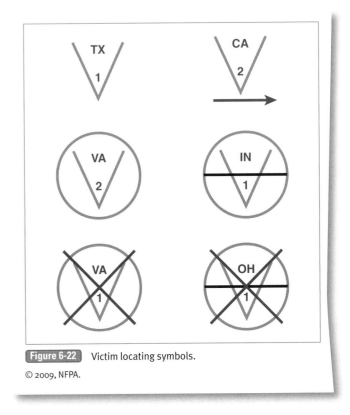

Figure 6-22   Victim locating symbols.
© 2009, NFPA.

Surface victims are the victims of a structural collapse who may be found on top of the collapse debris pile or along the perimeter of a collapsed building by the initial responders. These victims are generally not trapped or are only lightly trapped by loose rubble and can easily be accessed and removed to safety.

Victims who are only lightly trapped by easily removed loose or small debris may also be found. In these situations, the goal for responders is to rapidly access and remove these victims to a safe treatment area. Before responders approach the collapsed building or climb onto a rubble pile, they need to perform a thorough hazard assessment to assure that approach has only minimal risk and can be accomplished in a safe manner. Overhead hazards that may drop down on responders should be avoided. Personal protective equipment (PPE) must be utilized, and a rapid approach to and removal of the victim is needed to limit exposure of the rescuer to the hazard zone.

Rapid retrieval of a readily accessible victim from a structural collapse site requires the following steps:

1. Assess the location of the victim and confirm that a rapid retrieval from the collapse site can be completed relatively safely.

2. Assess and confirm the existence of structural collapse hazards. Avoid hazards during approach to the victim if possible.

3. Acquire and prepare rapid extrication/victim removal equipment.

4. As part of a team of two to four rescuers, approach, package, and remove the victim from the collapse site while one team member continues to assess the environment for any change in hazardous conditions.

Victims who are trapped by loose debris and small material that can be moved easily by hand or with small hand tools can be freed and removed during the initial stages of the collapse event. This effort may involve no more than moving bricks and broken masonry by hand to uncover a victim. Alternatively, it may involve using a pry bar as a lever to lift a small beam off a victim while another responder moves the victim out from under the entrapment. Any difficult rescues that require a long extrication process should not be attempted until a hazard control assessment is completed and trained structural collapse rescuers arrive.

Victim removal techniques may include the use of spinal immobilization devices and victim transfer devices. This equipment includes backboards, baskets, and flexible stretchers. If it is imperative that surface victims be removed quickly from a hazardous environment, even standard fire fighter drags and carries can be used to do so. To the extent possible, rescuers should consider the need for spinal immobilization and care for other injuries that may be present given the mechanisms of injury related to a building collapse.

Wrap-Up

Chief Concepts

- The types of buildings involved in structural collapse incidents can vary from a small, one-story house to a multistory, high-rise building. Although collapse events can vary in size and complexity, all such events will be difficult to manage.
- Resource requirements needed to manage a collapsed building incident effectively are significant and include search, rescue, EMS, heavy equipment, structural engineers, utility companies, and other specialized resources.
- Effective planning, which includes risk assessment, resource assessment, and development of written procedures, is important to manage a structural collapse emergency successfully. The implementation of the Incident Management System is a key factor in controlling this type of emergency effectively.
- The collapse characteristics differ for the various types of buildings that may be found within the community. The types of collapse voids and the potential victim locations may change depending on the construction features of the building.
- Hazards are significant at a collapsed structure incident. A thorough hazard assessment must be completed initially and hazards reassessed throughout the course of the event.
- Initial search measures by first responders may include physical search of safe areas and the implementation of a hailing system.
- First responders must be prepared to manage a large amount of walking wounded.

Hot Terms

Approach hazards Hazards that rescuers may encounter as they travel from their current location, such as a fire station, to the actual collapse site.

Canine search team Specially trained dogs, dog handlers, and, in some cases, personnel who direct the canine search teams.

Entrapment A condition in which a victim is trapped by debris, soil, or other material and is unable to perform self-extrication.

Extrication The removal of trapped victims from an entrapment.

Incident action plan (IAP) An oral or written plan containing general objectives reflecting the overall strategy for managing an incident. It may include the identification of operational resources and assignments. It may also include attachments that provide direction and important information for management of the incident during one or more operational periods.

Operations level The level of rescue training that allows for limited entry into a confined space with very few hazards, in such a way that a person could enter with self-contained breathing apparatus (SCBA) with no obstructions, and where the victim can be seen from the entryway of the confined space.

Physical search team Personnel who have been trained to perform a physical search in a collapse environment and do not require specialized equipment.

Resource typing Categorization of the resources requested, deployed, and used in incidents according to their capability to perform given tasks.

Secondary collapse A partial or total failure of structural members from a building that had been damaged from an earlier collapse event.

Shoring Tools and equipment used to support damaged buildings and damaged structural members, so as to stabilize the area involved and to avoid further (secondary) collapse.

Surface hazards Unstable debris piles, loose and uneven rubble, and sharp materials found at a collapse site.

Surface victims The victims of a structural collapse who may be found on top of the collapse debris pile or along the perimeter of a collapsed building.

Technical search The use of mechanical search equipment, including visual, thermal, and acoustic devices, to locate victims trapped within a collapsed structure.

Technician level The level of rescue training that allows the rescuer to be directly involved in the rescue operation itself, including difficult and hazardous confined-spaces entries.

Victim packaging The process of preparing a victim for movement as a unit, often accomplished with a long spine board or similar device.

Weapons of mass destruction (WMD) Hazards that include, but are not limited to, four main categories of threats: chemical, biological, radiological/nuclear, and explosive (CBRNE).

Rescue Responder *in Action*

A tornado has passed through your response district without warning. In its wake, your truck company is dispatched to a report of a house collapse with a report of people trapped. Along with your truck company, two engine companies, a rescue company, an ALS medical unit, and a chief officer have been dispatched.

1. What are some of the approach hazards you might encounter during this type of response?

   A. Downed power lines

   B. Trees blown down and blocking the roadway

   C. Structure fires

   D. All of the above

2. What are the potential hazards that might be present upon your arrival at the collapse site?

   A. Hazardous materials

   B. Secondary collapse

   C. Walking wounded

   D. Both A and B

3. Which actions can an awareness-level responder take prior to the arrival of the collapse rescue team?

   A. Establish scene control and limit further entry into the building

   B. Make entry into the collapse area to assist victims

   C. Deliver patient care to walking wounded

   D. Both A and C

Confined-Space Search and Rescue

NFPA 1670 Standard

7.1 General Requirements

7.1.1 Organizations operating at confined space incidents shall meet the requirements specified in Chapter 4. [p. 149]

7.1.2* The requirements of this chapter shall apply to organizations that provide varying degrees of response to confined space emergencies. [p. 149–151]

7.1.3 All confined space rescue services shall meet the requirements defined in 7.1.3.1 through 7.1.3.12. [p. 147–157]

7.1.3.1 Each member of the rescue service shall be provided with, and trained to use properly, the PPE and rescue equipment necessary for making rescues from confined spaces according to his or her designated level of competency. [p. 149–150]

7.1.3.2 Each member of the rescue service shall be trained to perform the assigned rescue duties corresponding to his or her designated level of competency. [p. 149–150]

7.1.3.3 Each member of the rescue service shall also receive the training required of authorized rescue entrants. [p. 149]

7.1.3.4* Each member of the rescue service shall practice making confined space rescues once every 12 months, in accordance with the requirements of 4.1.10 of this document, by means of simulated rescue operations in which he or she removes dummies, mannequins, or persons from actual confined spaces or from representative confined spaces resembling all those to which the rescue service could be required to respond in an emergency within their jurisdiction. [p. 149]

7.1.3.5 Representative confined spaces should—with respect to opening size, configuration, and accessibility—simulate the types of confined spaces from which rescue is to be performed. [p. 148–149]

7.1.3.6 Each member of the rescue service shall be certified to the level of first responder or equivalent according to U.S. Department of Transportation (DOT) *First Responder Guidelines*. [p. 147]

7.1.3.7 Each member of the rescue service shall successfully complete a course in cardiopulmonary resuscitation (CPR) taught through the American Heart Association (AHA) to the level of a "Health Care Provider," through the American Red Cross (ARC) to the "CPR for the Professional Rescuer" level, or through the National Safety Council's equivalent course of study. [p. 147]

7.1.3.8* The rescue service shall be capable of responding in a timely manner to rescue summons. [p. 147]

7.1.3.9 Each member of the rescue service shall be equipped, trained, and capable of functioning to perform confined space rescues within the area for which they are responsible at their designated level of competency. [p. 149–150]

7.1.3.10 The requirements of 7.1.3.9 shall be confirmed by an annual evaluation of the rescue service's capabilities to perform confined space rescues in terms of overall timeliness, training, and equipment and to perform safe and effective rescue in those types of spaces to which the team must respond. [p. 149]

7.1.3.11 Each member of the rescue service shall be aware of the hazards he or she could confront when called on to perform rescue within confined spaces for which the service is responsible. [p. 151–157]

7.1.3.12 If required to provide confined space rescue within U.S. federally regulated industrial facilities, the rescue service shall have access to all confined spaces from which rescue could be necessary so that they can develop rescue plans and practice rescue operations according to their designated level of competency. [p. 151]

7.1.4 A confined space rescue team shall be made up of a minimum of six individuals for organizations operating at the technician level, and a minimum of four individuals for organizations operating at the operations level. [p. 149–150]

7.2 Awareness Level

7.2.1 Organizations operating at the awareness level for confined space search and rescue incidents shall meet the requirements specified in Sections 7.2 and 5.2 (awareness level for rope rescue). [p. 150]

7.2.2 The organization shall have the appropriate number of personnel meeting the requirements of Chapter 4 of NFPA 472, *Standard for Competence of Responders to Hazardous Materials/Weapons of Mass Destruction Incidents*, commensurate with the organization's needs. [p. 151]

7.2.3 Organizations at the awareness level shall be responsible for performing certain nonentry rescue (retrieval) operations. [p. 158]

7.2.4 Organizations operating at the awareness level for confined space search and rescue incidents shall implement procedures for the following: [p. 147–159]

(1) Recognizing the need for confined space search and rescue

(2) Initiating contact and establishing communications with victims where possible

(3)* Recognizing and identifying the hazards associated with nonentry confined space emergencies

(4)* Recognizing confined spaces

(5)* Performing nonentry retrieval

(6)* Implementing the emergency response system for confined space emergencies

(7)* Implementing site control and scene management

Additional NFPA Standards

NFPA 1001, *Standard for Fire Fighter Professional Qualifications*
NFPA 1006, *Standard for Technical Rescuer Professional Qualifications*

Knowledge Objectives

After studying this chapter, you will be able to:

- Identify the need for confined-space search and rescue operations.
- Recognize and identify confined-space hazards.
- Identify various types of confined spaces.
- Describe the resources needed to conduct a confined-space search and rescue operation.
- Describe response planning and incident management requirements related to a confined-space search and rescue incident.
- Describe site control operations at a confined-space incident.
- Describe non-entry rescue considerations at a confined-space incident.

Skills Objectives

After studying this chapter, you will be able to perform the following tasks:

- Perform a non-entry retrieval of a victim at a simulated confined-space event.

You are a first responder assigned to a fire department engine company. Your unit is dispatched on an emergency medical call for a report of a man who has fallen off a ladder at a construction site. The initial report by radio states that the man is conscious but incoherent. He is reported to be lying on his back, and the caller indicated that co-workers planned to move the victim to a more accessible area prior to your arrival on the scene.

As you approach the scene with your vehicle, you notice a construction worker waving his arms frantically and motioning for you and your team to come to his location. As you walk up to the site, it becomes apparent that this is not a simple medical emergency due to a fall off a ladder. The original victim fell off an access ladder while trying to climb out of an underground electrical utility vault. He dropped 20 feet back down into the vault and was unconscious. Subsequently, two co-workers entered the utility vault in an attempt to remove the original unconscious victim. These two co-workers are also now unconscious at the bottom of the utility vault. It is not known why the two co-workers became unconscious during their attempt to remove their friend.

As you start to perform a size-up of this event, you realize that a confined-space incident requires specially trained responders and specialized tools and equipment to operate effectively.

1. What should you be looking for during a confined-space size-up?
2. Which types of hazards might rescuers find at a confined-space event?
3. Which actions can first responders take at this incident to manage the event more effectively?
4. Which non-entry actions can you take at a confined-space event?
5. Which types of resources are needed to manage a confined-space rescue event effectively?

Introduction

In March 1984, a 20-year-old construction worker attempting to refuel a gasoline-powered pump used to remove water from a sewer line was overcome and died from carbon monoxide exposure. A state inspector attempting a rescue of the worker also died, and 30 fire fighters were injured during the rescue operation. The following December, a 22-year-old worker was overcome by toxic vapors while attempting to clean the inside of a chemical storage tank and died. One fire fighter was killed and 15 others injured during the rescue effort.

These stories are but two examples of situations in which confined-space rescue incidents turned into nightmares for the rescue agencies involved. Federal and state officials of the Occupational Safety and Health Administration (OSHA) have established written rules governing safe operating procedures for confined-space entry. Private-sector organizations have begun to heed the many warnings being published about the dangers of this type of operation, and most are now working to achieve compliance with the written regulations.

Public safety organizations should not consider themselves exempt from these rulings. They should take action to adopt standard operation procedures (SOPs) (or suggested operating guidelines [SOGs]) that effectively safeguard rescuers who must expose themselves to these hazards while simultaneously complying with written law. In fact, public safety organizations (fire/EMS/police) that commit to performing confined-space entry and rescue must comply with the federal OSHA Standard 1910.146, Permit-Required Confined Spaces. They must also be able to respond to these incidents in a timely manner, must be certified to the first responder level according to the U.S. Department of Transportation's (DOT's) *First Responder Guidelines*, and must have successfully completed a course in cardiopulmonary resuscitation (CPR).

Information gathered during the development of OSHA's Permit-Required Confined Space standard reveals that a variety of circumstances may cause confined-space rescue operations to be dangerous, difficult, and confusing. The potential for confined-space incidents is high, especially given the many new utilities being placed underground, existing utilities, and confined spaces in need of repair and maintenance. Currently, more than 250,000 confined-space entries occur each day, carried out by construction workers, industrial plant operators, painters, electricians, welders, utility company personnel, and public works employees. There is no way to determine how many unauthorized or accidental entries occur every day by mischievous, adventuresome, or unlucky individuals.

Formal training in confined-space hazards and safe work practices may not exist in industry or the private sector. OSHA statistics show that 95 percent of victims of incidents in these areas had no confined-space training. In addition, the infrequency of confined-space situations encountered means that responders often have very little experience on which rescue teams can base their decisions.

Even relatively simple rescue situations may become much more complex when they occur within a confined space. For example, a simple medical emergency that can be easily handled on a living room floor becomes much more complex when it occurs 50 feet below ground in a sewer system. When the federal OSHA Permit-Required Confined Space Final Standard was adopted in 1993, statistics indicated that 98 percent of all industrial organizations were planning to depend on the local fire department for rescue operations; unfortunately, at that time, only 2 percent of fire departments were trained in confined-space rescue.

The other characteristic that makes confined-space entry and rescue dangerous for responders is the deceptive nature of the hazards encountered. The confined space may appear harmless. Hazards may not be obvious, to the point that a false sense of security can lure a rescuer into becoming a victim. Various safety organizations, including the National Safety Council (NSC) and the National Institute of Occupational Safety and Health (NIOSH), have published statistics over the years stating that anywhere from 50 percent to more than 60 percent of confined-space deaths result from secondary entries by personnel attempting to rescue the initial entrants into the space.

Safety Tips

In some cases, the patient may be conscious and talking to you from within the confined space. This situation may give an untrained responder a false sense of security and possibly lure him or her into making an entry into the confined space to render aid. Awareness-level responders do not have sufficient training to make entry into a confined space.

Confined Space

An enclosure must meet all of the following criteria to be considered a confined space:

- Any space having a limited means of entry or egress
- A space not designed for continuous worker occupancy
- A space that contains or has the potential to contain a dangerous air contamination or oxygen deficiency resulting from the accumulation of hazardous dusts, mists, gases, or vapors that cannot be removed or prevented by natural ventilation

In addition, an enclosure that meets these additional criteria is considered a Permit-Required Confined Space:

- Any enclosure that presents the potential for engulfment of persons by the product. **Engulfment** is a type of victim entrapment that may occur when a liquid product or a granular solid product flows into or moves within a confined space and then surrounds and/or buries a victim. An example would be a farmer who is working inside a grain silo and suddenly becomes trapped when the grain surface collapses; the victim is pulled down into the surface material, and grain then surrounds and engulfs the victim.
- Any enclosure that has an internal configuration such that an occupant could suffer an entrapment by inwardly converging walls or sloped and tapered floors. **Entrapment** refers to a condition in which a victim is trapped by debris, soil, or other material and is unable to extricate himself or herself.

Confined-Space Problem Sites

The following areas should be considered as having the potential for presenting the hazards of a confined space as outlined in the preceding definitions Figure 7-1 ▶ :

- Manholes
- Utility vaults
- Tunnels
- Storm sewer systems
- Sanitary sewer systems
- Wells and cisterns
- Trenches and excavation sites
- Open pits more than 4 feet deep
- Sump pits and sump rooms
- Silos
- Storage bins
- Hoppers
- Brewer's vats
- Septic tanks
- Tank cars
- Reaction vessels
- Industrial smokestacks, chimneys, furnaces, and boilers
- Cold storage facilities
- Large industrial transformers
- Ship's hold
- Auto repair lift pits
- Water treatment plants—sludge diffusers, pits, carbon tanks
- Sanitary sewer pumping stations

Rescue Tips

Awareness-level responders should learn to identify potential confined-space sites within their response areas. After doing so, they will not be so surprised if they respond to the site for a medical emergency or other type of emergency that is a result of exposure to the confined-space hazards.

A.

B.

C.

Figure 7-1 Examples of confined spaces include **A.** utility vaults, **B.** tank trucks, and **C.** sewage pump stations.

Applicable Standards

Emergency responders can find guidance on management of these types of incidents by referring to the National Fire Protection Association (NFPA) Standards 1006, *Standard for Technical Rescuer Professional Qualifications*, and 1670, *Standard on Operations and Training for Technical Search and Rescue Incidents*.

NFPA 1006 establishes the job performance requirements that are needed for responders to perform technician-level rescue skills. These skills include preplanning a confined-space site,

performing a site assessment, hazard recognition and control, atmospheric monitoring, entry team preparation, victim assessment, packaging and transfer techniques, and incident termination procedures.

NFPA 1670 establishes organizational requirements to operate safely and effectively at confined-space emergencies. The standard identifies the requirements for risk and hazard assessment, written procedures, safety protocols, personal protective equipment (PPE), establishment of operational response level, and resource considerations.

Federal OSHA Standard 1910.146, Permit-Required Confined Spaces, establishes regulatory requirements for all personnel who are required to enter confined-space work sites as part of their job. The standard establishes program requirements including confined-space site identification, confined-space classification, permitting systems, documentation requirements, permissible entry conditions, safety procedures, and training requirements.

Resource Requirements

Rescue Resources

Confined-space rescue that requires rescue teams to make entry into various enclosures is a high-risk event that requires the services of specially trained responders who use specialized equipment. All responders who may encounter a confined space during emergency response should be trained to the NFPA 1670 awareness level. All responders shall be equipped and trained to use equipment for confined-space entry based on their designated level of competency as described later in this section. All responders who will make confined-space entry must be trained to at least the NFPA 1670 operations level for very basic entry-type rescues; for more difficult entry-type confined-space rescues, responders must be trained to meet the technician-level job performance requirements as outlined in NFPA 1006. All responders who make a confined-space entry must be trained to meet the requirements of an authorized entrant as defined in federal OSHA Standard 1910.146, Permit-Required Confined Spaces. Training conditions should simulate potential confined space incidents. In addition, rescuers who make confined-space rescues must practice making these entries at least once every year. An annual team performance evaluation to confirm the team's capability to perform competently is required by NFPA 1670.

The three levels of competency are the awareness, operations, and technician levels. All must meet NFPA 1670 Chapter 4 requirements.

- **Awareness level**: the first level of rescue training provided to all responders, which emphasizes recognizing hazards, securing the scene, and calling for appropriate assistance. There is no actual use of rescue skills at the awareness level.
- **Operations level**: a level that allows for limited entry into a confined space with very few hazards and in such a way that a person could enter with self-contained breathing apparatus (SCBA) with no obstructions, and where the victim can be seen from the entryway of the confined space. An operations-level team should have a minimum of four team members.

- **Technician level**: the level that allows responders to be directly involved in the rescue operation itself, including difficult and hazardous confined-space entries. Training includes use of specialized equipment, care of victims during rescue, and management of the incident and all personnel at the scene. A technician-level team should have a minimum of six team members.

Safety Tips

Responders who are trained to make confined-space entries must be able to use atmospheric monitoring equipment to confirm that a safe atmosphere exists before making entry.

Specialized equipment required for confined-space rescue includes atmospheric monitoring equipment that is used to determine the environmental conditions within the space. Equipment such as positive-pressure ventilation fans and utility blowers is also needed to ensure that adequate clean air is circulated throughout the space, thereby helping to maintain a safer work environment. Lighting systems with explosion-proof or intrinsically safe ratings are needed as well; these lighting systems may include helmet lights, hand lights, chemical light sticks, and general area lighting.

Special breathing equipment known as supplied-air breathing apparatus (SABA) is used in confined-space situations where traditional SCBA is too big and bulky for safe operation. SABA is an emergency breathing-air system that is similar to SCBA, but uses an air line running from the rescuers to a fixed air supply located outside the confined space. SABA systems include the breathing apparatus itself, air lines, manifold systems to supply multiple users, regulators, an emergency egress bottle of 5–15 minutes duration, and an air supply. Extra breathing air or extra air bottles are often needed for difficult or prolonged entry situations.

Competent communications equipment is necessary to maintain full-time effective communications between the entry team and the rescue team support personnel who are operating outside the space. Sometimes communications can occur face to face or with hand signals during very basic entry situations where the members of the entry team and the outside rescue team can always see one another. An example of this situation is a scenario in which a rescue team makes entry into a utility vault where a victim is unconscious and the victim is located directly below the opening to the space.

In more difficult situations, entry team personnel may be required to travel some distance inside the space, such as a sewer system. In these situations, portable radios and **hard line communications systems** may be required. Each system has its limitations, however. The portable radio system may not consistently work due to interference from steel and concrete within the space. Hard line communications systems, which consist of wire cables that run between a control panel on the outside of the space and are attached to the rescuer, will not be affected by this type of interference, but the rescuers' movement is limited by the length of the cable.

Figure 7-2 Confined-space rescue teams require access and egress equipment for rescues.

Access and egress equipment is used by confined-space rescue teams to enter and exit a confined space; it includes tripods or similar devices and rope rescue equipment **Figure 7-2**. This kind of equipment is required for most confined-space entries. Once rescuers are able to gain access and extricate entrapped victims, they are required to use various types of victim packaging and transfer devices such as SKED stretchers, Reeves Sleeves, or similar devices. If the victims are located above or below grade, the situation may require rescuers to have rope rescue skills (in line with NFPA 1670 Section 5.3), such as the ability to establish rope anchors, operate lowering systems, use mechanical advantage raising systems, and implement safety belay lines.

Emergency Medical Services Resources

Emergency medical services (EMS) resources are required so that rescue teams can hand off victims for treatment and transport to medical facilities once the victims are removed from the confined space. Victims who have been trapped in confined spaces may be suffering from oxygen deficiency, hazardous materials exposure, traumatic injuries, or medical emergencies. Advanced life support (ALS) capabilities are required to treat these victims effectively once they are removed from the space.

EMS resources are also required at a confined-space rescue so that medically trained personnel will be available to assist with the rehabilitation of rescue team personnel. Difficult confined-space rescues can be time-consuming and very physically taxing for rescue personnel, especially if rescue team members are required to make multiple entries into the space to remove victims. EMS personnel can assist with the rehabilitation process by monitoring team member vital signs, ensuring adequate hydration, and administering medical care to team members as required.

Hazardous Materials Resources

Hazardous materials are materials or substances that pose an unreasonable risk of damage or injury to persons, property, or the environment if not properly controlled during handling, storage, manufacture, processing, packaging, use and disposal, or transportation. In some situations, hazardous materials resources are needed to assist with atmospheric monitoring, research the potential hazardous materials that may be present, determine the proper level of protective gear required, and assist with or manage decontamination of victims and entry team members. These hazardous materials resources should include personnel who meet the requirements of Chapter 4 of NFPA 472, *Standard for Competence of Responders to Hazardous Materials/Weapons of Mass Destruction Incidents.*

Special Logistical Support Resources

Confined-space events can present themselves in a variety of ways. Some will be more difficult than others. Very difficult entries and extrications may require the use of powered digging equipment such as a backhoe or another construction tool to dig down into sewer systems, tunnels, or collapsed earthen enclosures to gain access to entrapped victims. This option might be used if normal entry into the space where the victim is located is not feasible due to collapse, obstacles, size of the space, or other difficulties. Utility companies, public works, tunneling companies, or other community resources may all be required to assist rescuers when confined-space situations are unique and unusual. Keep in mind, however, that removal of victims using utility or power equipment should be the last option because of the potential for injuries to victims and rescuers alike. In these situations, power equipment should be used only to gain access to an area close to the victim or to the exterior of a confined-space enclosure, such as a pipe. The final access to the victim, such as by breaching a pipe, should always be accomplished with small tools to limit the potential for injury to the victim.

Incident Management Requirements

The strategic objective when operating at a confined-space incident is to control and manage the event in such a way as to evaluate the scene and identify potential victims and their locations, to initiate operations to minimize hazards to operating personnel and trapped victims, to initiate operations to search the confined space effectively, to initiate operations to rescue and remove trapped victims, and to minimize further injury to victims during search, rescue, and removal operations.

The Incident Command System (ICS) positions are staffed as determined by the scope of the event. In most cases, at a minimum, management personnel are required to fill the positions of incident commander (IC), technical rescue group supervisor, planning section chief, and safety officer.

The technical rescue group supervisor is responsible for implementation of tactical rescue decisions to support the strategy established by the IC. These tactical benchmarks will include hazard mitigation, entry-team readiness, rapid intervention capabilities, and emergency medical care for the victim. In cases where the IC is not well versed in technical rescue, the technical rescue group supervisor may be the person who makes decisions about tactical benchmarks and rescue methods.

Incident management personnel are responsible for strategic and tactical management of the confined-space emergency response. One of these responsibilities is the development of an incident action plan (IAP), which identifies the overall control objectives for the emergency. Because a confined-space emergency may involve unique industrial facilities, the incident management personnel may need to gather information from knowledgeable people from the private sector. These individuals may include industrial hygienists, plant or facility maintenance personnel, plant safety staff, and/or utility company personnel.

Rescue Tips

Pitfalls of Confined-Space Rescue
- Lack of understanding of confined-space rescue difficulties
- Lack of proper training
- Improper equipment
- Failure to perform an adequate situational analysis
- Inadequate staffing
- Failure to understand and recognize the hazards
- Failure to properly mitigate or reduce the hazards
- Improper personal protective equipment
- Lack of an incident management system

Response Planning

Responders must know how to initiate the emergency response system to ensure that appropriate resources are deployed to the event and that operational guidelines are followed. Many response organizations perform needs assessments within their response area to determine their vulnerability to confined-space events, including those in local federally regulated industrial facilities. As part of this assessment, the organization should consider the types of confined spaces they may encounter and the type of hazards that might be encountered, and then determine the appropriate emergency response resource needs. The emergency response system should include written procedures to request agency resources and resources that will respond through mutual aid agreements, contracts with private-sector companies, and any memoranda of agreement with regional, state, or federal assets. Private-sector agreements may be established to acquire resources that cannot be supplied by the response agency (e.g., extra breathing air, hazardous materials assets, dive resources, and industrial hygienists).

The capability of agency personnel should be determined as well. This capability assessment should include identification of the level of performance to which agency personnel are trained, such as the operations or technician level for confined-space rescue. As part of the emergency response system, the location, capabilities, and response times for mutual aid, regional, state, and private-sector confined-space rescue teams should be known, and procedures should be established to request these assets. Likewise, dispatch procedures should be in place to ensure that adequate and proper resources are deployed when a confined-space event occurs, thereby ensuring that effective search and rescue operations can begin.

Hazards and Hazard Assessment

Each confined space has its own specific problems and hazards. If personnel are to be protected against injury or death, they must be well trained to recognize and understand the hazards they encounter.

Scene Assessment

Always research the space to identify the hazards present, if any, prior to entry. As part of this effort, rescuers should look for an on-site confined-space permit. If a permit is available for reference purposes, it may provide valuable information about the space to be entered, the purpose for the entry, the hazards of the permit space, the type of work being performed, atmospheric testing results, hazard control actions taken by the company (ventilation, purging, control of energy sources), the number of entrants inside the space, personal protective gear needed, and the communications procedures used by the entrants. Such permits will vary from company to company, however, and they may be difficult to interpret for information-gathering purposes.

Figure 7-3 Atmospheric hazards pose a real risk to confined-space rescuers. An atmospheric monitor can help determine the contents of the atmosphere.

Safety Tips

Confined-space hazards can be very deceiving. The primary hazard at such events is an atmospheric hazard: This hazard may not be seen, smelled, or felt by the responder.

Additional assessment information can be gathered from co-workers on the scene; plans, drawings, and schematics of the space; industrial plant personnel; site safety personnel; on-site material safety data sheets (MSDS); or other hazardous materials references. The information gathered during the scene assessment should answer the following questions:

- What is normally stored in the confined space?
- Which operations are normally performed in the confined space?
- Which type of atmosphere might potentially be formed in the confined space?
- Are the responders properly trained and prepared to perform this rescue?
- Do the responders have the necessary resources to perform this rescue?
- What is the history of work or rescues that have occurred in the space in the past?

Atmospheric Hazards

Atmospheric hazards are not easily seen, smelled, heard, or felt—yet they can represent deadly risks to those who must work in confined-space areas. Approximately 90 percent of fatalities

in confined spaces involve a hazardous atmosphere containing carbon monoxide, hydrogen sulfide, oxygen deficiency, carbon dioxide, or a combustible gas **Figure 7-3 ▲**.

Many of these atmospheric conditions can create a dangerous environment known as an **immediate danger to life and health (IDLH)**. An IDLH environment contains an atmospheric concentration of any toxic, corrosive, or asphyxiant substance that poses an immediate threat to life or could cause irreversible or delayed adverse health effects. Three general types of IDLH atmospheres are distinguished: toxic, flammable, and oxygen deficient.

The federal OSHA Permit-Required Confined Space regulation states that a hazardous atmosphere is present when flammable gas, vapor, or mist is present in excess of 10 percent of its lower flammable limits. Flammable/combustible atmospheres may occur as a result of decaying vegetation or other organic matter that produces methane. Methane can also be produced during fermentation activities within wine vats, storage bins, and grain elevators. Anaerobic bacteria systems found in sewer systems, sewage treatment plants, and sanitary landfills can produce methane gas as well.

Flammable atmospheres can also be found as a normal product of storage, such as in a fuel storage tank. In other cases, flammable atmospheres may be created during maintenance activities within a confined space, such as when paints, solvents, or cleaning chemicals are used within the space.

Confined spaces such as storm sewers have sometimes been found to contain flammable atmospheres as a result of illegal dumping of chemicals. In addition, underground storage tank leaks and pipeline leaks can allow flammable materials to migrate through the soil, creating flammable atmospheres in utility vaults and sewer systems.

Airborne combustible dust is considered an atmospheric hazard when it is present at a concentration that meets or exceeds its lower flammable limit. This situation may be approximated as a condition in which the dust obscures vision at a distance of 5 feet or less. Airborne combustible dust hazards may occur in grain silos, agricultural storage buildings, and other industrial settings where finely divided dust particles may collect. When these combustible dust particles are disturbed and become airborne, they may rapidly ignite when an ignition source is encountered.

Oxygen Deficiency

Oxygen deficiency is another major confined-space atmospheric hazard of which responders must be aware. Oxygen deficiency can occur when the oxygen within a confined space is consumed, displaced, or absorbed. OSHA considers an atmosphere to be hazardous when oxygen is less than 19.5 percent of the atmosphere or exceeds 23.5 percent of the atmosphere.

Oxygen consumption within a confined space may occur during combustion operations such as those encountered in furnaces, smokestacks, or boilers, and when internal combustion engines are operated within the confined space. When responders enter these types of spaces, they must recognize that the exposed area may remain oxygen deficient long after fire and combustion operations are over.

Consumption of oxygen within a confined space may also occur due to bacterial action in sewer systems where aerobic bacteria consume oxygen; during fermentation activities in brewers' vats, grain silos, and feed storage bins; and during oxidation (rusting) of steel storage tanks.

Displacement of oxygen levels may occur within a confined space when another gas is produced within the confined space or introduced into the confined space from another source. For example, carbon dioxide and methane production within a space may displace oxygen in sewers, storage bins, and hoppers. Gases introduced from outside the space as a consequence of leaking storage tanks/pipelines or illegal dumping may displace the oxygen as well. Purging operations, such as the injection of nitrogen or argon gases into storage tanks and tank cars for cleaning purposes, may likewise create an oxygen-deficient atmosphere.

Oxygen deficiency due to absorption of the oxygen is a rare problem associated with the use of carbon storage tanks and areas within water or sewage treatment plants where activated charcoal filtering systems are used. These systems can act as a large oxygen "sponge," absorbing oxygen from the atmosphere of the space.

Toxic Atmospheres

Toxic atmospheres within a confined space may be present as a result of a natural occurrence, a normal product of storage, or an improper presence as a result of leakage or illegal dumping.

Naturally occurring toxic atmospheres may be the result of biological decay. For example, hydrogen sulfide is recognized as a major hazard in the oil refining and sewage treatment industries. This chemical compound results from the natural decomposition of sulfur-containing organic matter. The concentration of hydrogen sulfide in raw sewage may be very high—to the point that it is virtually as toxic as hydrogen cyanide (the chemical used to execute prisoners in gas chambers). Hydrogen sulfide can be recognized by its classic "rotten egg" odor. Sometimes, however, this odor may not be detectable, even when the gas is present at high concentrations, because of hydrogen sulfide's ability to cause paralysis of the olfactory nerve (which controls the sense of smell).

Other naturally occurring toxic atmospheres that may be found in confined spaces include excessive carbon dioxide, which is produced in some sewer systems due to bacterial action, and excessive methane, which is odorless and colorless and acts to asphyxiate people at high concentrations. Methane may be found in sewers, storage bins, grain elevators, caves, wells, and mines.

In some cases, toxic atmospheres may arise within confined spaces simply as the result of normal storage or normal operations within the space. Petroleum/chemical storage tanks, for example, will certainly contain toxic materials. Carbon monoxide, which is one of the most common asphyxiates encountered in industry, is often found in smokestacks, furnaces, and incinerators. Carbon monoxide is not created naturally in a confined space, but rather results from a process or is introduced into the space from the outside. Build-up of this gas within a confined space has been known to occur when responders' vehicles are parked where exhaust gases from the vehicle are picked up by intake fans and circulated into the confined space. Responders must be aware of this potential when positioning emergency vehicles at a confined-space incident.

Federal OSHA Permit-Required Confined Space guidelines refer to the concentration of any substance for which a permissible exposure limit is published in Subpart G, Occupational Health and Environmental Control, or in Subpart Z, Toxic and Hazardous Substances, as being of concern for confined-space entrants.

Atmospheric Changes

Atmospheres that initially test as or appear safe can suddenly change for a variety of reasons. The disturbance of the natural air flow into the confined space during ventilation or initial opening of a manhole cover can change the atmospheric mixture, for instance. Humidity and temperature changes will affect the atmosphere as well. Oxygen levels within the confined space can become fatally low in a brief period of time. Responders should not be lulled into a false sense of security by the presence of a viable and conscious patient; the oxygen level may not support the needs of any additional persons entering the confined space, especially when carbon dioxide from exhaled breath is displacing the available oxygen.

Maintenance operations, such as scaling operations to remove loose rust and sediment from pipes and containers, can release vapors and gases. The petroleum industry has experienced instances where storage tanks were rinsed, vented for several days, and tested with monitoring devices that showed no hazardous atmosphere. When workers then began scaling operations, oxygen deficiency occurred or flammable vapors were released and ignited.

The disturbance of sludge or sediment by maintenance or rescue workers as they move around on the floor of a storage vessel may release flammable or toxic vapors. In addition, stratification of gases due to various vapor densities may result in multiple gases being present at different levels within the space. This layering effect directly influences the ventilation and monitoring techniques required for safe entry into the space.

Voices of Experience

The information supplied by the public safety communications center was worrisome: "The caller states there is a man stuck in the sewer pipe. They lifted the lid but can't get to him and can only hear him yelling for help." A fire department engine company and ambulance that were dispatched arrived on the scene and were met by citizens and police who stated that there was a man in the storm sewer.

Additional fire department personnel specially trained in confined-space rescue began to initiate standard confined-space operations, which included assigning an entry team, a backup team, a rescue officer, and a safety officer. Initial confined-space operations began at the point where citizens met the first-arriving fire department units. This entry point was a standard storm sewer culvert that could be accessed through a standard-size manhole cover.

> **"Rain was falling heavily, and the vault in which rescuers were working, as well as the pipe in which the victim was located, was taking on significant water."**

A two-person entry team made the initial sewer system entry at an alternate location, where it was reported the victim had made his entry. At this point, the units were in a search mode and were still not certain that a victim was still in the sewer system. This initial search required the entry team to walk through a 48-inch pipe for a distance of approximately 125 feet to the original storm sewer culvert where citizens first reported the victim missing.

The entry team then began moving north through the pipe to continue their search. While this was occurring below ground, other rescue team personnel were moving forward above ground, opening manholes, checking for any signs of a victim, using a search camera for observation, and monitoring the atmosphere. Communication with the entry team was maintained utilizing portable radios.

As the entry team progressed through the pipe under a parking lot, they noticed water marks in the pipe that appeared to be footprints and could hear the victim making noise in the pipe ahead of them. At this point, the operational area had spread over a geographical area of about 400 feet. The personnel working above ground utilized police assistance to control traffic on a busy roadway to open another manhole, continue with atmospheric monitoring, and communicate with the entry team. At this point, the sewer pipe changed size from 48 inches to 24 inches.

Just as this portion of the event was transpiring, another culvert covered by a steel grate was discovered farther north in the parking lot; rescuers could hear the victim making noise at this location. This confined-space entry point was a 5-feet-deep vault and was a collection point for several smaller storm sewer pipes, including the 18-inch pipe that the victim had entered.

Rescuers entered this vault to determine the victim's location and found him to be approximately 75 feet down the pipe, south of their location. Rain was falling heavily at this time, and the vault in which rescuers were working, as well as the pipe in which the victim was located, was taking on significant water.

The rescuers could see that the victim's head was going under water at times as the 18-inch pipe started to fill with storm water runoff. Two rescuers entered the vault and used a water rescue rope, floating it down the pipe to the victim, in an attempt to pull him out. The victim was conscious but would not grab onto the rope to be pulled to safety. During their attempts to communicate with the victim, the rescuers realized that they had a language barrier.

As the victim's situation deteriorated, rescuers made the drastic decision to ask a Spanish-speaking citizen to enter the confined-space vault to give verbal direction to the victim. As a result of this action, the victim grabbed onto the rope and was pulled free of the pipe. Rescuers lifted the victim out of the vault, where he was treated by emergency medical personnel.

This example illustrates how dangerous confined-space rescue attempts can be and how rapidly a confined space rescue situation can deteriorate. Confined-space rescue requires effectively trained and equipped rescue teams to be successful.

Robert Rhea
Battalion Chief
Fairfax County Fire Department
Fairfax, Virginia

A.

B.

C.

Figure 7-4 Elevated confined-space structures include water storage tanks **(A)**, silos **(B)**, and hoppers **(C)**.

Physical Hazards

In addition to atmospheric hazards, responders must recognize the potential for physical hazards within a confined space. The types of physical hazards responders face depend on the specific space encountered and its primary function.

Energy sources such as electrical, steam, hydraulic, pneumatic, stored-energy, and gravity-fed systems may cause entrapment of a responder's arm or leg when operating machinery—such as gears, chains, augers, or vanes—is encountered within a confined space. Flowing product may enter a space and submerge or engulf a rescuer during confined-space entry operations; this scenario could happen in sewer systems or storage tanks, for example. Entry into sewer systems may also expose rescuers to high water conditions, hypothermia, raw sewage exposure, animals, slippery footing, and limited lighting, all of which can add to the complexity of a confined-space event.

> **Safety Tips**
>
> Always try to determine if a confined space has energized electrical equipment or some other type of energy source that could pose a danger to responders.

When responders are required to enter abandoned wells or mines, the potential for collapse is a key concern. The confined-space enclosures commonly found in industrial settings such as process vessels may expose rescuers to hot and cold surfaces, high noise levels, and vibration hazards. Elevated confined-space structures such as elevated water storage tanks, silos, and elevated hoppers expose responders to fall hazards and unstable work surfaces, especially if rescuers must enter a hopper or silo and work on top of granular materials such as gravel or agricultural feed products **◄ Figure 7-4** .

Basic Hazard Mitigation

The IC must determine which actions can be taken to eliminate or modify the hazards that are present. Competent atmospheric monitoring is extremely important in determining when safe entry conditions are present and in identifying the presence of hazardous atmospheres. Initial monitoring is always accomplished pre-entry and then conducted either continuously or at intervals after entry is made into the space. Confined-space rescue teams field calibrate atmospheric monitoring instruments and perform pre-opening testing of the space to limit the potential for ignition of lighter-than-air gases that may collect around manhole covers. All personnel who initially open and monitor the confined-space atmosphere should wear full protective gear and respiratory protection to protect themselves from unknown atmospheric contaminants. Continuous atmospheric monitoring should occur throughout the rescue operation to ensure that responders maintain awareness of any changes in atmospheric conditions.

Ventilation is another hazard control technique that is extremely important at confined-space emergencies. It involves the use of ventilation fans or blowers to introduce clean air into

Figure 7-5 Responders can use ventilation fans to reduce atmospheric hazards in confined spaces.

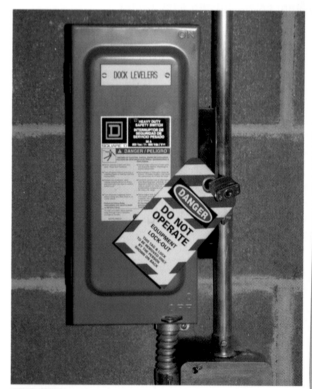

Figure 7-6 Lockout/tagout systems help keep responders safe when energy sources are present.

a confined space in an attempt to remove or reduce hazardous atmospheres **Figure 7-5 ▲**. This operation should be performed early in the rescue attempt to improve conditions for victims located inside the space. Responders can use on-site engineered ventilation systems, if available and in safe operable condition, for this purpose. Ventilation air flow should be applied at all levels within the space owing to the potential stratification of gases. Use positive-pressure ventilation whenever possible, in the form of electric ventilation fans or utility vault blowers.

The goal of ventilation is to ensure a nonhazardous environment for both rescuers and victims; to meet this goal, ventilation activities should not create an airborne combustible dust hazard. When combustible dusts are present, such as in a feed mill or similar enclosure, ventilation activities could cause combustible dusts lying on stable surfaces to become airborne, thereby increasing the potential for ignition or explosion. In these cases, ventilation may actually increase the hazards to rescuers and victims.

The control of energy sources that may be present within or next to a confined space is also necessary to keep rescuers from being harmed by moving machinery or flowing fluids or other storage products. Responders need to recognize the presence of various types of energy sources, such as electrical, pneumatic, hydraulic, gravity, momentum, stored energy, or steam sources. When assessing the presence of energy sources, always consider the possibility that multiple energy sources may be present—for example, battery backup systems that may be triggered when the main electrical switch is turned off.

When performing rescue operations at an industrial plant or other location where responders are not familiar with the process for controlling energy sources, ask industrial plant maintenance staff or on-scene industrial workers to assist with the lockout of energy systems. This form of energy control, often referred to as a **lockout/tagout system**, often involves a variety of special techniques such as applying multiple-lock hasps and multiple padlocks to a control valve, controlling the flow of a product through a pipeline using pipe plugs, installing blanks between pipe flanges, or performing a double block-and-bleed operation between control valves **Figure 7-6 ▶**. As part of the lockout/tagout operation, mechanical drive linkage or drive belts

may need to be removed from motors to eliminate movement of augers, vanes, or conveyor systems. Energy control valves such as quarter-turn valves, outside stem and yoke valves, and ball valves can be controlled using chains, padlocks, and special locking mechanisms.

If a positive lockout control is not possible, someone should be posted at the controls to guard against accidental activation of the energy source. Moving machinery parts within a confined space that have been locked out should be brought to a position where there is no movement in the machinery parts due to slack in the drive line (zero mechanical state) by blocking the moving parts to eliminate any rotation or movement from slack in the system.

The combustible atmospheres that may be found within a confined space have the potential to create fire or explosion hazards. When combustible atmospheres are identified, actions must be taken to position protective fire control hose lines near the opening to the space. It is also essential to identify and eliminate any ignition sources that may be located nearby to limit the potential for a fire or explosion.

Rescue Tips

Awareness-level responders should attempt to make verbal or visual contact with any victim in a confined space from *outside* the confined space to help determine the victim's location and the viability of rescue. However, responders who are not properly trained, staffed, and equipped should not make entry into a confined space or even stick their head into a confined space to look for the victim.

Non-entry Rescue Procedures

Responders at the awareness level should not consider making an entry into a confined space to initiate a rescue. At this level of training, the responder is not adequately prepared and equipped to perform an entry-type rescue safely and effectively.

Awareness-level responders may be able to perform a non-entry rescue in very specific circumstances. If the responder arrives on the scene of an emergency and finds a victim in a confined-space enclosure (such as a utility vault), and the victim is attached by way of a harness to a tripod and winch system, the responder can attempt to operate the winch by hand to remove the victim from the confined space without having to make entry. Under no circumstances should the awareness-level responder enter the confined space to attempt a rescue.

Rescue Tips

Responders should interview co-workers and witnesses to the confined-space accident to gain a better understanding of which operations occur within the confined space, which hazards are present, and how many victims are involved.

Another specific circumstance in which an awareness-level responder may effect a non-entry rescue would be a close horizontal retrieval scenario, such as a person located inside a tank, tunnel, or pipe with a horizontal opening. In this situation, the victim would need to be positioned close enough to the entrance that the responder could reach in with a tool and pull out the victim without having to make entry into the horizontal opening. For example, a responder on a fire truck might be able to use a pike pole or similar tool to grab hold of some part of the victim's clothing and pull the individual out of the confined space. Again, under no circumstances should the awareness-level responder place his or her head into the space or enter the space to perform a rescue.

Follow the steps in **Skill Drill 7-1 ▼** to make a non-entry confined-space horizontal retrieval:

1. Assess the location of the victim and confirm that non-entry retrieval is possible.
2. Position yourself next to the entry point of the confined space and use a long reaching tool to reach in and grasp hold of the victim's clothing. (**Step 1**)
3. Using the long reaching tool, pull the victim toward the entry point. At no time should you allow any part of your body to make entry into the confined space. (**Step 2**)
4. Continue to pull the victim to the entry point until the victim's body crosses the plane of entry point. At this point, you can use a standard victim drag or carry to complete the rescue. (**Step 3**)

Patient Care Considerations

The removal of the victim from within a hazardous confined space and out to a safe environment is the number one patient care priority. When dealing with confined-space emergencies, it is important to consider the potential mechanisms of injury—for example, falls, entrapment, or engulfment—and to package the victim accordingly. **Packaging** refers to the process of preparing

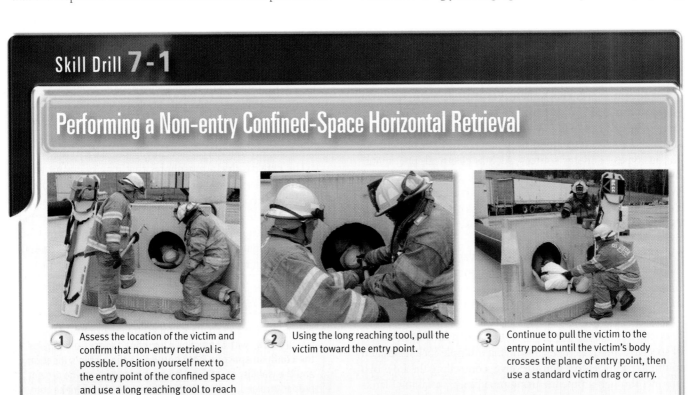

Skill Drill 7-1

Performing a Non-entry Confined-Space Horizontal Retrieval

1. Assess the location of the victim and confirm that non-entry retrieval is possible. Position yourself next to the entry point of the confined space and use a long reaching tool to reach in and grasp hold of the victim's clothing.

2. Using the long reaching tool, pull the victim toward the entry point.

3. Continue to pull the victim to the entry point until the victim's body crosses the plane of entry point, then use a standard victim drag or carry.

the victim for movement as a unit; it is often accomplished with a long spine board or similar device. The victim must be packaged for removal from the confined space based on your evaluation of any injuries, while simultaneously taking into account the current atmospheric conditions. If a hazardous atmosphere is present, responders must ensure that the victim is given respiratory protection and removed from the space as rapidly as possible. Responders must remember that the first victim care priority is removal to a safe environment.

Victim removal techniques used by confined-space rescue teams include drag-style devices that are used to package and move a victim through narrow horizontal spaces and openings such as sewer tunnels Figure 7-7 ▶ . By contrast, devices designed to package a victim and then hoist the individual vertically are used when victims must be removed from manholes, utility vaults, and other similar spaces. Extreme situations requiring rapid victim removal due to dangerous atmospheres may warrant the use of quick-application body harnesses or wristlets for removal.

During victim removal, it is important that responders always protect the victim's head (with a helmet, if possible) and guide the victim through narrow openings. Maintaining an open airway during a prolonged raising evolution may be required for an unconscious victim.

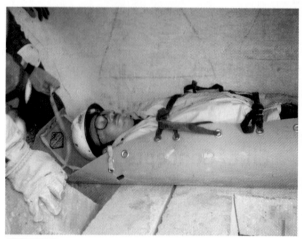

Figure 7-7 Devices such as the SKED stretcher facilitate victim removal from horizontal confined spaces.

Wrap-Up

Chief Concepts

- Rescuers may encounter emergency situations where victims are located within confined spaces—for example, sewers, storage tanks, utility vaults, railroad tank cars, elevated hoppers, or process vessels.
- Confined-space enclosures are found in every community. They come in various shapes and sizes, and responders must have a clear understanding of how to recognize what a confined space is.
- Confined-space emergencies require the use of various specialized resources. These unique rescue situations require the involvement of EMS personnel, hazardous materials specialists, and other specialized logistical support.
- Planning is critical for response organizations to operate effectively at confined-space emergencies. Needs and vulnerability assessments should be conducted to identify the potential threats in an agency's jurisdiction and the organization's capabilities for dealing with a confined-space emergency.
- Response personnel must have a clear understanding of written operational procedures, resource availability and capability, and procedures to initiate the proper response to manage confined-space events.
- Incident management requirements include trained personnel to implement command, safety, planning, and operational functions as the situation dictates. Specialized rescue resources are assigned to manage the tactical and task-level requirements of the confined-space event.
- Confined-space hazards can include oxygen deficiencies, combustible and toxic atmospheres, and various physical hazards. Hazard mitigation efforts are required before any responders attempt to make entry into a confined space.
- Awareness-level responders may be able to attempt non-entry retrieval of victims in uncomplicated confined-space situations. At no time, however, should they attempt to make a confined-space entry.

Hot Terms

Access and egress equipment Equipment used by confined-space rescue teams to enter and exit a confined space. This equipment may include ladders, tripods, davit arms, and rope rescue equipment.

Awareness level The first level of rescue training provided to all responders, which emphasizes recognizing hazards, securing the scene, and calling for appropriate assistance. There is no actual use of rescue skills at the awareness level.

Engulfment A type of victim entrapment that may occur when a liquid product or a granular solid product flows into or moves within a confined space and then surrounds and/or buries a victim.

Entrapment A condition in which a victim is trapped by debris, soil, or other material and is unable to perform self-extrication.

Hard line communications systems Wire cables used for communication purposes that run between a control panel on the outside of the space and are attached to the rescuer.

Hazardous materials Materials or substances that pose an unreasonable risk of damage or injury to persons, property, or the environment if not properly controlled during handling, storage, manufacture, processing, packaging, use and disposal, or transportation.

Immediate danger to life and health (IDLH) An atmospheric concentration of any toxic, corrosive, or asphyxiant substance that poses an immediate threat to life or could cause irreversible or delayed adverse health effects.

Lockout/tagout system Method of ensuring that electricity, gases, solids, or liquids have been shut down and that switches and valves are locked and cannot be turned on at an incident scene.

Operations level The level of rescue training that allows for limited entry into a confined space with very few hazards, in such a way that a person could enter with self-contained breathing apparatus (SCBA) with no obstructions, and where the victim can be seen from the entryway of the confined space.

Packaging The process of preparing a victim for movement as a unit, often accomplished with a long spine board or similar device.

Technician level The level of rescue training that allows the rescuer to be directly involved in the rescue operation itself, including difficult and hazardous confined-space entries.

Ventilation A hazard control method involving the use of ventilation fans or blowers to introduce clean air into a confined space in an attempt to remove or reduce hazardous atmospheres.

Rescue Responder *in Action*

Your engine company is dispatched as a first responder after the 911 system receives a report of a man who has fallen from a ladder and has a broken leg. An ambulance from a mutual aid station is also dispatched to assist your team. When your unit arrives on the scene, the officer gets out and meets with a member of the city public works department. According to this person, one of his crew members was installing a new valve on a water line in a utility vault and fell while trying to climb up the ladder to get more equipment. The co-worker states that he had walked across the street to a convenience store to get a cup of coffee when the accident happened; upon his return, he saw his partner lying at the bottom of the vault, holding his leg and asking for help. The officer climbs back up into the fire engine and uses the radio to request the assistance of the nearest confined-space rescue team and a ladder truck.

1. What are some of the clues that a confined-space rescue team is required to assist with this incident?
   A. The victim is located below grade.
   B. The victim is located inside a utility vault.
   C. The victim has a leg injury.
   D. Both A and B

2. What are the potential hazards within this confined space?
   A. Hazardous atmosphere
   B. Limited space for making entry, creating obstructions for rescuers
   C. Energy sources
   D. All of the above

3. What can an awareness-level responder do prior to the arrival of the confined-space rescue team?
   A. Establish scene control and limit further entry into the space.
   B. Make entry into the space to assist the victim.
   C. Put on SCBA and climb down the ladder to assist the victim.
   D. Allow the co-worker to make entry to assist the victim.

Trench and Excavation Search and Rescue

NFPA 1670 Standard

11.1 **General Requirements**

Organizations operating at trench and excavation search and rescue incidents shall meet the requirements specified in Chapter 4. [p. 169]

11.2 **Awareness Level**

11.2.1 Organizations operating at the awareness level at trench and excavation emergencies shall meet the requirements specified in Sections 11.2 and 7.2 (awareness level for confined space search and rescue). [p. 169]

11.2.2 Each member of the organization shall meet the requirements specified in Chapter 4 of NFPA 472, Standard for Competence of Responders to Hazardous Materials/ Weapons of Mass Destruction Incidents, and shall be a competent person as defined in 3.3.24. [p. 174]

11.2.3 Organizations operating at the awareness level at trench and excavation emergencies shall implement procedures for the following: [p. 164–177]

(1) Recognizing the need for a trench and excavation rescue

(2)* Identifying the resources necessary to conduct safe and effective trench and excavation emergency operations

(3)* Initiating the emergency response system for trenches and excavations

(4)* Initiating site control and scene management

(5)* Recognizing general hazards associated with trench and excavation emergency incidents and the procedures necessary to mitigate these hazards within the general rescue area

(6)* Recognizing typical trench and excavation collapse patterns, the reasons trenches and excavations collapse, and the potential for secondary collapse

(7)* Initiating a rapid, nonentry extrication of noninjured or minimally injured victim(s)

(8)* Recognizing the unique hazards associated with the weight of soil and its associated entrapping characteristics

Additional NFPA Standards

NFPA 1001, Standard for Fire Fighter Professional Qualifications

NFPA 1006, Standard for Technical Rescuer Professional Qualifications

Knowledge Objectives

After studying this chapter, you will be able to:

- Identify the need for trench emergency search and rescue operations.
- Describe various types of trench emergencies.
- List general hazards associated with a trench emergency incident.
- Describe the resources needed to conduct a trench collapse search and rescue operation.
- Describe response planning and incident management requirements related to a trench emergency search and rescue incident.
- Describe site control operations at a trench collapse incident.

Skills Objectives

There are no skill objectives for this chapter.

You are a first responder assigned to an engine company. Your unit is dispatched to a report of a person injured at a construction site. While responding to this emergency, you receive additional information: The person who is injured is having trouble breathing and is bleeding from the head because something fell on the individual.

When your engine company arrives on the scene, you grab the first-aid bag and start walking toward a crowd of people who are motioning for you to hurry up and help the injured worker. When you reach the crowd of bystanders, you suddenly find yourself next to an open trench that is about 4 feet wide and 10 feet deep. At the bottom of the trench you see the victim. It is the first time that you realize the victim is trapped by a collapsed trench wall that has buried the victim from the waist down. The victim is conscious but has suffered facial lacerations and is in obvious pain. The remaining trench walls that did not collapse appear to be very unstable. The crowd is telling you to jump in the trench and save their co-worker.

1. What should you be looking for during a trench collapse size-up?
2. Which hazards might rescuers find at a trench collapse event?
3. Which actions can first responders take at this incident to manage the event more effectively?
4. How should you search for victims at a trench collapse event?
5. Which victims should be treated first?

Introduction

A trench emergency includes any event(s) inside or outside a trench or excavation that could endanger people within the trench or excavation. In the construction industry, accidents during trenching and excavation work occur more frequently than other types of construction accidents in general. An **excavation** is any human-made cut, cavity, trench, or depression made in an earth surface that is wider than it is deep. A **trench** is a narrow (in relation to its length) excavation made below the surface of the earth. In general, the depth of a trench is greater than its width, but the width of a trench is not greater than 15 feet.

Excavation projects vary in complexity, and responders may encounter trench and excavation work in any community. This kind of work is performed when underground utilities are installed, during new building construction, during some plumbing repairs such as sewer line replacements, and during foundation waterproofing of homes and businesses.

The major occupational hazards of excavation work include cave-ins, underground utility exposure, atmospheric hazards, and material or equipment falling into the trench. A **cave-in** is the separation of a mass of soil or rock material from the side of an excavation or trench, or the loss of soil from under a trench shield or support system; the sudden movement of this material into the excavation, either by falling or sliding, occurs in sufficient quantity that it could entrap, bury, or otherwise injure and immo-bilize a person. A **shield** is a structure that is able to withstand the forces imposed on it by a cave-in, thereby protecting workers within the structure Figure 8-1 ▶ . When used in a trench, these structures are known as trench shields or trench boxes.

A cave-in results when one or more trench walls collapse onto workers due to a lack of proper support. Cave-in accidents are much more likely to be fatal to the victims involved than other types of construction-related accidents. The Occupational Safety and Health Administration (OSHA) estimates that the fatality rate for trenching work is as high as 112 percent greater than the fatality rate for the construction industry in general. According to OSHA, cave-ins account for the largest percentage of preventable fatalities in excavation work due to lack of (or inadequacies in) support systems. Some fatalities (a much smaller percentage) are related to problems during installation or removal of support systems that result in a collapse. In these cases workers are generally either asphyxiated or fatally injured by falling soil or rock.

Additionally, a smaller percentage of fatalities in excavation work are caused by disrupted utilities, which may result in drowning due to water main breakage, electrocution, or inhalation of toxic fumes. According to OSHA, failure to test atmospheres and properly ventilate is responsible for 2 percent of preventable accidents. In 1.4 percent of preventable fatalities, the stability of structures adjoining the trench was not properly ensured by adequate support systems, allowing the walls of the structures to fail and fall on the workers.

Figure 8-1 A trench shield protects workers from the potential of a collapsing trench wall.

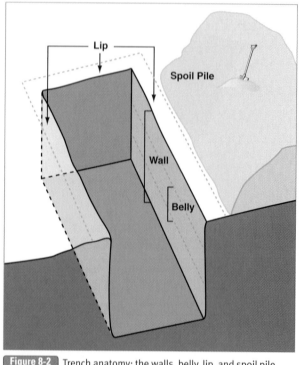

Figure 8-2 Trench anatomy: the walls, belly, lip, and spoil pile.

OSHA Standards 29 CFR 1926, 651, and 652 list precautions to be taken by workers to safeguard against these hazards, including requirements to protect workers from cave-ins.

Types of Trench Emergencies

Trench emergencies can occur in a variety of ways. The most likely cause is trench wall collapse; however, other emergencies can occur in the trench or excavation environment. For example, workers may be injured by loose rocks that fall from a trench wall or may become incapacitated by coming into contact with an underground electrical line. In some situations, rescuers may encounter a victim who is simply suffering from a medical emergency or trauma unrelated to a collapse. However, if the victim is located in an unprotected trench, rescue workers must first ensure that hazard control actions such as shoring are taken to make the trench safe before they attempt to enter it and deliver patient care.

■ Cave-ins

Trench and excavation emergencies present themselves in a variety of ways. As mentioned earlier, the cave-in of trench walls is the primary hazard of trench work. It is important for response personnel to know that several types of cave-ins are possible **Figure 8-2 ▶**. Cave-ins typically occur either from a shearing of one or both trench side walls, a slough in collapse of the trench belly, the failure of the trench lip, or a spoil-pile slide **Figure 8-3 ▶**. A **spoil pile** is a pile of excavated soil placed next to the excavation or trench.

■ Spoil-Pile Slide

The spoil pile of excavated soil and other materials that are removed from the trench and placed as a mound of material (spoil) next to the open trench can create a hazard if not properly handled. OSHA requires the spoil pile to be placed a minimum of 2 feet away from the edge of the open trench. If the spoil pile is placed too close to the open trench, the material may slide back in, causing a collapse.

■ Falling Rocks

Trench and excavation accidents can also occur when rocks or boulders break loose from a trench wall and strike a worker. This situation may or may not result in further collapse of the trench walls, but in many cases will render a victim unconscious and in need of medical assistance. If responders arrive to find a viable victim who is not trapped but still inside an unprotected trench, they must not enter the trench to administer treatment until the trench is stabilized and rendered safe by trained trench rescue personnel.

■ Shoring Collapse

Another situation that may result in a trench emergency is the collapse of existing shoring systems, such that workers become trapped. **Shoring** (also known as a **shoring system**) comprises a structure—such as a metal hydraulic, pneumatic/mechanical, or timber shoring system—designed to support the sides of an excavation and prevent cave-ins. A shoring system collapse may occur at construction sites where the trench or excavation has been open for a prolonged period. When shoring at trench sites remains in place for a long time, it may be exposed to temperature and weather changes that cause the shoring system to destabilize due to expansion and contraction of the surrounding soil.

At construction sites, OSHA requires a **competent person** to inspect the shoring system on a daily basis prior to workers entering the trench. According to OSHA 1926.32, a competent person is someone who is "capable of identifying existing and predictable hazards in the surroundings, or working conditions that are unsanitary, hazardous, or dangerous to employees, and who has authorization to take prompt corrective measures to eliminate them."

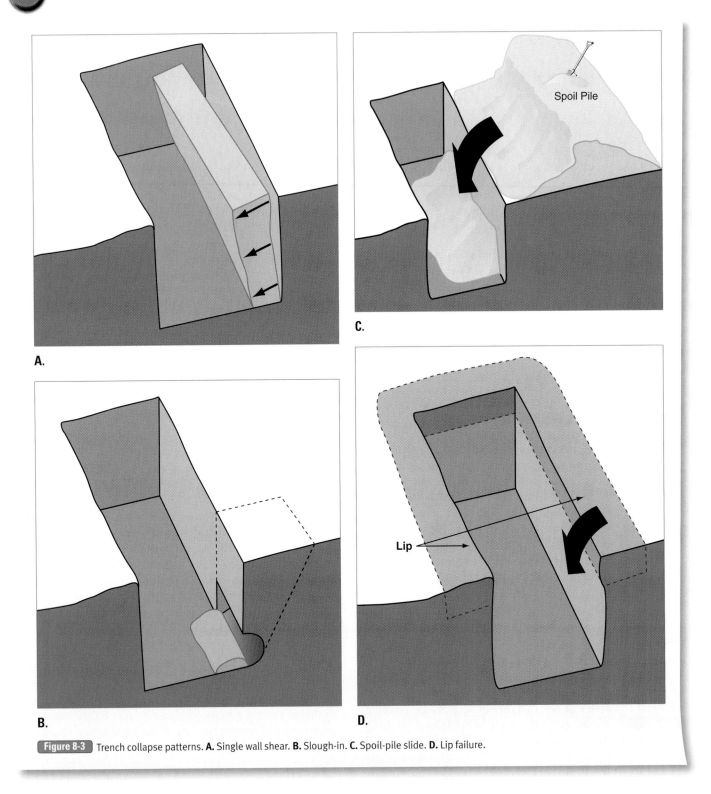

A.

B.

C.

D.

Spoil Pile

Lip

Figure 8-3 Trench collapse patterns. **A.** Single wall shear. **B.** Slough-in. **C.** Spoil-pile slide. **D.** Lip failure.

Additional Causes

In addition to the previously mentioned situations, trench and excavation emergencies may occur when workers become injured or trapped due to overturned construction equipment that falls into the trench; broken utility lines, such as those carrying electric, water, sewage, or natural gas services; or unanticipated atmospheric hazards such as build-up of methane gas from decaying vegetation. It is also important for responders to remember that they may arrive to find a construction worker who is simply suf-fering from an injury or medical emergency, but who is nonam-bulatory and located at the bottom of an unprotected trench.

Protection Against Trench Emergencies

OSHA requires that a trench be protected from collapse by instal-lation of shoring, shielding, sloping of the trench walls, or use of a benching system. A **sloping system** is a trench protection system that uses inclined excavating to form sides that incline away from the excavation so as to prevent a cave-in; the angle of the incline required to prevent a cave-in varies with factors such as soil type, environmental conditions of exposure, and application of surcharge loads. A **surcharge load** is any weight in the proximity of the trench or excavation that increases the site's instability or the likelihood of collapse. **Benching** (also known as a **benching system**) is a method of protecting workers from cave-ins by excavating the side of a trench excavation to form one or a series of horizontal levels or steps, usually with vertical or near-vertical surfaces between levels.

Phases of a Trench Collapse

Soil is extremely heavy, weighing approximately 100 pounds per cubic foot (depending on the moisture content and other factors). When a cut is made in the earth's surface, the trench walls can be visualized as a column of soil that weighs 100 pounds per cubic foot for each foot of trench depth **Figure 8-4 ▶**. If the trench is 10 feet deep, the 1-foot-square, 10-feet-tall column of soil that is the trench wall will weigh about 1000 pounds and thus exert a vertical force of 1000 pounds per square foot on whatever it rests on. This column of soil also exerts a horizontal force in all outward directions that is equal to one-half the vertical force. In the case of the 10-foot trench, the soil column exerts an outward force at the base of 500 pounds per square foot. The outward force increases with the depth of the trench, with soil blocks at the bottom of the soil column theoretically tending to become compressed and bulge outward.

Before soil is excavated, these columns are fairly stable, because they are held in equilibrium by the surrounding soil columns. When an excavation is made, the void (trench) created disrupts that equilibrium and destabilizes the trench walls. The weight of the soil and the increased outward pressure in the trench walls as the depth increases can lead to a situation in which the trench walls are no longer able to support the weight of the soil. The

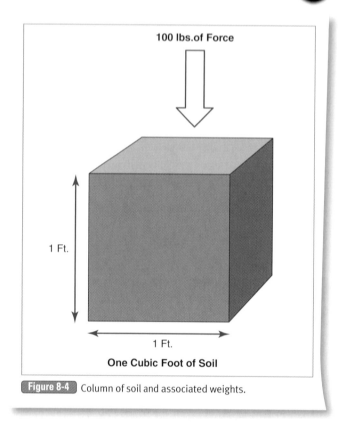

100 lbs. of Force

1 Ft.

1 Ft.

One Cubic Foot of Soil

Figure 8-4   Column of soil and associated weights.

resulting wall failure usually occurs in three phases. The first phase involves the trench wall moving into the trench at its lowest level near the trench bottom, which creates an overhang of soil at the top of the trench wall and destabilizes the wall **Figure 8-5 ▼**. The second phase of the failure occurs when much of the overhanging soil falls into the trench, which can result in a smaller, unsupported cantilever of soil remaining near the trench lip. The failure of the trench lip is the third phase of the trench failure.

Knowledge of these phases is important information for responders because there is typically some time lapse between

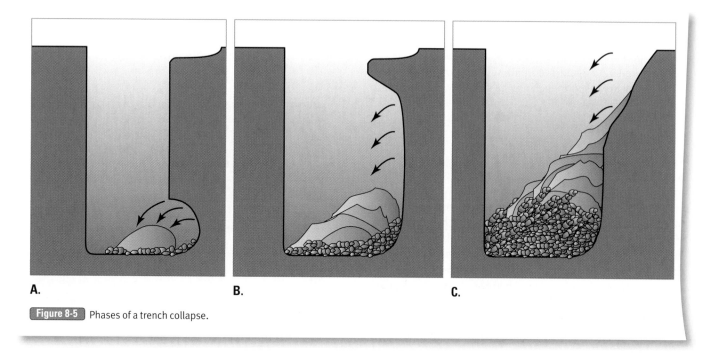

A.

B.

C.

Figure 8-5   Phases of a trench collapse.

phases. Thus, responders may arrive while the collapse is still in the first or second phase, potentially exposing them to the risk of a secondary collapse during shoring and entry operations. It also means that the victim who is partially buried when responders arrive may be totally buried before they have shoring systems in place.

Safety Tips

Under no circumstances should responders enter an unprotected trench! Rescuers and other response personnel should *never* enter a trench that has not been properly shored or protected by a trench shield or by proper sloping or benching systems.

Contributing Factors

A variety of conditions and situations can contribute to the cause of a trench collapse. It is important for rescue workers to have a good understanding of these contributing factors so that they can survey the rescue site to determine whether these conditions exist. If these conditions are found, additional collapses might endanger rescue workers during the rescue operations.

Narrow Roadways

Some excavations occur on very narrow roadways where traffic or construction equipment must work or travel close to the open trench. This narrow right-of-way issue creates a situation in which superimposed loads are placed on the trench walls, increasing the potential for collapse if the trench walls are not properly stabilized with shoring.

Vibration

Vibration sources from road traffic, heavy construction equipment, rail traffic, blasting, or other heavy industrial processes nearby may increase the risk of trench wall collapse **Figure 8-6**. It is important for rescuers and other responders to make sure that they park their response vehicles at least 100 feet away from the open trench to avoid becoming superimposed loads, and that unnecessary response vehicles be shut down to minimize their vibration effects.

Weather

Weather conditions and existing soil conditions can also serve as contributing factors to a trench collapse. Heavy rain conditions can flood an open trench or create high groundwater levels. Seepage of subsurface water can undercut trench walls, and saturated soil conditions can cause instability of the trench walls. When trenches are left open for prolonged periods of time during drought conditions, the drying of the exposed trench walls will reduce the moisture content in the soil; the soil particles will then lose cohesion, resulting in soil particles that easily fall apart. During weather conditions where temperatures swing between freezing and above-freezing levels, soil particles may freeze and then thaw out over the course of a day. This freeze–thaw cycle

Figure 8-6 Construction equipment can create vibrations that may cause a trench collapse.

can cause any existing shoring systems to loosen and either fail or be less effective in supporting the trench walls.

Soil

One factor that increases the likelihood of a trench collapse is previously disturbed soil—that is, the condition in which the existing trench is dug in an area that has been excavated in the past. When the previous excavation took place, typically backfill soil, bedding rock, and other nonhomogenous materials will have been used to fill in the old excavation. Rescuers may be able to determine that the excavation was previously disturbed by evaluating the trench walls and the spoil pile. The mix of materials causes the soil to be less stable and more prone to collapse failure than virgin soil that has never been excavated.

Excavation work may also create unstable soil situations where two or more trenches intersect **Figure 8-7**. Often this type of site is found when underground utilities are being installed and the utility line trenches cross, run parallel, or intersect (usually in the shape of an L, T, or X). This situation creates many unstable open trench corners, which are susceptible to collapse.

The existing soil types and conditions will also influence the stability of the soil and its susceptibility to trench wall col-

Figure 8-7 Intersecting trenches.

lapse. The most important factor affecting excavation safety is soil. At a trench collapse, a mass of soil typically crushes victims. Soil conditions do not remain uniform with respect to depth or along the length of the trench or excavation, and each trench site is unique. More than 100 distinct soil types exist, but no soil can be called entirely safe for creating vertical trenches. In fact, few soils have enough cohesive strength to justify working in unprotected trenches greater than 5 feet deep.

OSHA classifies soils based on site conditions, environmental conditions, and the characteristics of the soil itself that determine the soil's stability and ability to resist movement or collapse. These classifications include stable rock, Type A, Type B, and Type C soil. **Stable rock** refers to natural mineral matter that, when excavated, has vertical sides that remain stable when exposed to a lack of support. **Type A soil** refers to cohesive soils and cemented soils such as clay, silty clay, sandy clay, and hardpan. **Type B soil** refers to cohesive soil with less compressive strength than Type A soils, such as angular gravel, silt, and silt loam. Type A soils that are fissured or subject to vibration sources would also be categorized as Type B soils. Sandy soils, sandy loam, and gravel soils, submerged soils, or soils where water is freely flowing from the trench walls are categorized as **Type C soil**.

Rescuers are expected to categorize soils at construction sites by using both visual and manual tests. These tests are typically performed by a trained and knowledgeable individual, known as a *competent person*, who is familiar with all safety aspects related to excavation work. Visual tests include looking at the walls of the trench and observing the spoil material as it is removed from the trench to help determine soil characteristics such as the presence of water, amounts of sand, or cohesion (i.e., how well large chunks of soil stay together). Manual tests include thumb penetration and plasticity tests to determine the unconfined compressive strength of the soil, which is an indication of the soil's stability.

One notable situation that can contribute to trench collapse is layering of soil types Figure 8-8 ▾ . All soils are layered. When a situation occurs where the layers of soil are inclined downward toward the floor of the trench, however, it can create a very dangerous collapse hazard. Where the various soil types make contact, each layer will act as a lubricant to the other nonhomogenous soil

type, creating a situation in which a large segment of soil can slide into the trench and entrap a worker or rescuer.

It is also important to recognize those areas where rock is the primary soil type. Even in these sites, it is not safe to work in trenches or excavations greater than 5 feet deep without collapse protection. There is no such thing as "solid rock." Rock may split at undetermined times and locations after being "shot"—that is, following the use of explosives by construction companies to create excavations.

Applicable Standards

Emergency responders can find guidance on how to manage these types of incidents by referring to the National Fire Protection Association (NFPA) Standards 1006, *Standard for Technical Rescuer Professional Qualifications*, and 1670, *Standard on Operations and Training for Technical Search and Rescue Incidents*.

NFPA 1006 establishes the job performance requirements for responders who perform rescue skills at the technician level. These skills include hazard recognition and mitigation, trench stabilization techniques, performing lifting operations to effect rescues, extrication of victims from entrapment, use of heavy equipment, **victim packaging** (preparing the victim for movement as a unit, often accomplished with a long spine board or similar device), and victim transfer techniques. **Extrication** includes steps taken by responders to remove a victim from an entrapment at a trench collapse, which may include removal of soil, lifting heavy objects, or cutting through entrapping materials. **Entrapment** is the condition in which a victim is trapped by debris, soil, or other material and is unable to extricate himself or herself.

NFPA 1670 establishes organizational requirements to operate safely and effectively at trench and excavation emergencies. This standard identifies the requirements for risk and hazard assessment, written procedures, safety protocols, personal protective equipment, establishment of operational response level, training, and resource considerations. **Awareness level** is the first level of rescue training provided to all responders, which emphasizes recognizing hazards, securing the scene, and calling for appropriate assistance. There is no actual use of rescue skills at the awareness level, but responders at the awareness level for trench and excavation rescue must all meet the awareness-level requirements for confined-space search and rescue as well as the general requirements listed in Chapter 4 of NFPA 1670 and in some cases may initiate nonentry extrication.

The OSHA Standard 1926.650, 651, and 652 identifies safety requirements for workers who perform excavation and trench work. This standard is an excellent resource for emergency responders. It identifies hazards, training requirements, hazard control requirements, trench stabilization options, and other safety requirements to which workers (including emergency responders) must adhere.

Resource Requirements

For responders to a trench collapse or trench emergency, it is important to know which resources are required to manage this type of event. Knowledge of the complexity of trench rescue

Figure 8-8 Layered soils can be a dangerous collapse hazard.

operations enables first responders to rapidly request the proper resources that match the problem. Search operations at a trench collapse may vary from easily identifying injured victims who were able to escape the collapsed trench, to extricating partially buried victims that you can see from a safe area outside of the trench, to locating a totally buried victim (a much more complex problem).

Response personnel who have been trained in trench collapse rescue techniques have also been trained to identify clues that may suggest the locations of total-burial victims. An early request for qualified trench rescue teams is important to ensure a safe and possibly successful outcome of the event. Clues useful for locating total-burial victims are discussed later in this chapter.

Rescuers must be ready to use a variety of tools to remove material entrapping the victim. Entrapment problems may involve various types of soil, ranging from those that are easy to dig and remove by hand to those requiring pneumatic chisels or similar tools for breaking and removal. Rocks and boulders can also create difficult extrication problems because lifting or chiseling will be required to remove them.

Construction materials such as shoring timbers, utility pipes, or concrete from unsupported sidewalks or curbing may also fall into the trench, creating extrication problems. These materials can originally help the victim by supporting the trench, but may become a hindrance after soil movement. Even hand tools and other construction equipment, if positioned poorly, can create problems with extrication. Heavy equipment could overturn into an open trench, further complicating a lifting operation.

> **Rescue Tips**
>
> Trench rescue events can be long in duration and physically demanding. Rescuers should ensure that adequate qualified resources are on hand to allow for relief and rotation of rescuers.

Soil Removal Resources

The most common entrapping material at a trench collapse is soil. In these situations, the operations- or technician-level rescuers need to uncover the victim's head and chest area as quickly as possible, administer patient care, and then uncover the rest of the victim by removing any additional soil. Awareness-level responders do not have adequate training to participate in this task.

Soil removal is primarily accomplished with hand tools. Entrenching shovels are often used for this purpose because they are small, work well in a trench environment, and allow for limited leverage. Rescuers can then work gently around a victim's body. Soil is removed from the trench via small buckets. In situations where the soil must be removed entirely from the trench, these buckets can be attached to ropes and hoisted out of the trench for disposal of the soil in a remote location. In some situations, other small digging tools (e.g., hand trowels or post-hole diggers) can help remove rock or soil from a confined area where a shovel will not fit.

> **Rescue Tips**
>
> Digging operations at a trench rescue incident may require rescuers to modify or improvise existing tools such as shovels and digging bars so that they can remove debris in very tight and confined areas.

As rescuers install shoring systems, remove soil, and install supplemental shoring, the area around a victim becomes more confined. In some situations, rescuers must dig out a larger area around the bottom of the trench to increase access to the victim's lower extremities. As they begin digging, rescuers must also determine the positioning of the victim's body. For example, if the victim is lying down, then rescuers should square out the bottom of the trench to make victim packaging easier. If the victim is found standing up, then rescuers will typically dig a hole around the victim. Removal of material around the lower extremities, however, becomes more difficult due to the limited access.

Chipping and Cutting Resources

Rescue of trench emergency victims can be complicated by the presence of tools, ladders, timbers, concrete, pipes, or other construction materials in the trench. In these situations, rescuers may need to use tools such as pneumatic chisels, electric rotary hammers, and demolition hammers to free entrapped extremities. For example, these tools can reduce rocks to smaller pieces that can then be removed from the trench in buckets.

Rescuers may also use hand or power saws (e.g., reciprocating saws) when a victim is entrapped. These tools are small enough to be used in a trench, close to a victim. Cutting resources include knives or EMT scissors, which can be used on a victim's work boots, shoes, or tool belt if necessary.

Lifting Resources

Victims pinned by pipes, boulders, concrete, or other construction materials often require a lifting operation, which can be accomplished using air bags or bolting, rigging, and slings.

When using air bags in lifting operations, responders must first dig and remove material that would interfere with the correct application of the air bag. They must then build up the base under the air bag to distribute the load over the soft soil **Figure 8-9 ▶**. This keeps the air bag from sinking into the trench bottom and can be accomplished using a square of plywood, at least ¾ inch thick, that is larger than the air bag being used. If the ground pad and air bag continue to sink into the soil, use of a different lifting method must be considered.

Air bags can also be used outside the trench in combination with slings, timbers, rigging, and bolting devices to lift a pipe, boulder, or concrete material. For this application, responders place air bags on each side of the trench; they then place large timbers across the trench. Slings, bolting devices, and rigging materials can be used to attach the item to be lifted to the timbers laid across the trench. All slack is removed from the rigging prior to starting the lift, and then the air bags are inflated, raising the timbers. In this way, the entrapping material is lifted for rescue purposes.

Figure 8-9 When using an air bag for lifting operations, rescuers must first build up the base under the air bag to distribute its weight over the soil.

Bolting is a tool application in which rescuers drill a hole into the rock, stone, or piece of concrete that needs to be removed. A bolting device such as an expansion bolt or expansion anchor and eye bolt are then attached to the item (rock, concrete) to be moved. A chain, cable, or sling is attached between the bolting device and a rescue tool or a piece of heavy equipment, allowing for the rock or concrete item to be moved for rescue purposes. The use of bolting in this application is effective when the user is well trained and clearly understands the application guidelines from the manufacturer. Allowable bolting distances—for example, the minimum distance between bolts, the distance to the edge of the materials being lifted, and the minimum embedding depth—are all important requirements for effective use of bolting techniques.

Bolting, rigging, and slings can also be used in combination with heavy equipment for lifting. Rescuers considering this option must first evaluate the trench wall stability and determine how the increased loads and vibrations from the equipment might potentially affect the site. Regardless of the method of lifting employed, the material being lifted must be stabilized and the lift controlled. As the lift continues, cribbing can be inserted to help ensure that the material will not settle on the victim.

■ Specialized Resources

Vacuum Devices
Mechanical vacuum devices can be helpful in removing soil from around a victim in a process called **vacuum excavation**. Vacuum excavation consists of two phases: reduction and removal.

In the reduction phase, soil is reduced into smaller pieces using water, air, or mechanical means. The most effective reduction method is water; because it could cause hypothermia in the victim, however, this technique is not practical at a trench emergency. Reduction via air uses an air lance, a device that breaks the soil into small pieces using air flow from a high-volume air compressor. Mechanical reduction, which uses shovels and chisels to break up soil, is the slowest method.

Rescue Tips

A victim who is trapped under some type of obstacle such as a concrete pipe or concrete sidewalk slab presents a difficult lifting challenge for rescuers. The lifting operations can be complicated by the limited work area inside the trench and soft ground conditions, which may not allow for effective lifting with air bags from underneath the obstacle. Rescuers must be ready to develop alternative rescue plans in case the first plan proves to be unsuccessful.

The reduction phase is followed by the removal phase, in which responders remove the reduced material via a vacuum device. These vacuum devices can be hand-held devices powered by a separate high-volume air compressor, or they can be fully self-contained mobile units or skid load units that have their own air supply and a holding tank for the removed soil **Figure 8-10 ▾**. Vacuum units come in a variety of sizes and are capable of handling a variety of jobs.

Heavy Equipment
Significant trench collapse incidents can require rescuers to remove a great deal of soil during extrication, which can be very time consuming and labor intensive for rescuers. In such situations, responders may need to consider using heavy equipment to perform a **cut-back operation**. In these operations, heavy digging equipment is used to dig a parallel trench or a hole, thereby creating a void. The trench walls can then be pulled away from the trench to lessen its depth or the walls can be sloped back. A cut-back reduces secondary collapse hazards, allowing rescuers to walk into the trench area to complete the extrication by hand. Although cut-back operations are not typically the first option selected by trench rescue personnel, this tactical option can prove beneficial depending on the entrapment problems encountered. In body recovery scenarios, cut-back operations may be the best option to reduce the risk to rescue workers.

Rescue Tips

Although the use of heavy equipment around a victim is generally considered unsafe, cut-back operations can be very effective in victim recovery efforts.

Figure 8-10 A vacuum truck can be used during the soil removal phase of a trench rescue.

When rescuers are considering a cut-back operation using heavy equipment, it is important to perform a thorough risk–benefit analysis. The following considerations should be taken into account:

- The impact of the load and vibrations on the stability of the trench walls
- Information from experienced and knowledgeable heavy equipment operators
- The presence of any underground utilities

Emergency Medical Services Resources

In most situations, rescuers at a trench collapse incident will be dealing with fewer than five victims. Due to the mechanisms of injury associated with a collapsing trench wall and the potential for resultant trauma, advanced life support (ALS) emergency medical services (EMS) units should be requested to assist with this type of emergency. The number of units to request depends on the total number of victims. Local medical protocols should be adhered to in regard to medical unit requests.

EMS assets will also be needed to assist with the rehabilitation of rescuers. These operations are especially important at trench collapse sites because trench rescues are typically long-duration events. As rescuers are rotated out of the trench, they should be required to report to a rehabilitation area to be medically monitored, including evaluation of their vital signs, and given the opportunity to acquire food and drink for nourishment.

Other Resources

Hazardous materials resources may be needed if dangerous atmospheres are discovered within the trench or if underground utilities are broken and a release of natural gas, petroleum, or other pipeline products occurs. Assistance from local utility companies may also be required to deal with broken electrical, water, or sewage lines, all of which could affect trench stability and the environment in which rescuers must operate. Because most trench rescues are long-term events, incident command personnel should anticipate the need for acquiring long-term incident support resources such as portable toilets, food and drink, extra personal protective equipment, radio batteries, and scene lighting to support night-time operations. Law enforcement personnel may be needed for traffic and crowd control purposes, especially if co-workers, friends, or family members become distraught and begin demanding that responders take action and enter the unprotected trench.

Incident Management Requirements

The strategic objective when operating at a trench collapse incident is to effectively control and manage the event in such a way as to evaluate the scene and identify potential victims and their location, to initiate operations to minimize hazards to operating personnel and trapped victims, to effectively search for total burial victims, to effectively rescue and remove trapped victims, and to minimize further injury to victims during search, rescue, and removal operations.

The Incident Command System (ICS) positions are staffed as determined by the scope of the event. In most cases, at a mini-mum, management personnel are required to fill the positions of incident commander (IC), rescue group supervisor, planning section chief, logistics section chief, safety, and public information officer.

The IC or rescue officer must anticipate the needs of the rescue operation. As information relevant to the situation is gathered, the officer in charge of developing the rescue plan must anticipate tool needs. If these tools are not available, they should be requested as early as possible. The IC, rescue officer, and others should formulate multiple rescue options as part of the incident action plan (IAP). Once the action plan is devised, a team briefing to explain the plan of action to all team personnel should be conducted. Erasable marker boards can be utilized for this briefing to illustrate the rescue plan, identify command personnel assignments, and track the progress of the incident.

At least two functional work areas will need to be established:

- Logistics work area. This area is used to assemble needed tools and equipment, to cut timber shoring materials to required lengths, and to facilitate the modification of supplies as required for performing the rescue.
- Personnel staging/rehabilitation area. This area is used for the staging of personnel who will be used for various tasks throughout a long-term rescue effort and as a rehabilitation area for personnel who are rotated out of the rescue site work area.

Responders must coordinate their efforts with EMS personnel on the scene to initiate patient care as early as possible. After the trench has been shored and any hazards have been mitigated, rescuers may allow EMS personnel into the trench to administer initial patient care if local protocols allow. They should also continue to supervise the situation, ensuring the continued safety of the operation.

The rescue officer is charged with anticipating tool usage and allowing sufficient space (room for at least two rescuers) to perform tool operations and patient packaging. Additional shoring may become necessary as more trench wall is exposed during digging operations, requiring the rescue officer to reassess the competency of the shoring system throughout the rescue. It is also important, however, to take care not to disrupt the existing shoring system as responders, materials, tools, and victims move within the trench.

The officer in charge of the first-due trench rescue unit may recommend to the IC that the trench rescue resources be increased based on the situation report or dispatch information. The need for additional staffing and specialized equipment—such as heavy equipment, vacuum trucks, shoring equipment, and more trench rescue trained personnel—must be considered as well.

Rescue Tips

ICs must understand the need for tactical-level officers who are specially trained to manage trench and excavation emergencies. Specially trained officers who understand the challenges of trench rescue and know how to develop effective rescue plans for shoring a trench and performing an extrication within a trench environment are key personnel during these types of rescues.

Voices of Experience

In late January 2008, a 32-year-old man was assisting a neighbor with installing a water line on their property. The trench was 13 feet deep when it collapsed, burying the man completely. Local authorities called for the Louisville Fire Department's Heavy Urban Rescue Team for extrication of the victim. We sent a group of three members on the trip 30 miles northwest of Louisville while additional members of the team met to make the trip together.

When they arrived, it was obvious this was a unique and dangerous situation. The trench was still falling in and the spoil pile was positioned on top of the lip of the side opposite the victim. The soil was very dense, thick clay that was heavy and moist and that stuck together very well. The trench was 10 feet wide for a depth of 8 feet. At that level, the trench went to 3 feet wide for another 5 feet, so we had two trenches in one. There was also some heavy equipment in close proximity that needed to be locked and tagged out. The weather conditions were below freezing with light snow through the duration of the incident. Because of this, we also started a group of rescuers from the Jefferson County Trench Rescue Team that we train and respond with locally.

We started using local responders to remove the spoil pile while we got our game plan together. We staged all of our team members in a portable, heated tent to keep them out of the weather conditions. Locals also assisted us with transport of our equipment to its staging area. We decided we would treat this again like two separate trenches. This was unique because of the width of the trench. At one point, we had a member tied off on a straight beam ladder that bridged the trench from one side to another to assist with lowering our pneumatic shoring devices into place. Again, the trench was still falling in, so we needed to create a safe haven for our members to work inside before moving to where the victim was located. There was a shelf we could work from once this part of the trench was properly shored.

When we began to move where the victim was, we were in for an extensive digging operation. He was completely covered by the thick clay. We rotated the members every 20 minutes because of the weather conditions. As we dug deeper, we shored around the victim until he was completely uncovered. Another problem we encountered was that the soil around the victim's lower extremities was very wet, creating a vacuum when we tried to remove him. Once the victim was free, he was put in a Stokes basket and removed to the top of the trench. Many of the local responders knew the victim personally; they transported the victim to the ambulance.

Serious consideration was taken when removing our equipment due to the deteriorating conditions of the trench. Once everything was removed, a short debriefing was held before making the trip back to Louisville. A formal critique was held a few days following the incident.

Captain Edward J. Meiman III
Special Unit Coordinator
City of Louisville Division of Fire
Louisville, Kentucky

> **"The trench was still falling in and the spoil pile was positioned on top of the lip of the side opposite the victim."**

Response Planning

Responders must know how to initiate the emergency response system to ensure that appropriate resources are deployed to the event, and to ensure that operational guidelines are begun. Many response organizations perform needs assessments within a response area to determine its vulnerability to trench and excavation emergency events. As part of this assessment, the organization should consider which types of underground construction and utility repair or installation are common in the response area, and then determine emergency response resource needs in relation to those sites. The emergency response system should include written procedures to request agency resources as well as resources that will respond through mutual aid agreements, contracts with private-sector companies, and any memoranda of agreement with regional assets. Private-sector agreements may be established to acquire resources that cannot be supplied by the response agency, such as heavy equipment, lumber and construction tools, structural engineers, food, sanitation supplies, and housing.

In addition, the agency should identify the capabilities of its personnel. Topics covered by this capability assessment should include the level of performance to which agency personnel are trained, such as the operations or technician level for trench rescue. As part of the emergency response system, the location, capabilities, and response times for local, mutual aid, and regional trench rescue teams should be known, and procedures should be established to request these assets. Dispatch procedures should also be in place to ensure that adequate and proper resources are deployed when a trench emergency occurs so that effective search and rescue operations can begin as quickly as possible.

Hazards and Hazard Assessment

Initial Scene Assessment

Before even reaching the incident scene, all rescue responders should meet the competencies for awareness-level personnel outlined in NFPA 472, *Standard for Competence of Responders to Hazardous Materials/Weapons of Mass Destruction Incidents*. When responders initially arrive on the scene of a collapse event, they will often be confronted with a chaotic and uncontrolled situation. For this reason, responders should slow down as they arrive at the actual incident scene, thereby enabling them to assess the situation better.

As with any emergency incident response, rescue personnel must perform a thorough hazard assessment prior to initiating rescue operations. The initial scene assessment must begin with the dispatch information and the arrival of the first-responding units. Because there is a high potential for other building utilities to be disrupted with a trench collapse, and a chance for a secondary collapse may exist, responders should initially position their response vehicles at least 100 feet from the collapse area; additional responding vehicles should be directed to stage farther away until a situation assessment and immediate scene control operations are completed. This practice will help responders avoid locating vehicles and crews in an area where a gas leak or water main break may be encountered. At the same time, it is

important to ensure that any specialized vehicles—such as rescue squads, trench rescue supply vehicles, construction equipment, or vacuum trucks—can gain access to the scene if needed.

Rescue Tips

Responders must slow down their response vehicles as they approach a collapse site. Park at least 100 feet away from the collapse site, and walk toward the trench while carefully assessing the situation.

During the initial arrival and scene assessment, responders may be quickly overwhelmed by people requesting their assistance. Co-workers and/or family members may demand that responders enter the unprotected trench and take immediate action. At least initially, it may prove difficult to gain control of the situation.

As part of the initial assessment, the rescue site should be cordoned off by placing barrier tape on all sides of the area Figure 8-11 ▾ . All unnecessary spectators, responders, and nonessential personnel should be removed from the area by at least 100 feet for their own safety, to provide a free work environment for the rescue team, and to lessen the chance of a secondary collapse.

First responders should also attempt to locate a work-site foreman if the accident occurred at a work site or someone else who can supply information about the accident. Attempt to determine exactly what the problem is, which type of collapse has occurred, how many people are trapped, and whether the victims are trapped in a partial-burial or total-burial situation. Make note of or mark all known and approximate victim locations. In addition, try to determine how long ago the collapse occurred, whether victims are visible, and whether any underground utilities are broken or endangered.

During this information-gathering phase, rescuers should assess the immediate medical requirements of trapped victims as well as the care needs of any injured parties who may have escaped from the collapse. Nevertheless, it is very important that rescuers not approach the trench lip too closely to evaluate the

Figure 8-11 Trench scene controlled with barrier tape.

situation, owing to the instability of the trench lip and supporting walls. Likewise, they should not walk on top of a spoil pile to assess the situation—the spoil pile may shift and the responder could fall into the trench. It is safer to approach an unprotected trench from the narrow end, a point from which the rescuer can visualize the entire length of the trench yet be exposed to only a small area of unprotected trench wall.

Rescue Tips

Awareness-Level Responder Actions at a Trench Collapse

- Initiate the ICS.
- Set up control zones.
- Use barrier tape to establish control zones.
- Deny entry into the hazard zone.
- Identify the locations of trapped victims.
- Mark potential locations of total-burial victims.
- Gather incident information from bystanders or co-workers.
- If a victim is visible and has mobility, lower a helmet to him or her for protection.
- Lower an oxygen mask attached to oxygen tubing and an oxygen supply to the victim and direct him or her on how to use it.
- If there are broken utilities in the trench or the adjacent area, request a response from the appropriate utility companies.
- If a victim is mobile and is not trapped, lower a ladder into the trench so the victim can climb out, if this can be accomplished safely without creating a collapse hazard.
- Assist operations- or technician-level responders by serving as tool/equipment runners or helping to remove the spoil pile.
- Assist with the placement of tarps or tent devices to protect the trench from rain or excess sun exposure.
- Establish water diversion systems to limit rainwater from entering the trench.
- Assist operations- or technician-level rescuers with de-watering activities from outside of the trench.

Trench Collapse Hazards

Entry into an unshored trench or excavation is extremely dangerous due to the potential for a trench wall collapse. The hazards of being buried in a collapse include the crushing weight of soil. Given this risk, rescuers must assess the stability of trench walls and look for signs of potential secondary collapse Figure 8-12 . Evidence of trench wall distress should be marked and monitored throughout the rescue operation. Be vigilant for the following signs of impending secondary collapse and evidence of trench wall distress:

- Tension cracks (fissures) located parallel to or in the face of the excavation
- Subsidence of the edge of the trench
- Overhanging loose segments (cantilevers) of the trench wall
- Bulging of trench walls

- Proximity of the spoil pile to the trench
- Heaving or boiling of the trench floor
- Spalling (chipping) of the face of the trench
- Water in the trench

Safety Tips

Be alert for signs of an impending secondary collapse, and assess the presence of utility line involvement in the incident area.

The presence of water-saturated soil and flooding conditions in a trench from groundwater seepage or rainwater reduces the safety factor by 50 percent. Moisture in soil, when frozen, will expand 11 percent by volume—an action that can destroy or damage existing shoring systems.

Additionally, mixed layers of soil and loose, running, sand-like soils will be very susceptible to movement and collapse, so rescuers must be very cautious as they approach the trench. Superimposed loads created by heavy construction equipment located next to the trench can also lead to secondary trench wall collapse and should be removed if possible. Rescuers should also assess the surrounding area to identify the presence of any vibration sources, such as roadways, train tracks, or operating heavy construction equipment. Excessive or repeated vibrations can lead to trench wall collapse, so these sources should be controlled if possible.

Utility and Atmosphere Hazards

Underground construction work typically involves the installation or repair of utility lines, including those for electrical power, natural gas, steam, chemical pipelines, water lines, or sewage lines. These utility lines may represent a hazard to rescuers if they become broken during a trench collapse and release their product. Rescuers may be electrocuted or exposed to natural gas or another chemical that may result in a fire. Broken water and

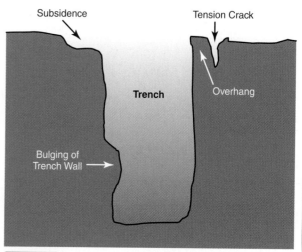

Figure 8-12 Responders should recognize the signs of potential secondary collapse.

sewage lines can flood a trench and cause a secondary collapse. If first responders arrive to find that utility lines pose a threat in the trench emergency, the assistance of the appropriate utility company resource should be requested. Rescuers may be able to crimp small broken water lines to control water flow or plug sewer lines with an air bag to limit the flow of sewage into the trench.

Rescuers may also encounter unanticipated atmospheric hazards during a trench emergency. For example, decaying organic matter in the soil where the excavation or trench was dug may create methane gas or other atmospheric hazards.

In addition, the trench may have been dug in an area where hazardous materials had previously been dumped, which may release hazardous gases. Another commonly encountered hazardous situation results from leakage of hazardous materials from underground storage tanks or pipelines. When this situation occurs, hazardous vapors, gases, or liquids can flow through the soil into an open trench, creating dangerous work conditions and requiring a hazardous materials team response.

Initial Hazard Mitigation

Awareness-level responders can take some basic actions to minimize their risk at trench collapse incidents. One of the most important things responders can do is to recognize and avoid trenching hazards. When they have a clear understanding of the types of hazards found at a trench site, responders can maintain a safe distance and avoid these potential problems. Upon recognizing the presence of trench site hazards, responders should notify other responders about the danger and mark the area with barrier tape or other marking devices to identify the unsafe condition or site. Responders can also limit exposure to hazardous areas by denying entry into those sites. Specifically, establishing operational zones (such as hot, cold, and warm zones) and marking these zones with barriers or barrier tape will help to limit exposure to other responders Figure 8-13.

As mentioned earlier, responders must also control vibration sources (such as road traffic or heavy construction equip-

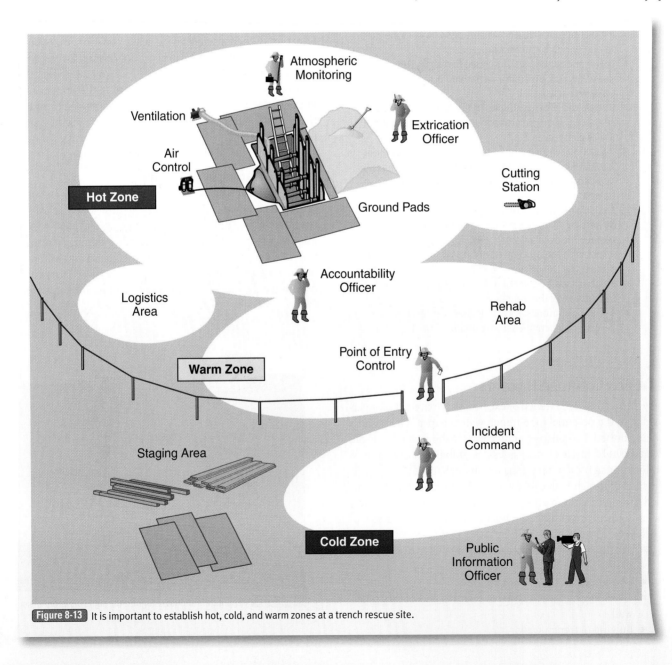

Figure 8-13 It is important to establish hot, cold, and warm zones at a trench rescue site.

ment) that could cause a secondary collapse. Heavy vibration sources located within a 500-feet radius of the trench should be shut down.

Patient Care Considerations

Awareness-level responders can anticipate that they encounter multiple injured people in need of patient care at a trench collapse emergency. The walking wounded may suffer from injuries including fractures, traumatic injuries, soft-tissue injuries, respiratory distress, or other related injuries, including hazardous materials-related injuries and the need for decontamination. Initial responders must ensure that adequate EMS resources are requested to assist with patient care. If responders are trained to deliver some level of emergency medical care, local triage and treatment protocols should be followed in regard to care for victims. As part of this practice, walking wounded should be directed away from the collapse hazard zone to a safe treatment area.

Responders should assess and mitigate immediately life-threatening hazards and provide initial medical and protective support for accessible victims without entering an unprotected trench area. Immediate care from outside the trench can be initiated for a conscious, partially buried victim by lowering oxygen supply tubing and an oxygen delivery mask and giving the victim directions on how to place the mask on himself or herself. As part of this effort, responders must identify the need for oxygen tanks and have them delivered to the scene; the oxygen mask and tubing can then be supplied by an oxygen cylinder located outside the trench. In addition, the victim's location should be marked and a rope lowered and placed adjacent to the victim. Pictures can be taken to document the victim's location in case of another collapse. Responders should also lower a hard hat or helmet to the victim, who can wear it to help protect the individual from additional falling debris.

Locating total-burial victims is very difficult. Responders should interview co-workers or other witnesses to help determine the last known location within the trench of an individual who is totally buried in a collapse. Possible burial areas should be marked with chalk or construction-type, high-visibility paint. If the location of a victim is not known, rescuers can look for clues to the victim's location:

- Results of interviews with eyewitnesses to determine worker locations when the collapse occurred. (This information is not always accurate.)
- The location of the end of the **pipe string**, the line of concrete pipe positioned along the length of the trench and ready for installation. Workers will normally be working in the trench near the end of the pipe string.
- The location of the lubrication grease can and brush used to lubricate concrete pipe ends during installation. These items will normally be located next to where workers were installing the next section of pipe. Beverage bottles and drinking cups are also typically found in this area.
- Laser targets and grade poles, which are used to align pipe and ensure proper pipe angle during installation. These items will typically be located next to where the workers were installing the next section of pipe.
- Cat tracks. If a trench wall collapse occurs while construction equipment is dumping bedding rock into a trench, a worker may be found near where the tire or cat tracks are located running perpendicular to the trench wall.
- The inside of large-diameter pipe. During the course of the collapse, workers may have been able to jump inside a large-diameter pipe (pipe larger than 24 inches). Rescuers should check whether they can gain entry to the end of the pipe line and determine whether this situation occurred. If rescuers can enter the pipe, they must follow confined-space entry procedures. This operation should be attempted only by technician-level rescuers.

Wrap-Up

Chief Concepts

- Trench rescue events can vary in size and complexity: Rescue workers may arrive at a trench emergency to find a worker with a back injury lying in an unprotected trench, or they could be faced with a total-burial situation.
- Responders should not enter a trench until all hazard control actions are completed; even then, only those trained in trench rescue should enter the excavation area.
- Resource requirements for managing a trench collapse incident will be significant and may include search, rescue, EMS, heavy equipment, utility companies, and other specialized resources.
- Effective planning—to include risk assessment, resource assessment, and development of written procedures—is important to successful management of a trench collapse emergency. The implementation of an incident management system is an absolute requirement to be able to control this type of emergency.
- Significant hazards may be encountered at a trench emergency incident. A thorough hazard assessment must be completed initially and throughout the event.
- Initial hazard control by first responders may include removal of unnecessary personnel from the trench work area, establishment of control zones, installation of barrier tape, and control of vibration sources.

Hot Terms

Awareness level The first level of rescue training provided to all responders, which emphasizes recognizing hazards, securing the scene, and calling for appropriate assistance. There is no actual use of rescue skills at the awareness level.

Benching (benching system) A method of protecting employees from cave-ins in which the side of a trench is excavated to form one or a series of horizontal levels or steps, usually with vertical or near-vertical surfaces between levels.

Cave-in The separation of a mass of soil or rock material from the side of an excavation or trench, or the loss of soil from under a trench shield or support system. The sudden movement of a sufficient quantity of this material into the excavation, either by falling or sliding, could entrap, bury, or otherwise injure and immobilize a person.

Competent person An individual who is capable of identifying existing and predictable hazards and is authorized to take measures to eliminate them.

Cut-back operation Use of heavy digging equipment to dig a parallel trench or a hole to create a void.

Entrapment A condition in which a victim is trapped by debris, soil, or other material and is unable to perform self-extrication.

Excavation Any human-made cut, cavity, trench, or depression made in an earth surface.

Extrication The removal of trapped victims from an entrapment.

Pipe string The line of concrete pipe positioned along the length of a trench, ready for installation.

Shield A structure that is able to withstand the forces imposed on it by a cave-in and, therefore, protects workers within the structure. When used in a trench, these structures are known as trench shields or trench boxes.

Shoring (shoring system) A structure such as a metal hydraulic, pneumatic/mechanical, or timber shoring system designed to support the sides of an excavation and prevent cave-ins.

Sloping system A protection system that uses inclined excavating to form sides that incline away from the excavation so as to prevent a cave-in.

Spoil pile A pile of excavated soil placed next to the excavation or trench.

Stable rock Natural mineral matter that, when excavating the vertical sides of a trench, remains stable when exposed to a lack of support.

Surcharge load Any weight in the proximity of a trench or excavation that increases its instability or likelihood of collapse. Also called overpressure load or superimposed load.

Trench An excavation that is relatively narrow in comparison to its width. In general, the depth of the trench is greater than its width, but the width of a trench is not greater than 15 feet.

Type A soil Cohesive soils and cemented soils such as clay, silty clay, sandy clay, and hardpan.

Type B soil Cohesive soil with less compressive strength than Type A soils, such as angular gravel, silt, and silt loam.

Type C soil Sandy soils, sandy loam, and gravel soils, submerged soils, or soils where water is freely flowing from the trench walls.

Vacuum excavation A two-phase process (reduction and removal) in which mechanical vacuum devices are used to remove soil from around a victim.

Victim packaging The process of preparing a victim for movement as a unit, often accomplished with a long spine board or similar device.

Rescue Responder *in Action*

The ambulance you are assigned to is dispatched to a report of a man who has fallen at a construction site and is complaining of a back injury. When your unit arrives on the scene, you are unable to drive your vehicle into the construction site due to the muddy conditions there. You and your unit officer grab your first-aid equipment, don protective helmets, and start to enter the construction site to find the patient. As you are walking into the site, a construction worker tells you that the victim tripped over a pipe, fell about 15 feet, and is at the bottom of a trench. As you approach the trench, the victim's co-workers point to where the victim is located and you see that one of the co-workers is climbing down a ladder into the trench to help the victim. Your officer uses the portable radio to request more assistance.

1. Which additional resources should be requested to help manage a trench emergency?
   A. Water rescue teams
   B. Command officer
   C. Specially trained rescue company
   D. Both B and C

2. Which initial actions should awareness-level responders take at this event?
   A. You should use the existing ladder to gain access to the victim to determine if the victim is breathing.
   B. You and your partner should rapidly enter the trench and remove the victim using a fire fighter's drag technique.
   C. You should secure the scene, move bystanders away from the trench, and attempt to communicate with the victim.
   D. You should attempt to install shoring to secure the trench.

3. What is the best away to approach the trench to assess the situation?
   A. Approach the narrow end of the trench.
   B. Climb to a high vantage point on the top of the spoil pile to get a better view of the site.
   C. Slowly walk up to the edge of the trench on the side closest to where the victim is located.
   D. Locate the ambulance about 25 feet away from the trench and get a high vantage point by climbing on top of the unit.

Vehicle and Machinery Search and Rescue

NFPA 1670 Standard

4.1 **General.**

4.1.1* The authority having jurisdiction (AHJ) shall establish levels of operational capability needed to conduct operations at technical search and rescue incidents safely and effectively, based on hazard identification, risk assessment, training level of personnel, and availability of internal and external resources. [p. 183–196]

4.1.4 The AHJ shall establish written standard operating procedures consistent with one of the following operational levels for each of the disciplines defined in this document:

(1)* Awareness Level. This level represents the minimum capability of organizations that provide response to technical search and rescue incidents. [p. 186–191]

4.1.9* At a minimum, medical care at the basic life support (BLS) level shall be provided by the organization at technical search and rescue incidents. [p. 186]

8.1* **General Requirements.**

Organizations operating at vehicle search and rescue incidents shall meet the requirements specified in Chapter 4. [p. 184]

8.2 **Awareness Level.**

8.2.1 Organizations operating at the awareness level for vehicle emergencies shall meet the requirements specified in Section 8.2. [p. 184]

8.2.2 All members of the organization shall meet the requirements specified in Chapter 4 of NFPA 472, *Standard for Competence of Responders to Hazardous Materials/Weapons of Mass Destruction Incidents*, commensurate with the organization's needs. [p. 191]

8.2.3 Organizations operating at the awareness level for vehicle emergencies shall implement procedures for the following: [p. 185–196]

(1) Recognizing the need for a vehicle search and rescue

(2)* Identifying the resources necessary to conduct operations

(3)* Initiating the emergency response system for vehicle search and rescue incidents

(4)* Initiating site control and scene management

(5)* Recognizing general hazards associated with vehicle search and rescue incidents

(6) Initiating traffic control

12.1* **General Requirements.**

Organizations operating at machinery search and rescue incidents shall meet the requirements specified in Chapter 4. [p. 184]

12.2 **Awareness Level.**

12.2.1 Organizations operating at the awareness level for machinery emergencies shall meet the requirements specified in Section 12.2. [p. 184]

12.2.2 All members of the organization shall meet the requirements specified in Chapter 4 of NFPA 472, *Standard for Competence of Responders to Hazardous Materials/Weapons of Mass Destruction Incidents*, commensurate with the organization's needs. [p. 191]

12.2.3 Organizations operating at the awareness level for machinery emergencies shall implement procedures for the following: [p. 185–196]

(1) Recognizing the need for a machinery search and rescue

(2)* Identifying the resources necessary to conduct operations

(3)* Initiating the emergency response system for machinery search and rescue incidents

(4)* Initiating site control and scene management

(5)* Recognizing general hazards associated with machinery search and rescue incidents

Additional NFPA Standards

NFPA 472, *Standard for Competence of Responders to Hazardous Materials/Weapons of Mass Destruction Incidents*

NFPA 473, *Standard for Competencies for EMS Personnel Responding to Hazardous Materials/Weapons of Mass Destruction Incidents*

NFPA 704, *Standard System for the Identification of the Hazards of Materials for Emergency Response*

NFPA 1006, *Standard for Technical Rescuer Professional Qualifications*

NFPA 1500, *Standard on Fire Department Occupational Safety and Health Program*

NFPA 1521, *Standard for Fire Department Safety Officer*

NFPA 1561, *Standard on Emergency Services Incident Management System*

NFPA 1994, *Standard on Protective Ensembles for First Responders to CBRN Terrorism Incidents*

Knowledge Objectives

After studying this chapter, you will be able to:

- Describe your role at vehicle and machinery search and rescue incidents.
- Recognize the need for a vehicle and machinery search and rescue.
- Identify the resources necessary to conduct operations.
- Initiate the emergency response system for vehicle and machinery search and rescue incidents.
- Initiate site control and scene management.
- Recognize general hazards associated with vehicle and machinery search and rescue incidents.
- Initiate traffic control.

Skills Objectives

After studying this chapter, you will be able to:

- Initiate the emergency response system.
- Recognize the resources necessary to conduct operations at an incident scene.
- Establish control and management of an incident scene.
- Recognize and control general hazards associated with a rescue scene.
- Provide traffic control.

Because of redistricting, your fire district has just been assigned an additional 50 miles of highway system that includes three industrial plants—a chemical plant, an aluminum production plant, and a steel mill. In the past, your department has been responsible for providing only basic fire and rescue service to a small municipality within your county. As a result of the redistricting, however, your department's areas of responsibility have grown significantly. Recognizing the likelihood that the department will have to deal with industrial incidents in its expanded territory, the fire chief requests that you form a committee for the purpose of conducting a thorough hazard analysis and assessment. This analysis should result in recommendations for additional equipment and resources to provide the necessary fire and rescue coverage for all residences, businesses, churches, schools, daycare centers, and industry facilities located within the newly acquired territory.

1. What are some possible operational issues or concerns that you might want to bring to the chief's attention?

Introduction

Vehicle and machinery search and rescue can include complex and unusual machinery, vehicles, and locations. To deal with these emergencies appropriately, responders must first be given the knowledge necessary to recognize the need for a vehicle or machinery search and rescue and to initiate the emergency response system, which can mean the difference between a successful response and a failed response. The emergency response system for vehicle and machinery search and rescue includes rescuers trained at the operations and technician levels who are capable of responding to these incidents, as well as local, state, and national resources.

Today, most public safety agencies receive emergency response system training as a part of their awareness-level training, regardless of whether that education consists of fire training, emergency medical services (EMS) training, or CPR training. Several methods of activating the emergency response system are feasible, with the simplest tactic being to dial 911 to access emergency dispatch. Other methods include use of portable radios, ham operators, mobile telephones, and direct lines to the local EMS or fire/rescue squad.

Awareness of the emergency response system is important in vehicle and machinery incidents, because almost every such incident is unique. Elements such as terrain, speed, type of vehicles involved, location, type of equipment involved, material(s) being transported, atmospheric concerns, scene control issues, utilities, energy sources, and the number of casualties involved combine to create a wide array of potential scenarios Figure 9-1 ▶.

In certain jurisdictions, municipal or city governments may choose to remain at awareness-level preparedness, calling upon mutual aid departments whose personnel are trained to the technical rescue level to respond to all incidents that involve vehicle- and machinery-type emergencies. Alternatively, the municipal or county government may elect to contract these services out to a private concern because of budgetary constraints. Whatever coverage plan is selected, it is essential that emergency responders realize their own internal capabilities and limitations and prepare a list of all local, county, and state resources that are available (and the amount of time necessary to activate those resources should they be needed). As you jointly train with surrounding departments and departments that provide mutual aid, you will quickly realize their capabilities and limitations. More importantly, however, you will find ways to reduce your own limitations and enhance your overall capabilities at an incident.

Vehicle and machinery incidents are sometimes predictable; at other times, they are not. Personnel who have been involved in vehicle and machinery rescue for a number of years know that if it rains on certain roadways in their response area, they can expect an incident. In many cases, even something as minor as an increase in traffic on an old two-lane highway where 18-wheelers travel may cause a spike in the number of incidents. Vehicle and machinery incidents do not always take place on two-lane roads or the interstate, however; they also occur on secondary

Figure 9-1 A wide variety of elements come into play during vehicle incidents.

roads, farmland, mountains, and other areas. Responders must take this factor into consideration when planning for all potential hazards associated with their response areas.

Many vehicle and machinery incidents happen across the United States annually. Whether the incidents are caused by human error, faulty equipment, alcohol, drugs, or some other issue, vehicle and machinery incidents continue to happen daily and are the number one cause of fatalities and serious injuries across the country. These fatalities are not limited to business people, service workers, teenagers, and taxi drivers; they involve rescuers as well. The number of rescuers killed or injured each year in the United States is increasing. The majority of the incidents involving rescuers and other responders occur while these personnel are responding to a false alarm or returning to the station. Responders must always remain focused on the task at hand; attention to detail will ensure a safe and successful rescue operation.

Specialized incidents also occur that involve vehicles and machinery, and responders must be prepared to handle these situations. Examples of specialized incidents include the following events:

- Vehicle coming to rest in a river or lake
- Vehicle and train collision
- Vehicle collision involving heavy rigs
- School bus incident
- Farmer caught in farm equipment
- Structural collapse
- Explosion
- Vehicle entering a residence or business

Applicable Standards

Several National Fire Protection Association (NFPA) standards and Occupational Safety and Health Administration (OSHA) regulations are relevant to vehicle and machinery awareness, operations, and technical rescue curricula. The importance of these standards and regulations at the awareness level is that they stress the importance of safety first and compel personnel to conduct not just risk/hazard assessments but also detailed risk/hazard assessments. They also outline the requirements for wearing (donning) the required personal protective equipment (PPE) for the tasks at hand and emphasize the importance of atmospheric measurement. In addition, these standards and regulations provide guidance to the student who wishes to advance to the operations and technical level by providing job performance requirements (JPRs) for specialized tasks, and the necessary training or certification requirements as the student advances his or her education and training.

NFPA 1670, *Standard on Operations and Training for Technical Search and Rescue Incidents*, establishes the requirements that organizations must follow to operate safely and effectively during an incident. This standard also includes requirements for risk and hazard assessment, written procedures, safety protocols, and PPE. It establishes operational response levels and guidelines for resource consideration and management as well. Vehicle and machinery responders at the awareness level should meet the requirements in Sections 8.2 and 12.2. They should also meet the general requirements listed in Chapter 4 of NFPA 1670.

NFPA 1006, *Standard for Technical Rescuer Professional Qualifications*, establishes the minimum JPRs for rescue service in an emergency response organization. NFPA 1006 is directed at Level I (operations level) and Level II (technician level) rescuers and gives performance requirements for rope rescue, surface water rescue, vehicular rescue, machinery rescue, confined-space rescue, structural collapse, and trench rescue training that is not considered at the awareness level.

NFPA 1521, *Standard for Fire Department Safety Officer*, lists the minimum knowledge, duties, and responsibilities of fire department safety officers, which are required per NFPA 1500. This standard addresses the importance of providing a safety and health officer and an incident safety officer at the scene of an incident. All incidents, regardless of their size, have a tendency to become large—very large—if rescuers do not understand the appropriate controls and the importance of constant and continual assessments. For this reason, an incident safety officer should be in place, thereby ensuring that the incident scene is safe for the rescuers to operate. Many fire departments, rescue squads, and specialized responding agencies have outsourced these two very important positions to private contractors because of the vast education and knowledge base required for their fulfillment.

OSHA regulations are equally important as a part of responders' training and certification processes, especially because operations- and technician-level responders, at some point, will be called upon to enter into a confined space. Although this regulation appears in the technical-level curriculum, it is equally important that awareness-level students have a basic knowledge of the hazards associated with responding to a confined-space incident or any incident that may require the control of hazardous energy sources. Remember, energy control covers more than electrical controls. In fact, rescuers will face more gravity control situations than any other types of energy controls.

For example, the Permit Required Confined Space regulation (29 CFR 1910.146) is intended to protect employees and rescuers from the hazards that might be encountered when entering into areas such as tanks, vessels, silos, storage tanks or bins, hoppers, vaults, or pits. Again, this topic of training will be addressed as you advance to the operations and technical level; nevertheless, the student at the awareness level of training should at least read about and be instructed regarding the dangers associated with confined spaces and confined-space entry.

The Lockout/Tagout regulation (29 CFR 1910.147) addresses the servicing and maintenance of machines and equipment for situations in which starting the machines or equipment unexpectedly or releasing their stored energy could result in injury or death to employees and rescue personnel. Energy can be deadly if not controlled. Earlier chapters of this book have addressed—and some readers may have witnessed— what happens when energy is not controlled. No rescuer, at any level of training or certification, should make entry into suspect locations before ensuring that all energy sources have been controlled.

The Personal Protective Equipment regulation (29 CFR 1910.132) applies to protective equipment for the eyes, face, head, and extremities; protective clothing, respiratory devices, and protective shields; and barriers that protect against hazards of the environment.

OSHA's Respiratory Protection regulation (29 CFR 1910.134) contains the requirements for respiratory protection, user training, and maintenance. The Bloodborne Pathogens regulation (29 CFR 1910.1030) focuses on occupational exposure to blood or other potentially infectious materials that could result from the performance of an employee's duties. The Hazard Communications regulation (29 CFR 1910.1200) addresses the evaluation of potential chemical hazards and the communication of information concerning hazards and appropriate protective measures. The Fire Brigade regulation (29 CFR 1910.156) contains requirements for the organization, training, and PPE of fire brigades whenever they are established by the employer.

Resource Requirements

The many variations on rescue situations can make it very difficult for any public safety agency to prepare for these rescues, regardless of the level of training or equipment available to its employees. Very serious vehicle and machinery incidents may require a great deal of specialized tools and equipment, not to mention specially trained personnel. Of course, these specially trained personnel may already be living in your neighborhood. Physicians, heavy-equipment operators, safety and health and environmental technicians, mechanical engineers, FEMA or state

technical rescue personnel, and National Guard weapons of mass destruction civil support team (NG WMD-CST) members must all be identified, and must be taken into account as part of the agency's resource planning.

Clearly, ensuring that the necessary resources are available is a key part of the initial response planning process. It is equally important that responding agencies recognize the need for additional resources as soon as possible, whether those resources consist of additional personnel, engine companies, aerial equipment, specialized equipment, or other items—the appropriate action must be swift and deliberate.

Resources necessary during a large-scale vehicle or machinery incident must be available and ready to respond when called upon. Many fire departments and rescue squads, as a part of their mutual aid agreements with neighboring departments and rescue squads, have established arrangements that allow for simultaneous dispatch. These arrangements normally address equipment needs, personnel, and other resource requirements. Additional assistance would include personnel from utility providers to control their companies' services. It is essential that all energy sources be controlled before fire fighters and rescue personnel enter an incident site.

Personnel Resources

The size of an incident will dictate whether a large management staff is necessary for oversight and planning. Of course, as an incident grows, the need for additional resources may grow as well. Each fire department's or rescue squad's standard operating procedures (SOPs) should address the levels of response and indicate what positions will be staffed based on the complexity of the incident.

Facility Personnel

Rescues at industrial facilities can involve a wide variety of machinery. Whether responders are dealing with roll mills, presses, slitters, furnace pits, overhead cranes, electrical basements, or confined spaces, all will pose problems for the rescuers. Facility safety managers are a valuable resource in providing training to rescue personnel as it pertains to the equipment, and in explaining emergency procedures for controlling associated energy sources. Other resources that rescuers should certainly take advantage of during an industry incident include electricians, mechanical millwrights, welders, fabricators, and equipment operators.

Rescue Tips

Facility safety managers and workers can be a very valuable resource at machinery search and rescue incidents.

Law Enforcement Personnel

Uncontrolled incident scenes pose serious threats to everyone at the site. Law enforcement should be requested to control crowds and traffic and to establish perimeter control **Figure 9-2 ▶**. In many cases, however, law enforcement may not be available;

Figure 9-2 Law enforcement personnel should be requested for crowd and traffic control at vehicle and machinery rescues.

Figure 9-3 The number of victims and types of injuries can help you determine whether ALS or BLS providers are needed.

instead, fire fighters and rescue personnel may be responsible for performing these tasks. Traffic and crowd control can evolve into a major concern for responders during vehicle and machinery incidents. Many fire fighters, rescue personnel, police officers, and EMS personnel have suffered serious injuries from uncontrolled crowds and traffic. In incidents that occur in high-traffic areas, it may be necessary to stop all traffic in both directions until all patients have been extracted from the wreckage.

EMS Personnel

Basic life support (BLS) is the minimum level of patient care; in contrast, advanced life support (ALS) is the required level of training for most EMS agencies. Unfortunately, in some areas where rescuers respond, the only medical service available in the district may be BLS-certified service. This does not mean that the rescuers cannot function, but it does force the question of whether the injuries the rescuers have identified can be treated by BLS providers or whether ALS support will be required. Clearly, the types and number of injuries will help responders make this decision **Figure 9-3 ▶**.

For example, if responders face a scenario in which patients have suffered amputations, crush injuries, severe cervical spine injuries, or other serious trauma, it makes sense that the rescuers would request ALS units. If necessary, those resources may be requested from a neighboring community where a mutual aid agreement has been established. At the same time, BLS-certified personnel can certainly assist in providing care to patients who have experienced less traumatic injuries, because the basic care training curriculum does include victim packaging, spinal immobilization, bleeding control, bandaging, and control and maintenance of a patent airway. Without question, it is always best to obtain ALS medical support as early as possible when a serious incident occurs, but as rescuers we must be reminded that not all municipalities and rural areas provide ALS medical service. For this reason, it is very important that BLS service providers are trained to the standard. Even so, it may be best to

request ALS units early in a call; they can always be dismissed if their services are not needed.

Hazardous Materials Personnel

Hazardous material incidents could require specialized teams, depending on the complexity of the incident. In incidents involving hazardous materials, rescuers must perform a proper size-up and evaluation. All too often in incidents involving hazardous materials, rescue personnel are unnecessarily exposed to dangerous agents because they rush into the incident site before they have gathered the necessary information pertaining to the material or agent. Although awareness-level responders' activities are limited in such scenarios, they can certainly assist operations- and technician-level responders by looking through a set of binoculars and identifying leaks, placards, product labels, numbers, and other information. They can also assist in the response by preventing others from entering the incident site by sealing the site perimeter and referring to the *Emergency Response Guide* (ERG) to identify the product, evacuation distances, flammability, the product's incompatibilities, and other pertinent data.

Additional Personnel Resources

Other valuable resources whose assistance might be requested during major incidents are state and county emergency services, state and county health departments, area hospitals, emergency flight services, and the state's National Guard civil support team. Other resources include K-9 organizations, urban search and rescue (USAR) teams, FEMA task force teams, and industry hazardous material response teams.

■ Specialized Equipment Resources

When responding to vehicle and machinery incidents, it is vital to know which additional specialized equipment is available to assist in managing an incident. All rescuers realize that vehicle and machinery incidents can require a great deal of time on scene, depending on the complexity of the incident and the factors involved. Although the availability of specialized equipment sometimes expedites rescue operations, bringing specialized equipment into the operation also adds another layer of safety concerns for the response process. This, in turn, signals the need for constant and continual reevaluation of the operation.

Depending on the state's or county's geography, assistance from other resources may be necessary during an incident, such as heavy equipment providers or operators, portable lighting companies, hardware stores, building suppliers, farm equipment sales, farm equipment mechanics, wrecker services, 18-wheeler operators, physicians, 4-wheeler sales, portable generator sales, fast-food restaurants, fire equipment dealers, and rescue equipment dealers **Figure 9-4 ▼**. All rescuers and other responding agencies should train together. Training together enhances all parties' equipment familiarity, identifies equipment limitations, and provides for practical training. Rescuers and other responders should train with the equipment that is available to them and know the equipment's limitations.

The response to incidents involving machinery—whether farm equipment, heavy equipment, or industry equipment—requires technical expertise. Rescue personnel should take advantage of training offered by farm equipment sales, heavy equipment vendors, industry facilities, and other outside sources. It is important that rescuers understand how the equipment operates and which safety concerns and potential hazards could arise if the equipment is involved in an incident. Farm equipment is especially tricky: Rescuers must be aware of potential dangers of rotating equipment, flying projectiles, stored energy, conveyor gears, and hydraulics associated with the various types of farm equipment and their attachments.

Rescue responders must also be trained in controlling and stabilizing the energy and gravity of the equipment before attempting rescue operations. Accomplishing this goal may be as simple as turning off the ignition switch or turning off a power disconnect controlling the equipment—but it could also be very complex. Using chains and come-alongs (pulling devices), cribbing (large and small), or stabilizers to relieve weight, manually

Figure 9-5   A crane can be helpful in lifting or displacing vehicles or equipment.

reverse rotating equipment, or apply other controls is essential before rescue personnel are allowed to place any parts of their body in or around the equipment.

A variety of specialized equipment may be necessary when dealing with vehicle and machinery incidents. In some incidents, more than one wrecker may be required. A crane or cherry picker may also be useful in lifting or displacing larger vehicles when necessary, or in lifting heavy industrial equipment **Figure 9-5 ▲**.

The need for this type of equipment should be determined as a part of the department's hazard assessment and analysis of its response area. If this assessment reveals that the potential exists for heavy equipment incidents, then the need for heavy equipment and an operator should be a part of your department's resource plan.

Rescue Equipment Failures

With the assistance of federal funding, most fire departments and rescue squads have managed to upgrade or maybe even purchase for the first time much-needed equipment. Nevertheless, rescuers must realize that this very expensive equipment doesn't last forever; whether equipment is damaged by misuse or simply through everyday wear and tear, the potential exists for equipment failure at any given incident. Equipment abuse can be controlled with the proper training and instruction, but everyday wear and tear cannot.

In recent years, hydraulic- and pneumatic-driven equipment has replaced many of the simple hand tools of yesterday. The old hack saw lubricated with a little soap and water for easy cutting has been replaced by super-torque hydraulic shears. The pliers and crescent wrenches once used to remove doors and seats, as well as ratchet and socket sets, have been replaced by super-torque hydraulic spreaders. The pocket knife used in the past to remove mastic and rubber moldings from around windshields and rear windows has been replaced by the super-duty AC/DC reciprocating saw. At the awareness level, responders must learn to use hand tools in preparation for the day when the hydraulics/pneumatic equipment fail, but must also recognize the importance of not abusing rescue equipment so as to extend its life as long as possible.

Figure 9-4   Assistance from other resources can be crucial during an incident.

Incident Command System

The development of the Incident Command System (ICS) provided emergency response agencies with a framework in which to operate and function in a consistent fashion regardless of their location, the disciplines involved, or the size or complexity of the incident. Common terminology has allowed for clear and concise communications, bringing a much-needed unified command structure that improves working relationships with all emergency response agencies.

Response Planning

Response planning is something that every public safety agency must perform and reevaluate annually. It is essential that all fire departments and rescue squads assess factors related to potential hazards in their response areas—for example, added roadways, revised roadways, rivers, streams, traffic volumes, types of transportation, types of equipment, number of employees, office complexes, apartment complexes, hardware stores, pool supply companies, and schools. They must also assess any warehouses and the materials they contain; manufacturers and the products they manufacture; and industries and what they are producing, how those materials are stored, and what quantity is stored.

Response planning is a very important part of every department's ability to provide adequate coverage. Rescue planning must be approached from a practical standpoint, meaning that the department must determine whether additional equipment purchases will be necessary to deal with potential hazards or whether a neighboring department has the necessary equipment.

Awareness-level personnel can and should participate in response planning as a part of their training process. Notifying neighboring departments to assess equipment availability, thereby helping the agency avoid purchasing unnecessary equipment, is another task that can be handled by awareness-level–trained personnel. Likewise, as a part of the area assessment, it is equally important that vital information such as the following is shared:

- Needs assessment
- Written procedures
- Mutual aid agreements
- Dispatch protocols
- Community resource list
- Employee training and capabilities

Once the assessment has been completed, a directory containing the results of the assessment should be prepared. This document should include the resource management lists along with personal contacts, telephone numbers, and other pertinent information; it will prove invaluable should an incident requiring technical rescue expertise occur.

The directory document should also include a memorandum of agreement (MOA) noting equipment and other resources available. These resources could include additional services such as rental agreements and methods of reimbursement for those rentals.

Specialized services should also be a part of this document. These services might include K-9 organizations and their handlers, technical rescue specialists who may reside in the district, physicians, safety experts, industrial hygienists, and heavy equipment operators and suppliers.

Incident Size-Up

Size-up at a vehicle and machinery incident should include the following evaluations:

- Incident scope and magnitude
- Risk–benefit analysis
- Number and size of vehicles or machines involved
- Stability of vehicles or machines involved
- Number of known or potential victims
- Access to the scene
- Exposed utilities, water, mechanical hazards, hazardous materials, electrical hazards, explosives, and other hazards, including environmental factors
- Exposure to traffic
- Necessary resources and whether they are available

In regard to the last point, in many locations rescue squads and other responding agencies are dispatched simultaneously (mutual aid) to ensure that the necessary personnel, equipment, and other resources are available. If mutual aid agreements are not in place, be sure to provide enough lead time for the outside resources to arrive.

Scene size-up is a crucial part of every rescue operation, including vehicle and machinery incidents. The success of the operation requires not only awareness of the positioning of apparatus and consideration of other arriving units, but also appreciation of many other factors that might influence the outcome of the incident. Other considerations in scene size-up include establishing a water supply, observing scene features, and obtaining information about the fire location, time of day, weather, and nearby activities. In particular, if continual scene assessments reveal that controlling the incident may require a long-duration operation, you must consider the possibility of changing weather conditions.

When approaching a scene, consider the following issues:
- General safety considerations
- Potential hazards
- Size-up techniques applicable to the incident
- Operational needs while responding
- Vehicle positioning

Search and Rescue Perimeter

The search and rescue area is the area within a 20-foot radius of the vehicle or machinery **Figure 9-6**. Awareness-level responders can ensure search and rescue area safety through the following tasks, although the specific situation will dictate the exact actions to be taken:

- Establishing site security (rescuers trained to a higher level will establish hot, warm, and cold zones)
- Using tools to stabilize the vehicle or machinery and to isolate the incident

Figure 9-6 The search and rescue area is the area within a 20-foot radius of the vehicle or machinery.

- Disentangling and extricating victims using hand tools
- Ventilating the area (rescuers trained to a higher level will monitor the atmosphere)
- Supporting unbroken utilities
- Providing victims with protective equipment
- Protecting bystanders and responders from unsafe areas

Personnel Safety

When working in or around vehicle or machinery incidents, responders must remember the importance of conducting a thorough scene assessment, which includes selection of the appropriate PPE to perform the tasks. PPE includes both protective apparel (e.g., clothing, footwear, gloves, headgear) and personal protective devices (e.g., goggles, face shields, hearing protection, respirators). Adequate PPE should protect the wearer's respiratory system, skin, eyes, face, hands, feet, body, and ears **Figure 9-7 ▶**.

Whether they are operating at a fire scene, hazardous material incident, or confined-space incident, all responders must be trained to recognize the need for wearing appropriate PPE to ensure responder protection. Many fire departments allow their personnel to wear limited body protection during incidents because the responders believe that their PPE is too hot, is too cumbersome, or takes away from their dexterity. Don't fall into this mindset: An incident scene is subject to change at any time, and changing conditions can cause serious injuries to rescuers and bystanders alike. Look and listen for hazards that could harm you, other responders, and victims involved in the incident.

Having a safety officer present at the incident will ensure a safe work environment for operations. This officer's job is to make observations and conduct reassessments of the incident and rescuers on an ongoing basis. The safety officer continually assesses the incident for unsafe acts and unsafe conditions that could harm the rescuers and victims.

A responder may easily recognize an unstable vehicle, fluid runoff, sharp edges, and fire. In contrast, when responders are

Figure 9-7 Always wear adequate PPE to protect against both known and potential hazards.

faced with incidents involving machinery, a different approach is required that necessitates additional education and training. Additional training for industry incidents may consist of OSHA standards 29 CFR 1910.146 for confined spaces and 29 CFR 1910.147 for lockout/tagout.

Responding rescuers should immediately begin thinking of questions that must be answered before rescuers are allowed to enter the incident site, especially when the incident is an industrial-type incident that involves complex machinery. Examples of some of these questions are presented here:

- Is the machinery locked and tagged out?
- Are there exposed rotating parts?
- Is the machinery stable?
- Are there hazards associated with the potential release of chemicals, steam, or other fluids?
- Are there fall hazards?
- What are the atmospheric conditions?

At the awareness level and as a rescue responder, you must have the ability to recognize the existing or potential dangers associated with all incidents. You must be constantly aware of your surroundings, which means looking and listening for any unusual sounds and double-checking that all energy sources, regardless of the type, have been controlled. Do not take anyone's word that energy has been controlled; verify that all energy sources involved with the rescue have been controlled by checking them yourself. Do not be afraid to ask questions and double-check controls before you and other rescuers enter into areas of uncertainty. Otherwise, you could place yourself and others in a position to be seriously injured or killed.

Pinch Points

Controlling **pinch points**—that is, areas where rescuers can trap fingers, hands, feet, or legs—is a part of the scene or incident size-up activity. Ideally, pinch points should be addressed as a part of the **vehicle/machinery stabilization** process. These areas are normally recognized during the initial assessment of the vehicle or machinery incident. Because incident scenes inevitably change, however, continual reassessment of the scene, vehicles, machinery, environment, weather, and pinch point areas by the incident safety officer is essential.

Slips, Trips, and Falls

The potential for slips, trips, and falls is also a focus of the responder's initial assessment. Any slip hazards must be controlled using sand, dirt, or any other absorbent material. Trip hazards such as debris, fluids, and equipment must be controlled by their removal or another method of rendering the area safe. Once all hazards are recognized and either removed or corrected, the potential for fall hazards is eliminated.

Controlling slips, trips, and falls speaks to the importance of maintaining crowd control. Rescue personnel require a clear and safe work surface when operating rescue equipment, stabilizing vehicles or machinery, extricating victims, and providing medical care.

■ Hazards and Risk Assessment

Rescue responders will face many types of vehicle and machinery incidents that present different hazards. Therefore, it is absolutely essential that a thorough risk assessment be conducted. In a perfect world, each fire department or rescue squad would have members with intricate knowledge and training of equipment operations. When this resource is not available, however, it is important that the incident commander (IC) and rescuers have the knowledge necessary to conduct a thorough risk assessment and identify all hazards associated with the incident. Responders must also understand the importance of using any technical resources available when necessary.

A variety of hazards (e.g., water, chemicals, extreme height), failure of essential equipment, or severe environmental conditions can further complicate vehicle and machinery rescues. A hazard identification and risk assessment analyzes those factors influencing the technical rescue incident and determines how they might affect the responders' ability to respond to and safely operate at these incidents. The goal of this assessment is to increase awareness of factors associated with potential technical rescue responses. In analyzing potential hazards in a facility containing machinery, information related to the following issues should be obtained:

- The type of energy source(s) that power the facility (e.g., natural gas, electricity, steam)
- The presence of a sprinkler system or other method of fire suppression in the facility

- The presence of a fire alarm system in the facility
- The number and adequacy of fire hydrants inside and outside the facility
- Written mutual aid agreements with surrounding fire departments
- The contact persons and telephone numbers available to the fire department and rescue squad
- Hazardous material concerns with any of the businesses
- The total square miles and population during both the day and the night
- The water supply available and its source

These are only a few considerations that should be addressed when conducting a hazard analysis and assessment at newly acquired territories. This type of hazard analysis and assessment is similar to that conducted during a department's fire preplanning activities; however, the depth of the questions is much different in that they will address every possible hazard, not just fire. Such a hazard analysis and assessment is very important because, in incidents involving the added hazards, the fire department must be prepared to provide optimal protection for its fire/rescue members.

When they are working vehicle incidents involving new vehicle designs and engineering changes, responders must realize that some safety equipment designed to protect the vehicle driver and passengers may actually create additional hazards for the rescuers. Examples of such equipment include passenger air bag systems; air bag cylinders hidden in the A, B, and C posts; seat pretensioner cylinders; composite materials; hydraulic cylinders hidden under vehicle hoods; and van and SUV rear doors **Figure 9-8 ▼**.

Before attempting to rescue occupants or moving a damaged hybrid vehicle, for example, rescuers should don full protective equipment (i.e., coat, pants, boots, helmet with hood, goggles/ face shield, gloves, and SCBA) to protect against the hazards posed by the vehicle's heat and its nickel metal hydride batteries, which produce toxic fumes when they are burned, includ-

Figure 9-8 At vehicle rescue incidents, be aware of hazards such as undeployed air bags, which can seriously injure responders and victims if they are activated during rescue.

ing oxides of nickel, cobalt, aluminum, manganese, lanthanum, cerium, neodymium, and praseodymium. The responders must also be aware of and protect themselves against the high voltage associated with hybrid vehicles. Rescue squads should seek further information from local car dealers about strategies for dealing with hybrid vehicles.

Here are three methods of protection in hybrid-vehicle incidents, beginning with the most desirable:

1. **Turn off the ignition switch.** This simple action turns off the engine and the electric motor, preventing current flow into the cables. After you turn off the ignition switch, remove the key so the car cannot be accidentally restarted.

2. **Deactivate the 12-volt system and main fuse.** Remove the main fuse and disconnect the negative cable from the 12-volt battery. This is the second best method for preventing current flow and should be attempted only if you are unable to turn off the ignition switch.

3. **Turn off the intelligent power unit (IPU).** Use this method only if the rescuer cannot turn off the ignition switch or make access to the under-hood fuse box.

Identification of power cables, cable protection, strategies for dealing with hybrid vehicles submerged in water, and the dangers when cutting through the frames of hybrid-style vehicles are covered at the operations or technician rescue level.

When dealing with new technology, rescuers should be trained to recognize the added hazards associated with these vehicles and ensure that proper rescue techniques are used when operating hydraulic or manual rescue tools during the __disentanglement__ processes. Many of these techniques will be learned as you continue your training. For rescuers to become truly proficient in their use, however, they must spend lots of time in the salvage yards working to hone their practical skills.

Rescue responders could be involved in incidents involving 18-wheelers, trains, boats, passenger cars, minivans, industry equipment, mining equipment, off-road equipment, and farm tractors and attachments. Obviously, many of these incidents could involve multiple individuals with multiple injuries. It is important, therefore, that rescuers have the intricate knowledge and training necessary to expedite the removal of the victims involved.

Regardless of the situation, responders must understand the importance of the following activities: conducting a good scene survey, ensuring appropriate apparatus placement (150–200 feet upwind or uphill when possible), managing environmental concerns, wearing the proper PPE, protecting rescuers with a charged line, conducting an ongoing size-up, controlling all associated hazards, stabilizing the vehicle or equipment, recognizing compressed cylinder locations, recognizing air bag system component locations, controlling energy sources, creating entry and exit points, dismantling equipment, gaining access to victims, treating and packaging victims, extricating victims, releasing victims to EMS for medical treatment and transport, and terminating the incident. All of these elements must be in place to ensure a safe and successful rescue operation.

Numerous other hazards are likely to be associated with vehicle and machinery search and rescue incidents. The authority having jurisdiction (AHJ) should attempt to identify potential hazards within the jurisdiction before an actual incident occurs and should

Figure 9-9 Crowd control is essential to provide rescuers with adequate space to operate.

provide awareness of and training for operations against these hazards. This step will ensure that rescuers and other responders can perform their duties safely around these hazards.

Awareness-level vehicle and machinery rescue responders are limited to minimal involvement during an operation; however, all members of the organization must meet NFPA 472, *Standard for Competence of Responders to Hazardous Materials/Weapons of Mass Destruction Incidents*. Rescuers trained to the awareness level can make valuable contributions at a rescue scene by detecting hazardous materials, conducting a survey of the hazardous materials from a distance (to identify it by name, UN/NA number, or placard), implementing actions consistent with their local emergency response plan and their organization's SOPs, and using the *ERG* to initiate protective actions and notifications.

Crowd Control

Scene control is an absolute necessity in protecting responders from individuals who might attempt to enter into the emergency site. It is not uncommon for individuals at a vehicle incident, industry incident, or any mass-casualty incident to try to assist those who are injured. Adequate crowd control provides the necessary space for fire and rescue personnel to operate without being concerned about the prospect of individuals interfering with their work **Figure 9-9 ▲**.

Traffic Control

Today, the majority of fire departments and rescue squads use fire or rescue apparatus to protect rescuers, victims, and bystanders from oncoming traffic. Others continue to use traffic cones, fire fighters, road flares, and specialty lighting for this purpose, although these methods have proven too risky.

Fire

Fire scenes present numerous hazards and ever-changing concerns for fire fighters and rescue personnel. The potential always exists for an explosion, a collapsing wall or roof, electrical issues, natural or propane gas issues, or water supply, personnel, entrapment, and other concerns. Given the many different hazards possible at vehicle and machinery incidents, donning the proper

PPE is clearly important when operating at an incident involving fire. Fire is not the only incident for which appropriate PPE is donned, however; appropriate PPE must be donned on every incident because every incident contains unknown elements.

Electrical Lines

Electrical lines down present a serious hazard to fire fighters and other rescue personnel. Most vehicle and machinery incidents occur during the night, which creates problems for emergency response personnel because of the lack of visibility to identify downed electrical lines. In this situation, attention to detail is very important. Rescuers must recognize when street lights that are normally lit are not functional, and residents in the area without lighting should alert the emergency responders to proceed very slowly into the suspect location.

Once an electrical line has been identified, responders should contact the appropriate utility provider and control the hazard from contact with both rescuers and pedestrians Figure 9-10 ▾ . Several methods can be used in controlling electrical hazards, such as placing traffic cones around the perimeter of the electrical line, placing fluorescent snap lights around the perimeter, or isolating the electrical line using barrier tape. When enough personnel are present, positioning fire fighters around the perimeter to prohibit travel into the hazard area is the best method of control.

Some locations do not have suspended electrical supply lines or electrical transformers; instead, electrical supply lines are buried and electrical transformers are positioned at ground level. In such a case, the hazards are the same as with suspended lines, but a little closer to the responder and surrounding properties. Like overhead electrical lines, those positioned underground are vulnerable to damage during construction activities and during vehicular crashes. Similarly, electrical transformers positioned at ground or suspended levels may create a number of problems for rescuers, such as open high voltage, toxic smoke and gas, intense heat, potential for explosion from oil-filled equipment, explosion and flying debris from glass insulators and porcelain insulators, and, with some systems, release of pressurized gas. Rescuers are trained to be aware of both overhead and underground electrical lines when dealing with incidents involving utilities.

Figure 9-10   Safety is of primary importance when dealing with electricity.

Fuel Sources

The most common fuel sources today are gasoline and diesel fuel, though ethanol is also growing in popularity as a fuel. Liquefied petroleum gas (LPG) continues to be the most widely used alternative fuel to gasoline and diesel on a worldwide basis, primarily because of the environmental improvements that use of LPG brings. More than 500,000 vehicles in the United States use propane gas, most of which have spark-ignition engines that can operate on either propane or gasoline.

Gas explosions and leaks typically involve natural gas and propane gas. Natural gas incidents usually involve a ruptured supply line or a line that has failed due to corrosion or surface shift. Rescuers should evaluate any suspected releases with gas and air monitoring devices to determine the actual release point. Once the release point has been pinpointed, all buildings in the immediate area should be monitored before the area is considered safe.

Propane gas explosions normally involve cylinders; for example, propane may leak from cylinders used in cooking grills, home heating supplies, and industrial forklifts. These conditions can be corrected by separating the propane cylinder(s) from a heat source. In its natural state, propane is odorless. When it is sent through a distribution system, an odorant is added. If the propane is being transferred by pipeline, however, the odorant may not have been added, creating a major concern for rescuers when incidents involve pipelines. Propane is heavier than air, so it will lie close to the ground surface, seeking an ignition source.

As mentioned earlier, the majority of propane leaks or releases occur from leaking tanks, usually caused by the tank being overfilled. One of the concerns with dealing with incidents that involve propane is the potential for a BLEVE created when fire impinges on the tank, resulting in temperature and pressure increases within the cylinder. As both temperature and pressure increase, the relief valve activates (opens), allowing the pressure to be released, which is good news for fire fighters and rescue personnel. If the relief valve fails to activate, the tank will fail, resulting in a violent explosion.

Fuel runoff can be very dangerous if not controlled by fire fighters or rescue personnel. Several concerns arise when dealing with fuel runoff, including the presence of ignition sources, environmental concerns, and reactions of fuels when they mix with each other and other products involved in the incident.

Damming, diverting, diluting, and absorbing are all methods used to control fuel runoff. Fire departments most commonly use floor/oil dry (oil absorbent), soil, and foam agents to dilute the materials and slow runoff; this technique makes clean-up proceed much more quickly, especially following a vehicle incident.

Ignition Sources

Rescuers must be aware of all potential ignition sources when working a vehicle or machinery incident. All ignition sources must be eliminated, if possible, without additional exposure or risk, and all apparatus and other vehicles must remain in the staging area upwind beyond the isolation zone. All individuals not associated with the incident must be removed, and no apparatus or portable equipment is to be started. Likewise, no matches or lighters, telephones, or radios may be used. Lights should not be switched on or off, nor should anything that could possibly create a spark be used.

Hazardous Materials

Responders, no matter what levels of training, must constantly be aware of the complexity, impact, and potential harm that various types of hazardous materials can and do present. They must also know how to avoid exposure to such materials, whether by inhalation, absorption, ingestion, or injection. The most intelligent course of action is to contact appropriately trained and equipped personnel to handle any incident involving hazardous materials. Responders must be aware of the threats that such materials pose to health, property, and the environment. In today's world, responders must also be trained to recognize indicators of a potential terrorist incident as a part of their size-up.

Hazardous materials can come in a variety of forms: solids, liquids, and gas. Such materials can have radioactive, flammable, explosive, toxic, corrosive, biohazardous, oxidizer, asphyxiant, pathogenic, allergenic, or other characteristics that make them hazardous in specific circumstances. Among the hazardous materials that rescuers encounter almost routinely at incidents are gasoline, diesel fuel, kerosene, nitric acid, toluene, and acetone.

Vehicles that carry flammable and nonflammable pressurized gases (e.g., nitrogen, hydrogen, oxygen), as well as transporters that carry flammable and nonflammable cryogenic liquids, including liquid nitrogen, liquid hydrogen, liquid oxygen (LOX), and liquefied natural gas (LNG), will also be observed occasionally where they are involved in incidents. Solid materials, including explosives, flammable solids, oxidizers and organic peroxides, poisons and corrosives, are also sometimes involved in incidents. Common examples include fertilizers, pesticides, caustic powders, water treatment chemicals, and Class 9 materials, which are often observed around mining or construction activities.

There are two primary NFPA standards and one OSHA regulation that apply to hazardous materials and training for emergency responders:

- NFPA 472, *Standard for Competence of Responders to Hazardous Materials/Weapons of Mass Destruction Incidents*
- NFPA 473, *Standard for Competencies for EMS Personnel Responding to Hazardous Materials/Weapons of Mass Destruction Incidents*
- 29 CFR 1910.120, Hazardous Waste Operations and Emergency Response

The Department of Transportation (DOT) defines a hazardous material as "any substance or material in any form or quantity that poses an unreasonable risk to safety and health and to property when transported in commerce." The Environmental Protection Agency's (EPA) definition of a hazardous material is "any chemical that, if released into the environment, could be potentially harmful to the public's health or welfare."

Material Storage

Material storage should be an important component of your hazard analysis when conducting fire safety inspections and writing fire/rescue preplans. This information can be easily obtained from the safety manager or store manager. Fire fighters and rescue personnel can often obtain material safety data sheets (MSDS) for specific materials housed or used at a facility; these documents provide information pertaining to storage locations, types of materials, amounts of material stored, appropriate PPE, health concerns, medical first-aid treatment, and fire protection media do's and don'ts, among other things. Additional information that could prove very helpful to fire fighters and rescue personnel includes identification of any containment area around the storage location, drains leading directly to a spilled contaminate area, sprinkler systems or other means of fire suppression in the storage area, and smoke or fire alarms in the storage area.

Material Transportation

Material transportation comes in a variety of shapes and sizes. Transportation may occur by water, highway, rail, or air Figure 9-11 ▾ . Materials may be transported in atmospheric, low-pressure, or pressurized highway tankers; specialized totes; intermodal containers; high- and low-pressure railway tankers; or drums, glass containers, boxes, or pressure cylinders. Rescuers must be trained to recognize the various methods in which materials are transported, stored, and packaged.

Unstable Vehicles and Machinery

If not controlled, unstable vehicles and machinery pose serious threats both to rescuers and to persons who have already been injured in incidents involving them. The shape, size, and resting positions of vehicles and machinery after an incident can create stabilizing challenges for rescuers. Rescue personnel must, therefore, ensure that vehicles and machinery are stabilized before entering or approaching the incident scene Figure 9-12 ▸ . Numerous methods are available for cribbing and stabilizing vehicles and machinery, such as box cribbing, wheel chocks, step chocks, high lift jacks, stabilizer jacks, rope, chain, cable, winches, and tow trucks.

Questions to consider after conducting a size-up of the incident include the following:

- Which method(s) can I use to stabilize the vehicle or machinery?
- Which method(s) can I use to extricate the trapped victims?
- Which tools do I have available or which tools may be required for the extrication?

Figure 9-11 Materials can be transported by water, highway, rail (shown here), or air.

Voices of Experience

It was the summer of 1970, and I was an 18-year-old rookie fire fighter who was about to witness an incident that would forever remain in my mind. The incident involved a Kenworth DART DE2772 Quarry Truck, and a 1970 Chevrolet Monte Carlo medium-sized passenger vehicle that was occupied by six teenagers who were cutting school classes. The teenagers were traveling at a high rate of speed on a dusty mining road that ran off Route 61 near the community of Chelyan, West Virginia. The road was primarily used by large, off-road–type vehicles transporting coal from a strip-mining operation to a lower-level dump area. The coal would then be transported by an 18-wheeler to a nearby barge loading facility, where the coal was then transported by river.

On this particular day, the fire fighters in my department were sitting around the department telling stories of rescues and structural fires, like most fire and rescue personnel do. Being a rookie fire fighter, I was all ears, trying to absorb every bit of information I could possibly gather: which techniques they used in extricating the victims, which vehicles were involved, how long it took for the extrication, and so forth. Of course, after listening to the more experienced fire fighters, I was eager to respond to *the big one*.

At approximately 3:30 P.M., our department was dispatched to respond to an incident that involved a passenger vehicle that had been run over by an off-road (DART) vehicle. I remember the incident like it occurred yesterday—from my initial reaction when I stepped on the rescue and my arrival at the incident location.

I remember arriving and seeing the DART covering the entire passenger vehicle—from the very front to the very rear of the vehicle.

The question was how we could get this large piece of machinery off this passenger vehicle to render aid to the victims. Even though fire departments from three different areas were on the scene of the incident, no one had a clue how to remove the large mass of steel from the vehicle. Finally, a decision was made: We would dig under the large machinery and try using railroad jacks to lift the machinery off the vehicle. Each of the rescuers worked through the slow process of lifting the machinery inch by inch, driving cribbing for every inch gained. Finally, rescuers gained enough clearance on the driver's side to allow rescuers access to check for any sign of pulses. Rescuers did obtain a pulse in the foot of the driver but were unable to do much else because of limited access. Rescuers continued to dig, raising and cribbing to gain every inch of additional access. Finally enough access was gained to check femoral pulses in the driver's legs—faint pulses, but pulses.

A medical decision was made by a WV Air Guard nurse to apply tourniquets to both legs and force what blood volume remained in the lower extremities to the heart and brain. Another concern was what would happen when the weight was lifted from the victim.

As darkness came and hours elapsed, any hope for the driver and other victims surviving was slowly diminishing. Suddenly, the superintendent of the mining company arrived at

> **"Even though fire departments from three different areas were on the scene of the incident, no one had a clue how to remove the large mass of steel from the vehicle."**

the incident location and suggested the use of another piece of heavy machinery: a huge end-loader. The superintendent said the end-loader could lift the bed or material hauling section off the vehicle, allowing rescuers access to the victims. The fire chiefs assembled briefly to develop a game plan and discussed their concerns with using the huge end-loader. In the meantime, rescuers continued digging, raising, and cribbing the huge piece of equipment, trying to gain access to the victims.

The fire chiefs agreed that the end-loader was the only option available if we had a chance of rescuing the victims. One of the fire chiefs was concerned with the end-loader dropping the load back onto the vehicle, but quickly agreed that this was our only chance of gaining access and rescuing the victims.

The end-loader arrived at the location in a matter of seconds. The area was cleared of all equipment and rescuers, and the end-loader lifted the bed off the passenger vehicle with relative ease. The vehicle appeared to have been dropped from a 100-story building; all the tires, shocks, and springs had burst and the bottom portion of the vehicle was positioned on the ground. The vehicle stood no taller than one foot high.

Rescuers used hydraulic shears to remove the top of vehicle, which allowed the remaining parts of the vehicle's roof to be peeled back. Five of the six individuals lay dead and the driver was seriously injured. The driver was flown to Charleston Memorial Hospital, where numerous surgeons were standing by, via West Virginia National Guard helicopter. After multiple surgeries and months of rehabilitation, the driver recovered from his injuries and regained his ability to walk.

Without question, this incident taxed our department and surrounding departments. During this era, hazard analysis, preplanning, and mutual aid agreements didn't exist. Granted, rescue equipment wasn't plentiful during the 1970s—but then again, no one ever thought an incident such as this would happen, even though the potential certainly existed. This incident also emphasized the importance of developing a good resource list: The end-loader was instrumental in saving the young man's life. The incident also demonstrated the importance of having mutual aid agreements. Although three fire departments responded, no mutual aid agreements existed. There were never joint training exercises, and the fire departments involved had no idea which equipment the other departments had on hand.

Today, proper training, preplanning, hazard assessment, established resources, upgraded equipment, mutual aid agreements, and the willingness to work and train together has certainly improved response capabilities. Would the result be different today if a similar incident happened?

Chief Jerry L. McGhee
Marmet Fire Department
Marmet, West Virginia

Figure 9-12 Ensure vehicles are stabilized before you enter the incident scene.

Agitated Victims

Agitated victims are not uncommon at an incident scene. Whether the cause of their distress is the incident, an injury, or an illness, it is important that all victims receive proper attention. In certain situations, EMS personnel may not be present at an incident scene; in such cases, rescuers may be required to render basic emergency care. Although this chapter does not address emergency medical care, responders should be prepared to provide basic first aid or CPR.

Agitated victims should be removed from the immediate incident location and moved away so that they do not see the aftermath of the incident. Responders should ensure that victims involved in the incident are not injured, and that their families have been notified. These steps will assist the rescue personnel in calming victims who are notably upset.

Sharp Edges (Metal and Glass)

During rescues, and especially in the course of vehicle <u>extrication</u>, responders may need to navigate around sharp edges and glass. Responders must be aware of all objects that could affect or endanger rescue personnel, and proper guarding must be applied to prevent injuries to everyone involved in the incident. Sharp edges can be controlled or eliminated by cutting back those edges with hydraulic shears, a reciprocating saw, or another metal tool. Sharp edges can also be controlled or eliminated by applying half hose covers, duct tape, or canvas. These controls and use of the appropriate PPE will limit the potential for injuries to rescue personnel.

Twists, Strains, and Sprains

Twists, strains, and sprains are injuries that can occur when stepping onto, down from, or across an uneven surface of some type. Obviously, these types of injuries can occur while performing many physical tasks. When they occur at a rescue incident, however, the injury is typically the result of unused equipment that is positioned at the feet or behind rescuers or vehicles or machinery parts that are not properly stowed away from the rescuers' work area.

Weather

Weather conditions can play a very important role in decision making as it relates to not only rescue operations, but also incidents involving hazardous materials. All rescuers are aware that a large majority of incidents occur during rainy, snowy, or icy weather; however, the importance of being aware of or requesting weather condition reports does not always relate to the potential for slippery roads. For example, current weather conditions or a five-day forecast can be very significant to rescuers when dealing with hazardous materials incidents. In fact, hourly weather reports could be very important when dealing with a hazardous materials incident, where changing winds and temperatures could lead to devastating results.

Wrap-Up

Chief Concepts

- Regulations and standards are intended to ensure that rescuers have the appropriate information needed to do their jobs more safely and efficiently.
- The many variations on rescue situations can make it very difficult for any public safety agency to prepare for these rescues, regardless of the level of training or equipment available to its employees.
- Facility safety managers are a valuable resource in providing training to rescue personnel as it pertains to the equipment, and in explaining emergency procedures for controlling associated energy sources.
- Vehicle and machinery incidents can require a great deal of time on scene, depending on the complexity of the incident and the factors involved.
- It is important that rescuers understand how the equipment operates and which safety concerns and potential hazards could arise if the equipment is involved in an incident.

Hot Terms

disentanglement The removal and/or manipulation of vehicle/machinery components to allow for proper victim removal.

extrication The removal of trapped victims from an entrapment.

pinch points Areas where rescuers' fingers, hands, feet, or legs may become entrapped.

vehicle/machinery stabilization The process of securing a vehicle or machine to prevent unexpected movement during a rescue operation, thereby avoiding injury to rescuers and preventing further injuries to the entrapped victims.

Rescue Responder *in Action*

On a hot day in June at approximately 3:00 P.M., your department is dispatched to a vehicular incident on a two-lane road. The incident is reported as involving a single vehicle: A tractor-trailer has flipped on its side on the roadway, resulting in the removal of all roadside guardrailing in its path, with the vehicle dropping 30 feet down an embankment. No further information is available at this time. The incident does not sound terribly bad or especially uncommon for this stretch of roadway until you receive the follow-up information from dispatch: An eyewitness observed a father and his two young daughters walking along the roadway when the truck flipped and left the roadway. The witness reports that the father and two daughters were struck by the truck and by falling and flying debris during its path of travel over the embankment.

1. Which additional resource equipment will be necessary at this incident?
   A. EMS
   B. Law enforcement
   C. Heavy equipment and operator
   D. All of the above

2. Which of the following teams should be considered for this incident?
   A. Technical rescue team
   B. Critical incident stress team
   C. Both A and B
   D. None of the above

3. What other concern exists for this incident?
   A. Hazardous materials
   B. Energy control
   C. Crowd control
   D. All of the above

Water Search and Rescue

NFPA 1670 Standard

9.1 **General Requirements.**
Organizations operating at water search and rescue incidents shall meet the requirements specified in Chapter 4. [p. 205]

9.2 **Awareness Level.**
9.2.1 Organizations operating at the awareness level at water search and rescue incidents shall meet the requirements specified in Section 9.2. [p. 203–226]
9.2.2 Each member of an organization operating at the awareness level shall be a competent person as defined in 3.3.24. [p. 205]
9.2.3 Organizations operating at the awareness level at water search and rescue incidents shall implement procedures for the following: [p. 203–226]

(1) Recognizing the need for water search and rescue
(2)* Implementing the assessment phase
(3)* Identifying the resources necessary to conduct safe and effective water operations
(4)* Implementing the emergency response system for water incidents
(5)* Implementing site control and scene management
(6)* Recognizing general hazards associated with water incidents and the procedures necessary to mitigate these hazards within the general search and rescue area
(7) Determining rescue versus body recovery

Additional NFPA Standards

NFPA 1006, *Standard for Technical Rescuer Professional Qualifications*
NFPA 1925, *Standard on Marine Fire-Fighting Vessels*
NFPA 1983, *Standard on Life Safety Rope and Equipment for Emergency Services*

Knowledge Objectives

After studying this chapter, you will be able to:

- Identify the need for water search and rescue operations.
- Identify various types of water and ice environments.
- Recognize and identify hazards common to all types of water.
- Recognize and identify specific hazards unique to ice, swiftwater, surf/marine, and underwater environments.
- Describe the resources needed to conduct a water search and rescue operation.
- Describe response planning and incident management requirements related to a water search and rescue incident.
- Describe site control operations at a water rescue incident.
- Describe non-entry rescue considerations at a water rescue incident.
- Describe the NFPA 1670 objectives for the operations and technician water-rescue levels, including how they apply to water rescue resources.
- Explain the modified RETHROG concept.
- Explain each component of the modified RETHROG mnemonic.

Skill Objectives

There are no skill objectives for this chapter.

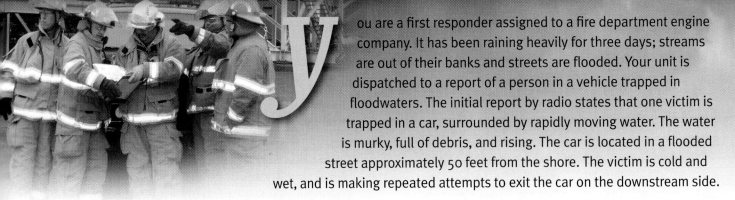

ou are a first responder assigned to a fire department engine company. It has been raining heavily for three days; streams are out of their banks and streets are flooded. Your unit is dispatched to a report of a person in a vehicle trapped in floodwaters. The initial report by radio states that one victim is trapped in a car, surrounded by rapidly moving water. The water is murky, full of debris, and rising. The car is located in a flooded street approximately 50 feet from the shore. The victim is cold and wet, and is making repeated attempts to exit the car on the downstream side.

As you approach the scene with your vehicle, you see several motorists waving you toward the water and pointing at the trapped victim. As you start to perform a size-up of the event, you realize that a water rescue incident requires specially trained responders and specialized tools and equipment to operate effectively. Your company does not have that specialized training, tools, and equipment. What should you do?

1. What should you be looking for during a water rescue size-up?

2. Which hazards might rescuers encounter at a water rescue event?

3. Which actions can first responders take at this incident to manage the event more effectively?

4. Which non-entry actions can you take at a water rescue event?

5. Which types of resources are needed to manage a water rescue scene effectively?

6. How can you get appropriate resources to a water rescue incident in a timely manner?

Introduction

Drowning is the second leading cause of death in the United States, second only to motor vehicle accidents. On average, three fire fighters die in water rescue accidents each year. A few examples will suffice to illustrate the danger associated with such incidents:

- On April 8, 2001, a kayaker died from being pinned underwater while paddling in a flooded creek. Approximately 45 minutes later, two fire fighter/divers died attempting to retrieve the kayaker's body. The kayaker's body was retrieved without incident the next day after the floodwaters had receded.
- On May 28, 2006, a paramedic/swiftwater-rescue technician drowned while attempting to save two teenaged boys from a low-head dam hydraulic. The boys' bodies were recovered two days later.
- On June 22, 2006, a fire fighter/rescue diver died while attempting to rescue two teenaged boys from a flooded patch of woods. The current overpowered the fire fighter's rope tether and pulled him underwater. He was retrieved within minutes and resuscitation was started immediately, but to no avail. The original victims were rescued a few minutes later in a boat operated by park rangers.

Well-intentioned but inappropriate water rescue efforts often kill rescuers in addition to the original victims. Federal and state officials of the U.S. Coast Guard and the Occupational Safety and Health Administration (OSHA) have established written rules governing safe operating procedures for operating in and around water and in boats. These rules apply to fire fighters and other rescuers, as well as to the general public.

U.S Coast Guard statistics show that a majority of water victims had no water rescue training. Typical risk factors include a lack of hazard awareness, alcohol use, and improper or no safety equipment. These factors, with the exception of alcohol use, have been implicated in many rescuer deaths as well. Many fire/rescue agencies do not consider water rescue to be a problem until it's been raining for days and the water is rising. Unfortunately, that is too late to obtain the proper training, equipment, teamwork, and experience, which is why poorly executed water rescues often end with tragic results.

A lack of understanding of the water's tremendous power can lead to underestimating the forces involved. A fast-moving current can easily overpower several fire fighters holding a rope attached to another fire fighter in the water. Other common water rescue mistakes are overestimating an individual fire fighter's swimming ability, not providing personal survival gear for fire fighters operating in or near the water, not providing an alternative means of propulsion for rescue boats, and assuming that the rescue will go according to plan.

Rescue Tips

Water Rescue Pitfalls

1. *Being unprepared for the water environment.* Water is a special environment, and operating in or near it requires specialized knowledge, skills, abilities, and equipment.
2. *Going near the water without wearing a personal floatation device (PFD).* The single largest factor in open water, swiftwater, or floodwater drowning is the failure to wear a PFD.
3. *Assuming that all water environments are the same.* Each type of water has its own temperature, current, and special hazards. Don't send a dive rescue team to a swiftwater rescue, or vice versa, unless the team is trained and equipped for both rescue types.
4. *Wearing heavy or negatively buoyant PPE or clothing in or near the water.* Turnout gear, steel-toed boots, and other PPE designed for use on land may kill you if you wear it in the water or if you fall into the water while wearing it.
5. *Ignoring the water temperature.* The air temperature may be much warmer than the water temperature. Accidental immersion in cold water can quickly lead to hypothermia if you are not dressed for immersion.
6. *Driving vehicles into moving water or water of unknown depth.* Moving water that is only a few inches deep can move a heavy vehicle off the road or tip it over. If you can't see the underwater surface on which you are driving, you can drive off the pavement or into invisible obstacles that can trap or disable your vehicle.
7. *Waiting for the 911 calls to begin before you seek extra help during floods, hurricanes, and other large storms.* Many people die for lack of rescue resources. If you know your area is going to flood, position swiftwater and flood rescue resources close by to reduce water rescue response times.
8. *Assuming that the boat motor will always work.* Motors, especially two-cycle outboards, have a high failure rate under rescue conditions. Props may foul or become damaged by submerged obstacles or debris, jet pumps may clog, or the motor may simply quit. If you put rescuers in a position where only the boat motor can save their lives, you need to realize that a motor failure can kill them. As Jim Segerstrom wrote, "If you go out in a boat, be prepared to come back without the boat."
9. *Ignoring simple, quick rescue techniques while waiting to implement a more complicated, high-tech solution.* The victim may not have the time it takes to wait on a technical rescue team, a helicopter, or a world-class kayaker to arrive.
10. *Waiting until the flood starts to recognize the need for water rescue training and equipment.* It can be difficult to visualize a dry wash, a babbling brook, or a major city under tons of raging water. Once the flood starts, it's too late to get the training and equipment. A high percentage of water rescuer deaths occur when rescuers without adequate training, equipment, and/or support attempt rescues for which their agency never prepared them.

Water Rescue and Recovery

An incident must meet all of the following criteria to be considered a **water rescue**:

- The water must be deep enough to create a risk of drowning.
- The water and/or air may be cold enough to create the risk for hypothermia or heat stress injury.
- There may be enough current to create the risk of being swept away.
- There may be natural and/or human-made obstacles that create the risk of injury, entrapment, or incapacitation.
- There may be energized electrical equipment that creates the risk of electrocution in the water.
- There may be waterborne hazardous materials that create the risk of exposure to fire; corrosive, toxic, or reactive chemicals; biohazards; or radiological hazards.
- There may be enough water to destroy community resources and infrastructure, including transportation, communications, medical, shelter, sanitation, and food and potable water supplies, in a community or even an entire region.

A **water recovery** is defined as a water emergency that meets the water rescue definition, plus has one or more of the following characteristics:

- The victim is known to be dead.
- The victim has been trapped underwater for more than 90 minutes.
- The victim is trapped in water from which there is no reasonably safe way to accomplish a rescue.

Water Rescue Problem Sites

The following areas should be considered as potentially being the sites of water rescues:

- Rivers and streams
- Lakes and human-made reservoirs
- Dams, especially low-head dams
- Bridges
- Storm sewer systems, especially storm drains
- Rapids and waterfalls
- Farm ponds
- Swimming pools
- Oceans and beaches
- Tidal creeks and rivers
- Ports, piers, marinas, and harbors
- Water and sewage processing and pumping stations
- Water parks and human-made whitewater recreation parks
- Large public fountains or similar water features
- Preserved warships and water-related war memorials
- Swamps and salt marshes

Figure 10-1 The four water specialties. **A.** Ice rescue. **B.** Dive rescue/recovery. **C.** Surf rescue. **D.** Swiftwater rescue.

Applicable Standards

Emergency responders can find guidance on how to manage these types of incidents by referring to National Fire Protection Association (NFPA) Standard 1006, *Standard for Technical Rescuer Professional Qualifications*, and Standard 1670, *Standard on Operations and Training for Technical Search and Rescue Incidents*.

NFPA 1006 establishes the professional qualifications needed for responders to perform Level I (operations-level) and Level II (technician-level) rescue skills. These qualifications include preplanning potential water rescue sites, performing a site assessment, hazard recognition and control, access techniques, disentanglement techniques, victim assessment, packaging and transfer techniques, decontamination, and incident termination procedures.

NFPA 1670 establishes organizational requirements to operate safely and effectively at water rescue emergencies. This standard identifies the requirements for risk and hazard assessment, written procedures, safety protocols, personal protective equipment (PPE), establishment of operational response level, and resource considerations. NFPA 1670 also defines the authority having jurisdiction (AHJ) as having the responsibility for assessing its water rescue needs, hazards, training requirements, and operational and safety standards. Rescue responders must meet the general requirements in this standard regardless of the level of responder or the type of rescue. Additionally, at least one member of an organization must qualify as a competent person, which is defined in NFPA 1670 and NFPA 1006 as "One who is capable of identifying existing and predictable hazards in the surroundings or working conditions that are unsanitary, hazardous, or dangerous to employees, and who has the authorization to take prompt corrective measures to eliminate them."

Water rescue differs in many ways from other technical rescue types. Several different water rescue techniques are used with different hazard types and require different types of rescue equipment and training. Both NFPA 1006 and NFPA 1670 address different specialty tracks within water rescue, as well as general objectives common to all water rescue incidents. The 2008 edition of NFPA 1006 has added a surface water–rescue specialty track, where "surface water–rescue" is defined as a rescue taking place in water that is either still or moving slowly.

The four water specialties, as defined by NFPA 1670, are ice rescue, dive rescue/recovery, surf rescue, and swiftwater rescue Figure 10-1 ◀ . In reality, water rescue types sometimes overlap. For example, the ice rescue category includes rescues from water beneath the ice, which overlaps with the dive rescue category. Similarly, rescue from tidal creeks and rivers requires the use of skills related to both surf rescue and swiftwater rescue Figure 10-2 ▶ .

A limitation of both NFPA 1006 and NFPA 1670 is that these standards do not specifically address water rescue from common accident sites such as swimming pools. Each AHJ needs to develop its own plans, training, equipment, and procedures for water rescue problems not addressed by the NFPA.

OSHA's General Duty Clause establishes federal regulatory requirements for all job-related employee safety and health issues, which includes water rescue. OSHA does not have a specific standard for water safety and rescue.

Some state agencies have taken the step of defining all volunteer fire and rescue personnel as *unpaid employees*. This definition covers all personnel who are required to enter water rescue work sites as part of their volunteer duties. Volunteer fire departments and rescue squads are commonly regulated as *unpaid employers* despite commonly held beliefs to the contrary.

U.S. Coast Guard regulations include **personal floatation device (PFD)** standards and requirements, boating safety requirements, channel marking systems, and "rules of the road" for boat operations.

Rescue rope and hardware used for technical rope rescues from the water should be compliant with NFPA 1983. Water rescue rope is designed to float but typically lacks the strength associated with technical rescue rope. Water rescue rope is usually made from polypropylene or a similar artificial fiber that has low to medium stretch, but is buoyant Figure 10-3 ▶ . This rope works well as a throw bag rope and for tethering rescue swimmers, but is not strong enough to handle the forces involved in highlines or vertical rescue work.

Figure 10-2 Some incident locations—for example, tidal creeks—require multiple rescue techniques.

Figure 10-3 Water rescue rope floats but is not as strong as technical rescue rope.

Common Water Rescue Misperceptions

There are several misperceptions about water rescue. One is that divers who use self-contained underwater breathing apparatus (SCUBA) are the only people who can safely make water rescues. In reality, most water rescues are made at the surface and from water conditions where SCUBA equipment is either not needed or inappropriate and increases the chance of rescuer injury or death. In addition, diver response times are usually too long for an underwater victim to still be alive when the divers arrive on the scene. SCUBA gear can even be deadly in swiftwater rescues because SCUBA regulators typically free-flow in current and dive weights can entrap divers in undercut rocks or other submerged entrapment hazards.

Another misperception is that a lifeguard is capable of rescuing anyone from any water condition. In fact, standard lifeguard training is oriented to seeing the emergency happen in the controlled conditions of a public swimming pool. Lifeguarding rescue techniques appropriate for swimming pools often do not work in swiftwater, surf, flood, or ice conditions. The ability to swim in flat, calm, clear pool water does not translate into the ability to swim in rapids, waterfalls, low-head dams, surf, rip currents, murky water, or urban floodwaters.

The notion that SCBA and turnout gear can be used for surface or underwater rescue is another misperception. No SCBA manufacturer will certify a firefighting SCBA for underwater use. Most pressure-demand SCBA will free-flow when the second-stage regulator is submerged. While it is possible to survive a fall into calm water while wearing turnout gear and SCBA, these types of equipment are not designed for water rescue use. Fire fighters who fall into the water while fighting marina and boat fires typically require rescue by other fire fighters. Put simply, never attempt a water rescue while wearing turnout gear. (If wearing turnout gear in the water is a good thing, why don't Olympic swimmers wear turnout gear?)

Another misperception is that in the absence of a rescue boat, responders can use any boat around to make a rescue. That's a little like saying, "I don't have a neurosurgeon, so I'll just commandeer whatever doctor is around when I need brain surgery." A wide variety of boats can be found around typical waterways, but many of them are not appropriate for rescue under any conditions. Many others are not designed for the specific problems encountered in rescues from floods or ice, or for dive support. Commandeering boats often causes more problems than it solves, including an increase in liability for the incident commander (IC), boat-based rescuers, and the boat owner/operator who may participate in the rescue attempt.

Yet another misperception is that any technical rescue team can effect a swiftwater rescue. This idea is problematic for several reasons. First, technical rope rescue involves a single force vector—gravity. By comparison, swiftwater rescue always involves two or more force vectors—gravity and at least one horizontal current vector. Swiftwater can involve multiple current vectors and helical flows such as eddies and hydraulics. Second, technical rope rescue is based on gravity tensioning the rope. This is not the case when the rope is pulled in multiple directions or swirled by the current. Third, rope rescue teams are trained to

use redundant belay systems. Redundant belay systems in the water may create additional entanglement hazards rather than providing an additional safety margin.

Rescue Resources

Water rescue requires rescue teams to operate in, around, over, or under water. A water rescue is a high-risk event that requires specially trained responders using specialized equipment and teamwork. All responders who will perform non-entry water rescue work near the water or operate as part of a rescue boat crew must be trained to at least the NFPA 1670 operations level. Responders should also be trained to meet the technician-level job performance requirements as outlined in NFPA 1006 prior to performing rescues that require water entry.

The specialized equipment required for water rescue includes several general categories of equipment. These categories can overlap or they can include specific requirements for one of the water subspecialties—ice, dive, surf, or swiftwater rescue.

Water rescue equipment consists of PPE, basic water rescue equipment, technical rope rescue equipment, swim aids, boats and related equipment, search equipment, emergency medical services (EMS) equipment, and decontamination/clean-up equipment. Underwater rescue/recovery equipment includes SCUBA systems that allow divers to breath underwater.

Personal Floatation Devices

PPE includes a minimum of a PFD for any responder who works from a boat, enters the water, works from bridges or structures over the water, or works on shore within 25 feet of the water. Additional PPE can include specialized helmets, eye protection, wetsuits or drysuits, special water rescue footwear and gloves, whistles, and rescue knives. For night work, personal lighting is needed. Personal lighting systems may include helmet lights, hand lights, PFD-mounted strobe lights and flashlights, and chemical light sticks.

PFDs used for water rescue should meet the U.S. Coast Guard's PFD standards **Figure 10-4 ▶**. Type III work vests are suitable for most water rescue work. Type V rescue PFDs with a quick-release tether are a good choice for swiftwater rescue and for marine rescues outside the surf zone. Type V inflatable PFDs are a good choice for surf rescue work where there is a chance that the rescuer may be swept away. Type I PFDs tend to be too bulky for rescue work, while Type II PFDs generally do not stay secured to the responder's body during the physical activity required for rescue work. Type III PFDs are intended for survival in calm water only.

Basic water rescue equipment includes throwable rescue rope and devices. The most commonly used types are throw bags and Type IV (throwable) PFDs.

Throw bags typically contain 50 to 75 feet of buoyant rope. The rope is attached to the bag and then stuffed into the bag. The rescue is made by opening the mouth of the bag, holding the rope end, and throwing the bag over the victim so the rope lands within the victim's grasp. Throw bags are useful for virtually any

A.

B.

C.

D.

E.

Figure 10-4 U.S. Coast Guard–approved personal floatation devices. **A.** Type I. **B.** Type II. **C.** Type III. **D.** Type IV. **E.** Type V.

type of water rescue. They can be thrown from shore, from boats, or from human-made structures such as piers.

While the Coast Guard recognizes three different Type IV PFDs, the one most suited for water rescue is the **ring buoy**. Ring buoys, as their name describes, are essentially floating rings. They are typically constructed from closed-cell foam, are covered with a layer of waterproof fabric, and have a grab rope installed around the outside edge. Ring buoys may be used alone to provide floatation to a conscious victim; alternatively, they may be attached to a rope to provide both floatation and a method of victim retrieval.

A third throwable rescue device is the rescue disk **Figure 10-5 ▸**. This device resembles a large Frisbee but contains a spool of light-duty cord. The rescuer throws the rescue disk to the victim while holding the end of the cord. A well-trained rescuer can throw a rescue disk accurately almost twice as far as he or she can accurately deploy a throw bag.

Rope Rescue Equipment

The technical rope rescue equipment used for water rescue is essentially the same equipment used for vertical rope rescue

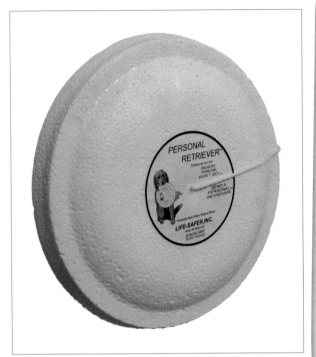

Figure 10-5 The rescue disk is a throwable rescue device that can be thrown much farther than a throw bag.

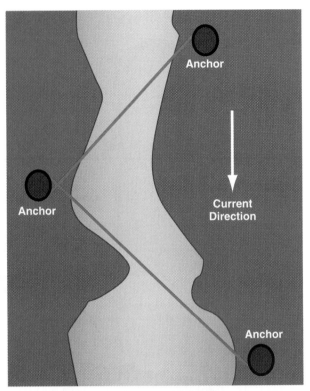

A.

B.

Figure 10-6 Tension diagonal. **A.** A tension diagonal system. **B.** A tension diagonal in use during a one-on-one rescue.

from cliffs or structures. NFPA 1670 specifies that technicians (or operations-level personnel who are supervised by technicians) are required to build and operate technical rope rescue systems for water rescue. Typical technical rope systems used in water rescue include tension diagonals, boat V-lowers and X-lowers, and boat telfer systems.

A **tension diagonal** is a rope stretched at an angle of at least 45 degrees across a stream. The rope is attached to anchor points at both ends, and tension is applied with a mechanical advantage rope system. Once the rope is tensioned, rescuers can attach a prussic or webbing loop to the tension diagonal with a carabiner. They can then hold the loop and use the current to assist in sliding the diagonal line from one stream bank to the other **Figure 10-6 ▶**. A small inflatable boat or raft can be ferried across a stream using the same system.

A **V-lower** uses two ropes attached to a boat or raft's upstream end, where the rescuers control one rope from each river bank **Figure 10-7 ▶**. The boat can be moved up and downstream using these control ropes. The boat can also be moved laterally by simultaneously applying tension to one control rope and slack to the other. An **X-lower** starts with the V-lower setup, then uses two additional control ropes attached to the downstream end of the boat. The X-lower allows for both upstream and downstream control in situations where the current swirls or forks in multiple directions.

A boat **telfer system** uses a highline stretched at 90 degrees across a stream **Figure 10-8 ▶**. A **movable control point (MCP)** system, also known as a *trolley*, is attached to the highline, and a multi-anchor plate or rigging ring is attached to the MCP. A tag line is attached to each side of the trolley to allow the MCP

to be pulled from side to side. An additional pulley is attached to the MCP to enable a boat control line to be run from shore to the upstream end of the boat. The boat control line moves the boat upstream or downstream to avoid obstacles and to be placed near the victim.

In water rescue, technical rope rescue systems are used predominately for swiftwater rescue, but they have applications in ice rescue and tidal marine rescue as well.

Figure 10-7   V-lower used to control a boat.

Figure 10-9   Rescue boards, fins, and can buoys.

Figure 10-8   Raft telfer system with a movable control point.

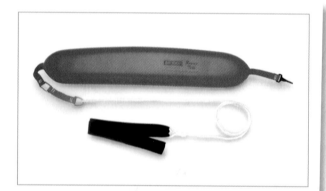

Swim Aids

Swim aids are items designed to assist rescuers and victims with floatation, vision, breathing, or propulsion. Surf/marine rescue and surface rescue from open water bodies such as lakes and ponds can require rescuers to wear a dive mask, fins, and snorkel. The fins increase the rescuer's power and control and reduce fatigue, the mask improves visibility beneath the surface, and the snorkel allows the rescuer to breathe while his or her face is submerged.

Boogie boards, which were originally used for recreational surfing, can be used in virtually any type of water, including swiftwater. Riverboards are a larger, two-person version of a boogie board. Rescuers can use boogie boards or riverboards to get to locations not accessible by boat, and can use them for rescue when access to river hydraulics is required. Special short fins designed for boogie boarding increase the rescue swimmer's propulsion while reducing the entrapment potential associated with using standard dive fins in whitewater or other confined waters Figure 10-9 ▶ .

Additional swim aids include rescue tubes and can buoys Figure 10-10 ▶ . Rescue tubes are standard lifeguard equipment at any public pool. They can be thrown to victims near shore, can be used as reach devices or arm extenders for contact rescues,

Figure 10-10   Rescue tube and can buoy.

and can be used to tow a victim. Can buoys, which were made famous by the television show *Baywatch*, are a hard plastic torpedo with a tow rope. They are used similarly to rescue tubes, but create an impact hazard to the victim if they are thrown and hit the victim.

Boats and Boat Equipment

Rescue boats can range in size from small inflatable rafts to sea-going rescue craft such as Coast Guard cutters. Many variables must be taken into account when selecting and using rescue boats. Agencies considering the use of one or more rescue boats should factor in the type(s) of water involved, the boat's size, its hull type, the propulsion method, and required boat and crew functions. All boats used for water rescue should have at least one method for retrieving unconscious victims from near the water level. Moving an unconscious victim even 18 to 24 inches vertically out of the water can be an almost impossible task. If a rescue boat will be purchased, the agency should ensure that it makes victim and rescuer retrieval as easy and efficient as possible. Hull doors, swimmer recovery platforms, parbuckling systems, and victim hoist davits are all very useful features for recovering victims and rescue swimmers. A dive door, dive platform, or dive ladder is essential if the boat will be used to support dive operations.

Two of the key variables in boat selection—the hull type and the propulsion system—are discussed next.

Hull Types

In motorized craft, there are three general hull types: inflatable/partially inflatable hulls, planing hulls, and displacement hulls Figure 10-11 ▶ .

Inflatable hulls are made from tubes of rubber or artificial materials that are sealed to keep the air contained. These hulls typically include two or more chambers, which helps the craft to maintain floatation if one chamber is punctured. Each chamber is inflated through an intake valve with a manual or powered air pump. Inflatable and partially inflatable hulls are commonly used as tenders for larger vessels, for dive support, for surf rescue, and for swiftwater and flood rescue if the water is relatively flat.

Planing hulls have flat bottoms and are intended for shallow-water use. The "john" boat is a typical planing hull, but most rescue planing hulls are larger and displace more water. Planing hulls are effective in shallow tidal waters, in swamps and bayous, and in flat flood water, providing that the engine provides adequate power.

Displacement hulls usually have a round or V-shaped hull bottom. This hull type is appropriate for deep-water ocean, lake, or other open water rescue. The hull cross-section allows for the boat to "bite" into the water and to track well in wave, tide, and wind cross-currents.

Human Propulsion

Boats are propelled either by motors or by human power. Human-powered boats include inflatable rafts and specialized raft derivatives such as the Shredder and Oceanid RDC. These boats are propelled either by a guide rowing an oar rig or by a team using paddles and captained by a guide. Paddle raft guides

A.

B.

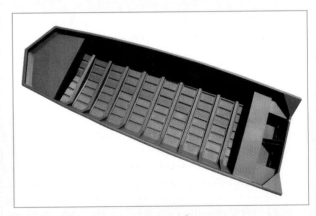

C.

Figure 10-11 Hull types. **A.** Inflatable hull. **B.** Planing hull. **C.** Displacement hull.

are usually positioned in the stern and use rudder strokes to turn the raft.

Most human-powered rescue boats fall into one of two hull types: nonbailing and self-bailing. Nonbailing rafts are based on World War II–era amphibious assault boats. Any water that enters these boats must be bailed out manually. An improvement on nonbailing rafts is the self-bailing raft. Self-bailers have a positively buoyant floor laced into the inflatable side tubes. Buoyant floors either are inflatable or are filled with a buoyant material such as closed-cell foam. This construction allows water that enters the boat to simply drain out through the drain holes in the floor.

Two additional categories of self-bailing rafts are distinguished: slow-bailing floors and fast-bailing floors. Slow-bailing floors use the water's weight to increase the boat's stability in whitewater containing large waves, souse holes, and hydraulics (discussed in detail later in this chapter). Fast-bailing floors seek to keep the craft light by quickly draining the water out of the boat.

Motorized Propulsion

Four categories of motorized propulsion are used in boats: outboard motors, inboard motors, stern drives, and specialized propulsion.

Outboard motors attach to the boat's transom. They are steered by hand or from a steering wheel linked to the motors by a wire or hydraulic steering system.

Inboard motors use automotive engines mounted inside the boat's hull and drive propellers with shafts that penetrate the hull. These systems are steered by a steering wheel linked to a rudder system separate from the propulsion system.

Stern drives combine both inboard and outboard features. Boats using such propulsion systems have one or more motors mounted inside the hull. Each of these motors powers a steerable external transmission and propeller unit. Small boats used for rescue should have permanently attached propeller guards; these propeller guards greatly reduce the chance of injuring victims or rescue swimmers operating near the boat.

Specialized propulsion systems are classified into three subcategories: jet pumps, airboats, and hovercrafts **Figure 10-12 ▶**. Jet pumps can be substituted for the propellers on outboard or stern drives and are the primary means of propulsion for personal watercraft. Jet drives add to victim safety by eliminating the risk of propeller injury to victims and rescue swimmers. They also require more horsepower than equivalent propeller engines and are vulnerable to being shut down by aerated water or debris clogging the engine intakes.

Airboats eliminate the need for propellers or jet drives by using aircraft-type engines and propellers to move the boat. They are typically based on large john boat hulls or similar planing hulls. The motor moves the boat by pushing air with an aircraft propeller mounted in a safety cage in the boat's stern. The operator steers the craft by moving rudders mounted behind the propeller. Airboats can operate in extremely shallow water and are seldom disabled by debris in the water or underwater obstacles. Disadvantages of airboats include their lack of a reverse gear, difficulty in operating in confined areas, noise, their tendency to project contaminated water onto other boats and rescuers, and their vulnerability to shipping water in large river or ocean waves. Airboats are most suitable for operating in tidal marshes, swamps, bayous, and other similar flat coastal waters.

Hovercraft operate on a pressurized cushion of air created by propeller systems that drive air through the boat's superstructure and under the craft. Sidewalls keep the pressurized air under the boat. A second propeller drives the hovercraft forward in much the same manner as airboat propulsion. Small hovercraft designed for rescue can operate both in the water and on land, increasing the possible number of launch and recovery points.

A.

B.

C.

Figure 10-12 Boats with specialized propulsion. **A.** Jet pump. **B.** Airboat. **C.** Hovercraft.

Hovercraft limitations include limited deck space for victims, the lack of a reverse gear, and vulnerability to sinking if the sidewalls are damaged by collision with objects such as rocks, piers, and other vessels.

Motorized Hull Types

Planing hulls are rigid, flat-bottomed hulls designed for use in shallow, protected waters. They typically have a square or sharply angled junction between the hull's sides and bottom.

Planing hulls work best in situations where the water is too deep for displacement hulls, or where the ability to maneuver smoothly across conflicting currents such as <u>eddy</u> (horizontal current reversal) lines is important. Boats with this design require less horsepower to skim across the water's surface than do boats with similar-size displacement hulls.

Displacement hulls typically have a rounded junction between the hull's sides and bottom. They can also have a keel, a V-shaped cross-section, or both. Displacement hulls are designed for open waters where the wind, wave action, tides, and other currents may simultaneously push the boat in three or more directions. The keel or V-hull carve the water and allow the <u>coxswain</u> (the responder steering the boat) to achieve better boat control.

Other Boat Considerations

If the boat will support dive operations, it must have plenty of deck space for divers, dive tenders, and gear. A dive platform or dive ladder is mandatory for recovering the divers. Likewise, if the boat will be used for surface rescue, a hull door, dive platform, and/or a davit and winch system will be very useful for recovering rescue swimmers and victims immobilized on spineboards or in rescue basket litters.

If the boat will be used for EMS response, ensure that adequate space is available to treat and transport at least one immobilized victim. Space for at least one EMT to provide life-saving care is a primary consideration for rescue boats.

Search Equipment

Search equipment includes two general categories: navigation equipment and victim location equipment.

Navigation equipment for water rescue includes topographical maps and **global positioning systems (GPS)** for land-based searches involving bodies of water Figure 10-13 ▶. The National Association for Search and Rescue's (NASAR) Fundamentals of Search and Rescue (FUNSAR) course teaches the basics of land-based searches. FUNSAR principles work very well in most searches from the shore, but are not designed for responses involving searches from the water. When searching from the water, GPS systems, radar systems, sonar systems, marine navigation charts, and navigational aids such as channel markers and navigation buoys are useful in helping responders determine where they are and in providing visual reference points for search coordination.

Victim-locating equipment can be subdivided into two types: equipment for locating lost rescuers and equipment for locating the original victim. GPS systems are one of the best ways to keep rescuers from becoming lost. They are typically not waterproof, however, and their batteries often die at inopportune times. For this reason, all water rescuers should carry at least one flashlight, one strobe light, and a few chemical light sticks at all times. Standard boating distress flare guns or hand-held flares are a good way for victims or lost rescuers to help themselves be found. Boating smoke flares or smoke grenades are the daytime equivalent of flares.

One of the best ways to locate victims in the water is to use a thermal imaging camera (TIC), like the ones typically used

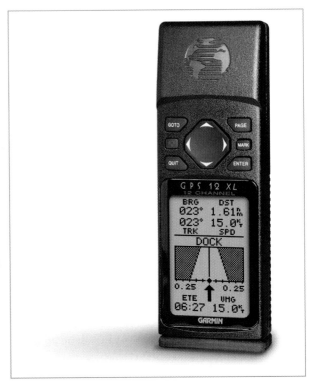

Figure 10-13 A hand-held GPS device can help responders during land-based searches.

by fire departments Figure 10-14 ▼. TICs can be especially effective at night, in fog, in heavy rain, and in other low-visibility conditions. To be effective, some part of the victim's body must be at or above the water's surface. That body part must be at least a few degrees warmer than the water for a TIC to discern the temperature difference. These cameras may not work in all water search situations, however: Specifically, TICs cannot see

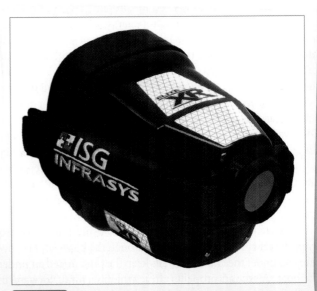

Figure 10-14 Thermal imaging cameras can help find victims in reduced-visibility situations.

below the water's surface, and they will not detect a victim or an object whose temperature is the same or very close to the water temperature.

If the victim is not quickly visible with the naked eye or a TIC, responders must attempt to establish the **point last seen (PLS)** as quickly and accurately as possible. This task can prove very difficult if there are no reliable witnesses at the scene. If bystanders are present, you can usually locate the general PLS area through a process known as triangulation. In triangulation, two or more witnesses stand as close as possible to the position from which they last saw the victim. Mark these locations with traffic cones, barricade tape, chemlights, spray paint, or other easily visible marking device. A rescuer then stands at each of these points and draws imaginary lines to where the victim was last seen. The point at which these imaginary lines intersect is the approximate PLS.

Emergency Medical Services Resources

Water rescuers need to have EMS support available. A minimum of one advanced life support (ALS) unit should be assigned to all water rescues. The rescue team should have basic life support (BLS) capability for any unit operating in an area not accessible by the ALS unit. This is especially true for rescues in remote wilderness areas and when operating boats. ALS intervention is ineffective or may even be impossible for victims in the water or in boats in extreme environments such as in surf or whitewater rapids. In contrast, larger boats used for marine or open water rescue may have the capability to function as ambulances with full BLS or even ALS capability.

In general, water rescue procedures call for "*Rescue first, then medical.*" This principle is based on the understanding that it is best to extricate the victim rapidly without regard for spinal **packaging** (applying a cervical collar, placing the victim onto a spinal immobilization device, and securing the victim) or other medical procedures if the victim is exposed to a life-threatening scene hazard. In general, it is impossible to safely and effectively package a victim in surf, rip currents, or swiftwater, or on thin ice. In less extreme conditions, it may be advisable to apply a spinal package before removing the victim from the water. Examples of such scenarios include dive injuries in shallow swimming pools, shallow eddy packaging after the victim is removed from swiftwater, and spinal packaging from boat accidents in deep open waters with minimal wave action.

Water rescue teams need the capability to provide CPR, spinal immobilization, rewarming, rehydration, bandaging, and minor environmental emergency treatment whenever their assistance is engaged in water rescue operations. Care of the airway, breathing, and circulation (ABCs) is the most immediate concern, but it may not be possible to perform ABC care in the water, particularly in big waves, swiftwater, cold water, or storms.

EMS resources are required so that rescue teams can hand off all victims for treatment and transport to medical facilities once they are removed from the water. Water rescue victims may be suffering from hypoxia from near-drowning, hazardous materials exposure, traumatic injuries, or medical emergencies. ALS capabilities are required to effectively treat water rescue victims once they are removed from the water and moved to

shore. In fact, the national resource typing system calls for Type I Swiftwater Rescue Teams to have ALS capability.

Access and egress equipment that includes rope rescue equipment will be required for some water rescues. Once rescuers are able to remove victims from the water, various types of victim packaging and transfer devices such as long spineboards, Stokes basket litters, or other similar devices may be required Figure 10-15 ▾ .

Spinal packaging devices that are negatively or neutrally buoyant—such as Stokes basket litters, metal scoop stretchers, and some spineboards—should have additional floatation kits applied prior to in-water use. An exception to this rule would be

Figure 10-15 After rescuers have removed victims from the water, they have several victim packaging and transfer options.

when immobilizing a victim in the shallow part of a swimming pool or calm river eddy where all rescuers have secure footing and there are adequate resources to ensure that the victim's airway remains above the water's surface at all times.

Floating spineboards may be used as floatation or swim aids as well as being useful for in-water spinal immobilization. These devices do not have enough positive buoyancy to float a victim without additional floatation, however, and they may not keep the patient's face out of the water. The primary purpose of a floating spineboard is to keep the board from sinking prior to being attached to the victim.

A new device, called the HydroSpine™, has been introduced as a means of in-water spinal immobilization. This device, which was developed at Virginia Polytechnic Institute, is designed to immobilize a victim who has a spine injury and is in the water while providing positive buoyancy and keeping the victim's face out of the water. This device is currently considered experimental by the water rescue community.

Water rescuers may be able to apply a spinal package in deep, calm water if the packaging system floats, if there are enough personnel to support the victim while other rescuers package the victim, and if a method of moving the victim into a boat or to the shore without risking the victim's or rescuer's safety is available. Shallow-water spinal packaging may be done in the shallow end of a swimming pool, on a flat shoreline or beach, or in a calm eddy at the river's edge. If the water's edge is located below grade, the situation may require rescuers to use rope rescue skills such as establishing rope anchors, operating lowering systems, using mechanical advantage raising systems, and implementing safety belay lines.

EMS resources are also required at a water rescue so that medically trained personnel are available to assist with the rehabilitation of rescue team personnel. Water rescues can be both time-consuming and physically taxing for rescue personnel, especially if rescue team members must make multiple entries into the water to remove victims. EMS personnel can assist with rescuers' rehabilitation process by monitoring their vital signs, ensuring adequate hydration, and administering medical care to team members as required.

Rescuers must be aware that EMS responders may lack the proper PPE and knowledge of water rescue techniques and must not be placed in situations in which they are untrained or unprepared to perform. They must be able to work safely without jeopardizing themselves or other rescuers.

Hazardous Materials Resources

In some situations, hazardous materials resources are needed to assist with water quality monitoring, research potential hazardous materials that may be present, and determine the proper level of protective gear required. Hazardous materials teams may also need to assist with or manage decontamination of victims, entry rescue team members, boats and boat crews, and rescue equipment.

If crews must operate in contaminated floodwaters, the local and/or state health department can assist with assessment of biohazard threats, immunizations for crew members who may enter the water, and treatment of water-acquired diseases.

Decontamination

Decontamination for all water rescues involves washing, soaking, rinsing, and drying all PPE, ropes, equipment, and boats to the manufacturer's specifications. This consideration is particularly important if the rescue is effected in saltwater or contaminated floodwaters. Gear that has been exposed to serious chemical contamination should be disposed of and replaced.

The following are a few basic water rescue decontamination rules:

- Decontaminate boats from top to bottom, then from bow to stern. Allow the bilge pump to return the decontamination runoff to the contamination source (the water) whenever practical.
- Most chemical contamination can be reduced to safe levels by using detergent soap and lots of potable water. Saltwater is corrosive, so items that come in contact with it should be decontaminated with large quantities of fresh water.
- Most waterborne biological contaminants are bacterial in nature. A 10% solution of hypochlorite solution (household bleach) in water will destroy most bacteria. Hypochlorite is corrosive, so rinsing the item with copious amounts of fresh water is required to complete this decontamination process.
- PPE and gear that cannot be decontaminated at the water's edge should be bagged in waterproof containers and transported to a suitable decontamination facility. Do not cross-contaminate clean gear by transporting it mixed with contaminated gear.
- It is permissible to return contaminants to their source. Floodwaters and other water rescue decontamination runoff do not have to be controlled, but rather may be returned directly to the water source from which they originated.
- Electronic equipment is virtually impossible to decontaminate. Radios, GPS units, depth finders, and other electronics should be mounted in water-resistant locations or carried in waterproof dry bags or dry boxes. It is a fairly simple task to decontaminate the exterior of most dry bags and dry boxes.

Additional Resources

Communications

Competent communications equipment is necessary to maintain full-time effective communications between the rescue team and incident command, between two or more rescue units or groups working the same incident, between divers and surface personnel, and between the rescue team and the rescue team support personnel operating from shore. When possible, rescuers should operate within the line of sight of incident command, the rescue group officer, and one another. In situations such as massive flooding or a dive operation, however, this goal may be impossible to achieve.

Verbal communications may occur on a face-to-face basis when the situation is limited and when ambient noise levels are

A.

B.

C.

Figure 10-16 General hand signals. **A.** I'm OK. **B.** Medical help needed. **C.** Help me, quick.

low. When working near surf, rapids, waterfalls, or motorized boats, it may be impossible to communicate verbally or by radio. In this situation, it may be necessary to resort to hand signals **Figure 10-16 ▲**.

When working underwater, rescue/recovery divers should use hard-wired or wireless underwater communication systems. This type of system will keep the divers in constant voice contact with their surface support personnel and backup diver(s). Wireless systems operate over only a set distance between the diver and the surface receiver. Likewise, hard-wired systems are limited by the length of the communications wire and, in addition, are vulnerable to entanglement on submerged obstacles.

Underwater communication systems come in wireless and hard-wired models. The wireless systems are the best choice in most situations, but are the most expensive options. They are only as dependable as their battery life permits. In addition, wireless communication systems may have reduced effectiveness over long distances, in certain water conditions, or when the diver is inside an underwater cave or sunken ship. By comparison, hard-wired systems are more dependable and tend to be less expensive, but they limit the diver's movement to the length of the communications cable and the cable can create an entanglement risk for the diver. Both wireless and hard-wired systems are typically monitored by a tender on the surface.

If voice communication between the diver and the surface fails, the diver's tether rope can be used as a communication method. The number of tugs indicates the action that crew members on the diver's tender should take. The OATH mnemonic describes how this manual system typically works **Table 10-1 ▶**. This system is not perfect, however. For example, if the dive tether becomes snagged or entangled, the surface crew may pull the diver into an underwater obstacle and complicate the entrapment. If personnel on the tender do not pay close attention, it is easy to miss a tug and misinterpret the intended message from the diver.

Every rescue boat should be provided with at least a hand-held fire/rescue radio. Some manufacturers actually make submersible emergency radios, although these devices are very expensive compared to standard fire/rescue hand-held units. For most motorized rescue boats, a covered/enclosed steering

Table 10-1 OATH System

	Meaning	Number of tugs
O	Okay	One
A	Advance the line	Two
T	Take up the line	Three
H	Help	Four (may also be continuous tugs)

position not only protects the officer and the coxswain from the weather, but also provides mounting positions for fire/rescue mobile radios, mobile telephones, marine radios, and roof-mounted antennae. Marine radios are essential for any rescue boat operating in ocean, marine, or tidal waters. Marine radios allow the rescue boat to contact shipping vessels, pleasure craft, and the U.S. Coast Guard.

◼ Special Logistical Support Resources

Water rescue events can present themselves in a variety of ways. Some will be more difficult than others. Utility companies, public works, salvage and recovery companies, or other community resources may all be required to assist the responders when water rescue situations are unique and unusual.

▌Incident Management Requirements

The strategic objective when operating at a water rescue incident is to effectively control and manage the event in such a way as to evaluate the scene and identify potential victims and their locations, minimize hazards to operating personnel and victims, effectively search the water rescue area, effectively rescue and remove victims, and minimize further injury to victims during search, rescue, and removal operations.

Incident Management System (IMS) positions are staffed as determined by the scope of the event. In most cases, personnel are required to fill the positions of incident commander (IC), water rescue group supervisor, boat group supervisor, and safety officer. If technical rope access and/or egress are needed, a rigging group supervisor is also required. If the responding agency has no operations- or technician-level capability for a specific type of water rescue incident, a planning section chief should be assigned.

In larger incidents, an operations branch, multiple rescue groups, and additional incident management components may be required. Several safety officers may be required for large incidents such as urban floods. In catastrophic incidents, area command, multiple Emergency Operations Center (EOC) activations, state and federal assistance, and long-term reentry and recovery plans may be needed.

The water rescue group supervisor is responsible for implementation of tactical rescue decisions to support the strategy established by the IC. Tactical benchmarks in such scenarios may include victim location, boat launch and recovery site identification, hazard mitigation, entry team readiness, rapid intervention capabilities, and emergency medical care for the victim.

Incident management personnel are responsible for strategic and tactical management of the water rescue emergency response. Among their responsibilities is the development of an incident action plan (IAP), which identifies the overall control objectives for the emergency. Because a water rescue emergency may involve unique conditions, including the need to identify local conditions, incident management personnel often need to seek input from knowledgeable people from the private sector in these scenarios. These people may include commercial raft guides, kayakers, dive shop personnel, fishing and ecotourism company personnel, harbormasters and marina facility maintenance personnel, and dam operators or other utility company personnel. Other government agencies such as the U.S. Coast Guard, the state's Department of Natural Resources, and health and environmental department personnel may also be very helpful.

Response Planning

Responders must know how to initiate the emergency response system to ensure that appropriate resources are deployed to the event and that operational guidelines are followed. Many response organizations perform needs assessments within their response area to determine their vulnerability to water rescue events. As part of this assessment, the organization should consider the types of water rescue its personnel may need to perform and the type of hazards that will be encountered, and then determine its emergency response resource needs. The emergency response system should include written procedures to request agency resources as well as resources that will respond due to mutual aid agreements; in the latter case, the organization should document the minimum training and knowledge levels of responders who will be answering the mutual aid call, contracts with private-sector companies, and memoranda of agreement with regional, state, or federal assets. Private-sector agreements may be established to acquire resources that cannot be supplied by the response agency, such as swiftwater rescue teams, rescue boats, hazardous materials assets, and dive resources including extra personnel, extra breathing air, and boat and personnel transport assets.

As part of the planning process, the organization should determine the capabilities of its own agency personnel. This capability assessment should include identification of the level to which performance agency personnel are trained, such as the operations or technician level for water rescue. As part of the emergency response system, the locations, capabilities, and response times of mutual aid, regional, state, and private-sector water rescue teams should be known, and procedures should be established to request these assets. Dispatch procedures should be in place to ensure that adequate and proper resources are deployed when a water rescue event occurs, thereby enabling effective search and rescue operations to begin as quickly as possible. The NIMS resource typing system has established minimum training and capabilities for some water rescue team types.

Preplan and Training

Preplanning a response to aquatic emergencies includes more than just risk and hazard assessments, specialized teams, and mutual aid departments: It also includes training with the initial response agencies, special rescue teams, public-sector resources, and the primary AHJ dispatch agency. Training must instruct the primary dispatch agency and their personnel about dispatch procedures, but must also include required scribe information that will later be used by the rescue teams, including terminology not often used during the everyday emergency dispatch. Dispatch personnel should also be exposed to the on-scene environment during training so that they develop a better understanding of the what, why, and how regarding the use of this information, terminology, and technology. Many emergency response agencies are now assigning a response team dispatcher to respond to the incident, serve as a scribe, and act as an aide to the IC.

Hazards and Hazard Assessment

Each water rescue has its own specific problems and hazards. To protect personnel from being injured or killed, responders must be well trained to recognize and understand the hazards they encounter.

Scene Assessment

Additional assessment information can be gathered from bystanders or other first responders on the scene. Aerial maps, topographical maps, marine charts, and agency-specific pre-emergency plans can all offer valuable data for planning the rescue. The information gathered during the scene assessment should answer the following questions:

- Which type of water or ice rescue is this?
- How many victims are involved?
- Is the victim visible?
- If the victim is not visible, what is the PLS?
- Can we communicate with the victim?
- Is the victim trapped or entangled?
- Is the victim under water or beneath the ice?
- Has the victim been swept away?

- Is the victim injured?
- Is any type of contamination present and, if so, does that level of contamination pose a hazard to rescuers?
- Are the responders properly trained and prepared to perform this rescue?
- Do the responders have the necessary resources to perform this rescue?

Natural Hazards

Natural hazards are those hazards that are not created by humans. The primary water hazard is drowning. Other naturally occurring hazards comprise current, natural obstructions such as trees or rocks in the water, rapids, waterfalls, bad weather, and darkness. Natural hazards may also include attacks by bears, poisonous snakes, and other predators that use the water to hunt or hydrate.

Marine hazards can include attacks by predatory fish such as sharks and barracuda, poisoning from jellyfish stings or stingray punctures, and lacerations from oyster beds, barnacles, and other marine life. An emerging threat to divers is the lionfish, which has multiple fins tipped with poisonous spines. Lionfish are native only in tropical waters, but may be found in colder port waters after they are discharged from seagoing ships pumping ballast water overboard.

Natural hazards also include poor footing on wet or slimy surfaces and the potential for being sucked down by quicksand or **pluff mud** (soft, semi-liquid material) in swampy areas or tidal marshes. In addition, divers may be lost in underwater caves, entangled in roots or vines, pinned against obstacles by currents, or sucked through underwater drainage or irrigation tunnels.

Human-made Hazards

Human-made hazards are those hazards physically created by humans. These water rescue hazards comprise primarily physical, chemical, and biological hazards. Obstacles such as bridges, piers, dams, shipping and sunken vessels, and loose fishing nets can prove lethal to unwary water rescuers. Chemical and biological hazards exist in waterways near chemical plants and sewer facilities and create a "toxic stew" in large, slow floods such as the 2005 Hurricane Katrina flood in New Orleans.

Physical Hazards

Responders must recognize the potential for physical hazards that may be present at a water rescue. The type of physical hazards that responders face will depend on the specific water area and the primary function of the physical hazards encountered. In particular, physical hazards may include dams, piers, flood control channels, culverts, storm drains, low-water crossings, flooded structures, bridges, piers, and sunken boats or ships.

NFPA 1670 requires that rescuers know how to mitigate these hazards to the greatest extent possible. Nevertheless, there may not be a way to eliminate them in a way that is both practical and safe. This is especially true when personnel are not trained and equipped for in-water rescue work. Mitigating large hazards such as a low-head dam, a pipeline crossing, or a sunken boat often requires major engineering work—which, in turn,

Figure 10-17 Flooded residences can present an electrical hazard.

typically requires a substantial infusion of public funds, public support, and political capital. That level of mitigation is beyond the scope of this text.

Electrical hazards can include flooded residences, businesses, and electrical production and transmission facilities such as power plants, substations, and transformers **Figure 10-17**.

Divers are especially vulnerable to underwater hazards such as net entanglements and risk being hit by boats or shipping vessels when they are surfacing. When divers are required to enter abandoned wells, flooded mines, or sunken boats or ships, the potential for collapse or other entrapment is also a concern. In some locations, divers perform inspections and routine maintenance in elevated water tanks and water towers; water tower dive rescues, while rare, are essentially an elevated/dive/flooded confined-space incident characterized by hazards from all three of these categories.

NFPA Type-Specific Water Rescue Hazards
Dive Rescue Hazards

Dive rescue hazards are classified into two types, based on the primary environment in which they occur: underwater and barometric hazards for technician-level divers, and surface hazards for operations- or technician-level surface support personnel such as dive tenders and boat coxswains.

Divers use SCUBA equipment so that they can breathe underwater. Public safety divers should be at least average swimmers, be in good physical condition, and have basic dive, open water dive, and full-face piece dive qualifications. Certification as a current diver and training in diving medical emergencies are desirable as well. Dive tenders ideally will also be divers, but may be specially trained operations-level personnel. Coxswains should be qualified at the technician level as boat driver/operators.

Hazards to divers include equipment failure, running out of air, entanglement or entrapment underwater, barotrauma, and medical emergencies that occur while they are underwater. Any equipment failure or diver error that causes the diver to run out of air is an immediate life threat. For this reason, all public safety divers should carry a second source of air as a backup.

Figure 10-18 A "diver down" buoy should be anchored to the dive site.

Divers are also vulnerable to being run over by boats when they are surfacing or operating at shallow depths. A "diver down" buoy should be anchored at the dive site during all dive operations Figure 10-18 ▲.

Entanglement or entrapment in any human-made or natural obstacle can vary in seriousness from an irritation to an immediate life threat to the diver.

Barotrauma results from the increasing water pressure as the diver's depth increases. It may be as simple as a "squeeze" or as serious as a fatal case of the bends. A **squeeze** is an episode of painful pressure in the sinuses and ear canals. It occurs when a descending diver is unable to equalize pressure in parts of the head that contain air. In contrast, the **bends** is a barotrauma emergency that occurs when a diver ascends from depth too quickly without adequate decompression. Inadequate decompression can lead to extreme pain as gas bubbles become trapped in body tissues, as well as incapacitation, heart attack, stroke, and other medical emergencies. A very serious case of the bends may be fatal. Divers who have the bends must be rapidly transported to a decompression chamber, preferably one operated by a medical facility.

The Divers Alert Network (DAN) is a nationwide network designed to provide pre-emergency information and dive emergency treatment information for divers. All participants in a water rescue should be aware that any medical emergency—such as

seizure or trauma—that occurs on the surface can also occur underwater. The seriousness of underwater medical emergencies is magnified, however, owing to the increased chances of diver incapacitation and drowning.

Hazards to surface personnel include falling or being pulled overboard, storms or other bad weather, collision with other boats, fires, and mechanical failure.

Ice Rescue Hazards

The primary hazard present in ice rescues is the risk of breaking through the ice into the water below and suffering exposure to extremely cold air, ice, and water temperatures. Ice fractures caused by the initial victims breaking through the ice surface can easily enlarge, and drop rescuers into the water. Both victims and rescuers who fall through the ice will quickly become hypothermic unless they are dressed in waterproof, insulated ice suits. In addition, people who fall through the ice can be swept away by underwater currents.

Vehicles and even aircraft may also plunge through the ice surface. Consider the crash of an Air Florida plane through the ice on the Potomac River in January 1982: At least one victim of that crash was seen struggling under the ice surface but drowned before he could be rescued. Vehicles that crash through the ice can result in several or even hundreds of simultaneous ice and under-ice victims requiring rescue or recovery.

The ice surface can present additional hazards that are not readily apparent to responders untrained in assessing those hazards. Ice thinned by thaw/freeze cycles, changing under-ice water levels, protruding plants or structures, changes in weather, or warming of the water may all produce weaknesses that make an ice surface very likely to break under the weight of even a single rescuer.

Surf Rescue Hazards

Surf rescue is defined as any rescue from saltwater or the brackish water in the tidal zone. It does not include just rescues made on beaches with breaking waves, but also encompasses incidents that occur in offshore areas, harbors, marinas, bays, coves, sounds, tidal rivers and creeks, salt marshes, and coastal mangrove swamps.

Surf rescue hazards come in several types. Notably, surf breaking directly on a beach, sand bar, or reef presents a danger to victims and rescuers alike. Rescuers wearing buoyant PFDs in a surf zone will typically float to the top of the wave, then be dashed the full wave height onto the sand or reef. For that reason, it is recommended that surf lifeguards and surf zone rescuers wear inflatable Type V PFDs in the deflated mode Figure 10-19 ▶. The responder can then dive under breaking waves rather than being lifted to the top of the wave. If the responder needs additional floatation outside the surf zone, he or she can then inflate the PFD and take advantage of the positive buoyancy thus created.

Surf rescuers in marine environments that lack breaking waves should wear Type III work PFDs or swiftwater Type V tether vests. Swiftwater Type V PFDs are particularly useful in the marine swiftwater environment found in tidal creeks and rivers, and for deploying tethered rescue swimmers from boats.

Marine rescue boat crews may wear the water-activated inflatable Type V PFD. This equipment allows for freedom of movement while a person is operating a boat or working as a

Figure 10-19 It is recommended that surf lifeguards and surf zone rescuers wear inflatable Type V PFDs in the deflated mode.

deckhand, but automatically inflates if the responder falls or is pulled into the water.

Human-made surf rescue hazards include ships and boats, drifting objects, piers, lost fishing nets, fishing lines, and pipe and power line crossings. All of these items can entangle, injure, or even kill the unwary responder.

Natural surf rescue hazards include tidal currents, rip currents, and marine animal life. Tidal creeks can have currents that flow as swiftly as 15 to 20 mph, particularly when the tide is receding. Swimmers, no matter how well conditioned, will be swept away by a current of that strength. If caught in a tidal current, the individual should swim diagonally across the current to the closest place of safety. The current pushing against the swimmer's up-current side will help move the swimmer laterally to the shore.

Rip currents form when surface currents sweep toward the shore, tumble down at the surf break, and then become subsurface currents as the water retreats. If the subsurface current is funneled between submerged sand bars or other obstructions, the outgoing current has an increased velocity compared to the incoming surface current. Many people drown each year while attempting to swim against rip currents. If a person is swept away in a rip current, the self-rescue technique is the same as the tidal creek method: Simply swim laterally across the current until you get outside the rip, then swim in toward the shore.

Marine animals that pose hazards to victims and rescuers include predators such as barracuda, sharks, and some whales.

In addition, shellfish such as oysters and barnacles have shells that can cause nasty, deep lacerations to unprotected parts of the responder's body. Other dangerous marine animals include poisonous snakes in some coastal marshes and swamps.

Responders can protect themselves from surf/marine hazards through pre-emergency planning, maintaining situational awareness, wearing appropriate PPE, and, if necessary, discontinuing rescue operations in areas where the hazards are not manageable.

Swiftwater Rescue Hazards

Swiftwater, including floods, is probably the most dangerous of all water rescue environments. All too often, planners, agency heads, and responders underestimate the power of moving water. It is also easy to underestimate the increasing frequency and severity of flash flooding, large river flooding, and coastal and inland flooding caused by hurricanes.

In many areas, the "asphalt county" phenomenon exacerbates urban flooding. Asphalt counties result from increasing percentages of streets, parking lots, and other nonporous surfaces in urban and suburban areas, which means that less porous earth is available to absorb rainwater. During rainstorms, any water that cannot be absorbed will run off to low areas. If drainage from the low areas cannot keep up with the rate of runoff, a flash flood will result **Figure 10-20 ▾**. A few communities are requiring that porous paving tiles be used instead of solid surfaces to help reduce stormwater runoff. This mitigation attempt can reduce—but will not prevent—urban flooding. Indeed, in 2006, two fire fighters and a paramedic drowned while attempting swiftwater or flood rescues in three separate incidents; all of these fatalities occurred in urban or suburban areas.

The primary swiftwater hazard is the current's force. That force can easily sweep a person off his or her feet and into whatever downstream hazards are present. Even a few inches of swiftwater can float a vehicle on its tires and move it into more dangerous and less accessible spots.

Figure 10-20 Nonporous pavement forces water to low areas, increasing the risk of flash flooding.

Swiftwater is the result of the weight of water falling downhill. This force can push large amounts of water across flat surfaces or even upward during flooding or tidal action. Swiftwater forces can be quantified by calculating the water speed and multiplying that speed by the approximate square footage of the river channel. For example, if a river channel is approximately 10 feet deep and 50 feet wide, the river has a capacity of approximately 500 square feet at the measurement point. If the current speed is 10 feet per second (FPS), the volume of water passing the measurement point is 5000 cubic feet per second. Water weighs approximately 62.5 pounds per cubic foot. Thus the weight of the water passing the measurement point in our example is 312,500 pounds per second. In other words, more than 156 tons of water passes the measurement point every second. It is no wonder that moving water can erode riverbanks, change a river's course, and destroy human-made structures and objects with ease.

Another way to quantify a river's force is to measure that force when it is applied to a person or an object. Suppose a person has a cross-section of 400 square inches from waist to ankles. A 10 FPS current initially results in hundreds of pounds of pressure being applied to the person standing in the water. That pressure will actually increase if the person continues to stand in the water because the water continues to pile up on the upstream side. Even a very strong person will not be able to withstand hundreds or thousands of pounds of pressure for long.

Swiftwater currents are categorized into several different flow types, including laminar, horizontal helical, vertical helical, confused, and tidal.

Laminar flow is a term used to describe differences in the current speed and force in different parts of a stream or channel. This type of flow is strongest at the river's surface, in the center of the channel **Figure 10-21 ▶**. Laminar flow at the riverbanks and river bottom is the slowest; at these locations, it interacts with the turbulent flow caused by friction. Laminar flow may be visualized in terms of an "inverted rainbow" model, which shows the fast water in the center of the channel followed by increasingly slower current layers as you move toward the river bottom or shoreline.

Horizontal helical flow is water moving in a circular motion, parallel with the river's surface. This flow is most commonly found in eddies at the river's edge or immediately downstream of obstacles that penetrate the water's surface **Figure 10-22 ▶**. Water at an eddy boundary tends to flow in the opposite direction as the laminar flow. The area created by these opposing currents is called an eddy line. Eddy lines tend to spin boats trying to cross them, and they can be difficult for swimmers to cross because the swimmer must fight through two different perpendicular currents.

Vertical helical flows are typically one of two types. The first type of flow, a **hydraulic**, occurs perpendicular to the laminar flow. Hydraulics are formed at the bottom of water flowing over a relatively short, shallow drop such as a **low-head dam** (a small dam that creates a hydraulic) or a smooth, wide river ledge **Figure 10-23 ▶**. This results in a circular current that flows upstream toward the obstruction. People and buoyant objects tend to be pulled upstream, then into the downstream current that flows under the hydraulic. If the person is lucky, he or she

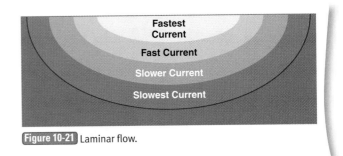
Figure 10-21 Laminar flow.

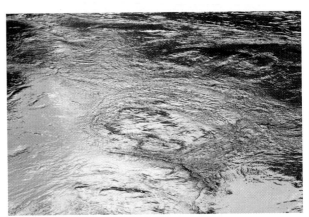
Figure 10-22 Horizontal helical flow is most commonly found in eddies.

Figure 10-23 Hydraulic vertical helical flows can be formed at low-head dams.

will be flushed downstream of the hydraulic. Usually the person floats to the surface and is then pulled back to the obstacle; this cycle may repeat itself until the person either drowns or succumbs to hypothermia.

The second type of vertical helical flow is an eddy fence. Eddy fences are a vertical exaggeration of a normal horizontal eddy line. Usually a high-water phenomenon, they are most commonly found on the outside of river bends or where streams are forced through human-made channels. An eddy fence is essentially a hydraulic facing parallel to the laminar flow—and it can be just as lethal as a dam hydraulic.

Confused flow is a term used to describe the chaotic water where two or more current flows come together **Figure 10-24 ▼**. The combined currents can cause surges, exploding and collapsing wave cycles, whirlpools, and very aerated whitewater. Confused flows are very difficult to navigate, whether you are swimming or boating. These flows are typically found in Class IV, V, and VI whitewater and floodwater areas. They can cause crosscurrents, underwater helical flows, and exploding waves.

Tidal flows are normal features in coastal rivers and creeks and may involve either saltwater or brackish water. (Brackish water is a mix of salt and fresh water.) Two primary features characterize tidal water.

First, the current flows in the reverse direction with the tide cycle, which can cause problems when referencing river orientation. Some swiftwater rescuers recommend using a "freshwater river right and river left" orientation scheme. This means that the river right and river left orientations will switch banks when the tide reverses. Given this fact, it is probably better to keep a single river right and river left reference pattern regardless of the tidal state, particularly in marked channels. For instance, the standard U.S. Coast Guard channel marker system employs red channel markers on river left and green channel markers on river right, regardless of the tidal state **Figure 10-25 ▶**. The mnemonic "Red–Right–Returning" helps boat operators remember that the red channel markers will be to the boat's right when returning to port from the open ocean.

Figure 10-25 The U.S. Coast Guard channel marker system places red channel markers on river left and green channel markers on river right.

The second characteristic of tidal flows—and one that can cause problems—is that current force and channel depth may vary widely within a few minutes, depending on local geography and the moon phase. Outgoing tides can easily reach normal speeds of 10 to 15 miles per hour, particularly when the tide is outgoing. Spring tides—which occur three days prior to, the day of, and three days after the full moon—cause extreme low and high tides and faster tidal current speeds.

Ebb tides can also expose obstacles that do not present a problem when the water level is higher. As a consequence, boats and swimmers may face hazards that are not a problem when the water level is higher. Those hazards can include sunken vessels, mangrove roots, and the razor-sharp edges of oyster shell beds and barnacle encrustations.

Swiftwater has other dynamic characteristics in addition to the basic flow types. These properties are usually associated with specific river features known as pillows, waves, and holes. Determining the presence of these features and understanding how to navigate them safely constitute a skill set known as "**reading the river**."

Pillows usually form on the upstream side of obstructions that penetrate the water's surface. Pillows are caused by water literally piling up against the surface of the object. Continued water flow increases both the water level and the force that pushes on the obstruction. Pillows typically form against large rocks, bridge piers, vehicles in the water, and people who are wading or pinned against an obstruction.

Figure 10-24 In a confused flow, two or more current flows come together.

Voices of Experience

It had been raining constantly for three days due to the remnants of a tropical storm. Area creeks and rivers were swollen past flood stage. Any street that crossed a creek or stream was flooded.

The information supplied by the public safety communications center was minimal: "There is a boy trapped in a car in the floodwater in a creek." Two fire department engine companies, a rescue company, a paramedic ambulance, an EMS supervisor, and a battalion chief were dispatched. Upon their arrival at the scene, they found a teenage boy in a car, trapped in floodwaters. The teen had attempted to drive a late-model sedan down a street covered with approximately two feet of rapidly flowing floodwater where the road crossed the creek. The water rose until it stalled the car's motor and trapped the victim. The creek was several feet out of its banks, full of debris, and rapidly rising; the car was stalled about 50 feet from the nearest shore. The victim was conscious, wet, and cold, and was trying to exit the car on the downstream side. The first-arriving units established command and immediately requested a tower ladder company, boats, and a swiftwater rescue team.

❝The creek was several feet out of its banks, full of debris, and rapidly rising; the car was stalled about 50 feet from the nearest shore.❞

Additional scene hazards included rising, debris-laden water; a 50-foot-wide, 4-foot-tall waterfall at the downstream edge of the road; barbed wire fences and tree strainers downstream of the scene; and an electrically energized, flooded pump house immediately downstream of the street.

Fire and EMS personnel specially trained in swiftwater rescue initiated standard water rescue operations, which included establishing command and assigning an upstream lookout, downstream safety personnel on both sides of the creek, a rescue group officer, and a safety officer. All personnel dressed in PFDs, water rescue helmets, and gloves. Later-arriving responders dressed in full swiftwater rescue PPE including wetsuits or drysuits.

The incident commander cleared the area where the street entered the water and had the tower ladder set up at the water's edge. A backup team suited up and was prepared to conduct a Wade water entry rescue if the Reach effort with the aerial ladder was not successful. A mechanical rope rescue system was rigged to the tip of the aerial behind the tower ladder bucket; a fire fighter was suspended from the rope system. The tower ladder was extended to the downstream side of the car. The rescuer was lowered into the eddy, where he contacted the victim and determined that the victim was conscious, alert, and uninjured, but cold. A webbing harness and PFD were placed on the victim. The rescuer and victim were then hauled up to the tower bucket, the ladder was retracted, and the rescuer and victim returned to shore without further incident.

The victim was assessed by EMS personnel, but did not require ambulance transport. The tower ladder operation was quickly terminated and the truck moved away from the water that was literally lapping at the rear wheels of the apparatus as the rescue was completed.

Battalion Chief/Paramedic Ben Waller
Hilton Head Island Fire & Rescue, South Carolina
Fire and Technical Rescue Instructor, South Carolina Fire Academy
Swiftwater Rescue Instructor, TARS and SRSG

River waves differ from ocean waves in an important way. In ocean surf, the water stays relatively immobile, but the waves come and go. In contrast, river waves are called *standing waves* because the water moves through the wave but the wave stays in one place. A series of standing waves is a common feature of many whitewater rapids. A series of three or more standing waves is called a *wave train* or *haystacks*.

Holes are the depressions in the water's surface found immediately downstream of a standing wave. They are also called *souse holes* because any boater navigating a hole will be "soused" with water.

Other swiftwater hazards consist of objects in the water. Moving objects are described as loads Figure 10-26 ▾. Positively buoyant objects floating on top of the water are known as **top loads**. These objects can be trees, propane tanks, or anything else that floats. Neutrally buoyant objects moving between the water's surface and the bottom of the channel are known as **suspended loads**. Suspended loads may include silt, waterlogged wood, and anything else moving in the current. Heavy objects with negative buoyancy will sink to the bottom. If the current is strong enough to move these objects, they are known as **bottom loads**. Bottom loads may include vehicles, boulders, slabs of asphalt or concrete, or other heavy objects.

Objects in the water that do not move are described as obstacles. The most common (and one of the most dangerous) objects in swiftwater are **strainers** Figure 10-27 ▸. Strainers are porous objects that let water through but trap people, boats, gear, and debris. For example, downed trees, fences, and log jams may serve as strainers. Strainers should be avoided whenever possible. A risky "last resort" strainer crossing technique is taught in technician-level swiftwater rescue courses.

Undercut rocks are another swiftwater hazard. These rocks are either angled upstream or have had the lower part of the rock worn down by the current. Either way, undercut rocks are a foot entrapment hazard if small and a body entrapment hazard if large.

Boulder sieves are another common swiftwater hazard. These areas form when rocks or boulders tumble together in a rapid, resulting in one or more funnel-like vertical passages. In boulder sieves, the vertical water flow can pull a person underwater and trap him or her against the narrowing sides of the sieve. The water's force will continue to hold the person underwater until he or she drowns.

The presence and locations of known strainers, boulder sieves, and other obstructions at a swiftwater rescue scene should be communicated to all responders.

A.

B.

C.

Figure 10-26 **A.** Top load. **B.** Suspended load. **C.** Bottom load.

Figure 10-27 Tree strainer.

Flood Hazards

An analysis of floods shows that they occur in four phases:

1. The pre-flood phase: The period between floods.
2. The emergency phase: The water rises rapidly and runs swiftly.
3. The stabilization phase: The water rises slowly, reaches its crest, and then recedes.
4. The recovery phase: The water recedes to normal levels and the flooded area works to return to normal conditions.

The primary hazard of the pre-flood phase is complacency. It can be difficult to visualize 10 or 20 feet of water in your house, the businesses on Main Street, or in City Hall when it hasn't been raining for three days. Even so, the potential for flooding exists everywhere—even in the desert. The pre-flood phase is the best time to educate public officials and the general public, to train and equip water rescue teams, and to develop flood response and recovery plans.

The primary hazard of the emergency phase is the rapidly rising water and its threat to human life, livestock, and pets. The general public and many responders tend to underestimate how rapidly flood waters rise and how short their escape window is. During the emergency phase, there can be tremendous demand for water rescue assistance. If your community and your agency are not prepared, people will die for lack of rescue. Unprepared but well-intentioned rescuers may die as well. Additional emergency phase hazards include the extremely turbid water, which hides underwater hazards, and the unusual loads picked up by floodwaters when they venture outside the riverbanks.

The primary hazards of the stabilization phase are chemical and biological contamination. Floodwaters near their crest often contain hazardous chemicals from chemical manufacturing and storage, hardware stores, lawn and garden stores, small businesses, and homes. Propane tanks; gasoline and diesel from underground storage; raw sewage; runoff from chicken, hog, and cattle farms; and almost any other imaginable contamination will pollute the water and anything in it.

The primary hazards of the recovery phase are the hazardous residues from the previous phase, physical hazards such as electrocution when utilities are restored, and illness from drinking contaminated water, eating unsafe food, and respiratory problems secondary to the mold that will grow on every previously flooded surface.

Basic Hazard Mitigation

The IC must determine which actions can be taken to eliminate or modify the hazards that are present. In most cases, water hazards such as current or floodwater contaminants cannot be eliminated or even reduced. An exception is that managers of dam-controlled rivers may be able to reduce the current flows by slowing the dam release. This action is possible only when dam integrity is not threatened by floodwaters, the reservoir is not full, and water diversion into a penstock, irrigation system, or floodwater diversion channel is an option.

Water-related risks can also be reduced by preventing untrained or inadequately trained personnel from attempting a rescue. Hazard modification may include curtailing access to the area and denying entry to non-rescue personnel.

In major floods and hurricanes, hazard reduction may be an impossible task during the response phase. These disasters are often so large and complicated that response resources may not be adequate, even when the response occurs on a national scale.

Non-entry Rescue Procedures

Awareness-level responders, and in some cases even responders at the operations level, should not consider making an entry into a water rescue to initiate a rescue. At this level of training, the responder is not adequately prepared or equipped to safely and effectively perform an entry-type rescue. Water rescuers should be trained and equipped to at least the operations level prior to attempting even shore-based water rescues.

Awareness-level responders may be able to perform a non-entry rescue in very specific circumstances. If such a responder arrives on the scene of a water rescue emergency and finds a victim has been able to self-rescue and is ashore or at the edge of a calm body of water, he or she may be able to assist the victim safely. Responders should avoid wearing turnout gear and fire helmets and should wear at least a PFD prior to assisting a victim near the water. This requirement acknowledges the possibility that the responder might slip, trip, fall, or be pulled into the water.

Another way in which an awareness-level responder may assist in a water rescue is to throw a buoyant object to a victim. A Type IV PFD, boat cushion, plastic picnic cooler, inner tube, or anything else that floats can provide floatation assistance to the victim until other rescuers arrive on the scene. Under no circumstances should the awareness-level responder enter the water to attempt a rescue.

A water rescuer responder who has been trained to the awareness level may be able to assist in a water rescue in a limited way if he or she is trained to the operations or technician level in rope rescue. In some swiftwater, marine swiftwater, and ice rescue situations, technical rope systems may be required. Non-water responders with rope rescue qualifications can assist more technically advanced personnel by rigging and operating a variety of technical rope rescue systems under the supervision of a rope rescue technician. The rope rescue technician should also be qualified in the specific type of water rescue for which the rope system is used.

The Modified RETHROG Concept

For many years, lifeguards have used the RETHROG mnemonic (reach, throw, row, go/tow) to describe a variety of rescue techniques. These techniques are ranked in order from the quickest and least dangerous to the most dangerous and time-consuming. Water rescuers have modified the original concept to add helicopter techniques and the "no go" decision, as follows:

- *Reach* techniques may be accomplished with any "arm extender," such as pike poles, tree branches, or other objects extended to victims near shore.
- *Throw* techniques include throw bags, Type IV PFDs, or other buoyant objects that may be thrown to a victim who is out of reach but within the rescuer's ability to make accurate throws.
- *Row* techniques are any rescue using a boat, regardless of its propulsion method.
- *Go/Tow* techniques include any wading or swimming technique that involves water entry rescue.
- *Helo* techniques involve the use of helicopters to rescue victims from the water.
- *No Go* is a conscious decision not to attempt a rescue because the victim is dead or the situation is hopeless.

This mnemonic has been successfully used as a guideline for water rescuers for many years. It is not a perfect model for any conceivable situation, and it does not imply that rescuers must try each technique in ranked order. For example, if the victim is 200 feet from shore, reach and throw techniques are not effective, so riskier techniques are required. This mnemonic was designed as a decision tree for responders new to swiftwater rescue, not as an absolute for all swiftwater and flood rescue situations.

A revision of the original mnemonic is based on separating wading rescues from swimming rescues and on recognizing that helicopter rescue techniques may actually be safer than boat-based rescues if done correctly. The latest mnemonic is recommended by Jim Segerstrom, co-founder of the Swiftwater Rescue Technician program. Segerstrom's system can be modified with the additions of the Techno and No Go options.

Talk
Reach
Throw/Wade
Techno
Helo
Row
Go
Tow
No Go

This new mnemonic is based on changes in technology, rescue helicopter availability, the many documented problems with two-cycle outboard engines used with many rescue boats, and new analysis of the relative risks inherent in the many different rescue techniques. The sequence can be modified depending on local resources and rescue team qualifications and capabilities.

Talk is the first option. You may be able to talk to the victim and direct him or her to swim to safety, or to remain in a place of safety rather than entering the water until additional resources arrive. This technique may be useful for awareness-level rescuers.

Reach techniques now include basic arm extenders such as pike poles or extremely advanced techniques such as using a fire department aerial ladder as a reach device.

Wade has been split from Go, and moved up in the sequence for rescues in shallow water in which it is safe to wade and when Throw techniques may not be available or effective. An example would be a wading rescue of a small child who is unconscious but floating in a swimming pool or farm pond.

Techno is the application of technical rope systems, primarily for swiftwater. These systems are complicated and take more time than simpler techniques to implement, but provide better safety margins in many situations.

Helo has been placed before Row in the revised sequence, but is pertinent only in areas where helicopters have rescue capability, their crews have been trained in water rescue, and the helicopter has a response time short enough to start the rescue in a timely manner. Helicopters with water rescue-trained crews are a terrific asset. The U.S. Coast Guard's thousands of helicopter rescues during Hurricane Katrina amply demonstrated the impressive effectiveness of this technique in water rescues.

Row techniques require experienced and well-trained paddle crews if human-powered craft are used. If motorized craft are chosen, then rescuers must remember that most small rescue boats are powered by unreliable two-cycle outboard motors. These motors often fail in the middle of a rescue attempt, slowing the rescue and putting the crew at higher risk. As Jim Segerstrom comments, "You shouldn't go out in a rescue boat unless you are prepared and equipped to come back without the boat."

Go techniques involving rescue swimmers are among the most dangerous to the rescuer, as are *Tow* rescues requiring a rescue swimmer to assist a panicky, disabled, or unconscious victim.

No Go is always an option when the level of risk is not reasonable given the conditions, the rescuer's level of training, or other available resources. This option (not entering the water) should be used by awareness-level rescuers.

Be aware that the modified RETHROG technique is designed for rescuers who are trained to the operations and technician levels because it requires work either in the water or nearby.

Victim Care Considerations

The removal of the victim from the water to a safe environment is the number one victim care priority. It is important to consider the mechanisms of injury associated with falls, entrapment, or other trauma and to package the victim accordingly. The victim may be packaged for removal from the water based on your evaluation of any injuries. This evaluation must be conducted in accordance with the conditions, however. Put simply, if the victim is in danger of drowning, responders must remove the victim from the water as rapidly as possible.

Victim removal techniques used by water-rescue teams include **parbuckling** (using lengths of rope, webbing, or netting to roll a victim from the water's surface into a small boat), rescue davits, hull doors, or dive platforms to assist victims from the water into a boat. Water rescue victims can be packaged on buoyant plastic spineboards or in rescue basket litters with floatation kits for their removal from deep, calm water. By contrast, spinal packaging in surf, current, and especially from whitewater is not appropriate. Do not apply a spinal package in these water types until the victim is removed from the surf zone or into a calm eddy. If it is necessary to move a packaged victim over swift water by boat or highline system, the victim should *not* be strapped to the spineboard or rescue basket litter Figure 10-28 ▶ . If the victim enters the water because the inflatable boat flips or pins the victim, the basket litter drags in the water, or another problem occurs, the victim will likely drown: He or she cannot swim while strapped to the spinal package.

Figure 10-28 Swiftwater spinal packaging.

If the rescue will involve moving the victim over water, the victim should be placed in a PFD whenever possible. For this reason, water rescue teams and boat crews typically need to carry victim PFDs in a variety of sizes, from infant models to PFDs that can accommodate obese individuals. When evacuating structures during urban flooding, large quantities of victim PFDs may be required.

Wrap-Up

Chief Concepts

- Rescuers may encounter emergency situations where victims are located in pools, rivers, lakes, tidal creeks and rivers, whitewater, floodwater, oceans, bays, ports, and a variety of human-made structures including water and sewer treatment plants, flood control channels, and low-head dam hydraulics.
- Potential water rescue sites will be found in every community. They come in various shapes and sizes, and rescuers must have a clear understanding of how to recognize what a water rescue is.
- Water rescue emergencies require a variety of specialized resources, including specialized rescue equipment and specially trained rescue teams. These unique situations require the involvement of emergency medical services personnel, hazardous materials specialists, and other specialized logistical support.
- Planning is critical for response organizations to operate effectively at water rescue emergencies. Needs and vulnerability assessments should identify threats in the area and the organization's capabilities to deal with a water rescue emergency. Response personnel must have a clear understanding of written operational procedures, resource availability and capability, and the procedure for initiating the proper response to manage water rescue events.
- Incident management requirements include trained personnel to implement the command, safety, planning, and operational functions as the situation dictates. Specialized rescue resources are assigned to manage the tactical- and task-level requirements of the water rescue event.
- Water rescue hazards can include drowning, being swept away, being swept under ice, divers running out of air, entrapment, trauma, various physical hazards, wildlife hazards, and waters contaminated with hazardous materials.
- Hazard-mitigation efforts are required before responders make entry into a water rescue, with the understanding that some water hazards such as strong currents are not controllable.
- Awareness-level responders may be able to attempt non-entry retrieval of victims from water's-edge rescue situations; however, at no time should they attempt to make a water rescue entry.

Hot Terms

Bends A type of barotrauma emergency that occurs when a diver ascends from depth too quickly without adequate decompression.

Bottom load A heavy object with negative buoyancy that sinks to the bottom of a water channel.

Coxswain The responder responsible for steering a powered boat.

Eddy A horizontal current reversal along the riverbank and/or immediately downstream of an obstacle that penetrates the water's surface.

Global positioning system (GPS) An electronic device that pinpoints a location on the earth's surface by triangulation with two or more special satellites designed for this purpose.

Horizontal helical flow Water moving in a circular motion, parallel with the river's surface.

Hydraulic (water flow) A form of vertical helical flow (current reversal) that occurs in rivers and streams and that has trapped and drowned many people, including water rescuers.

Laminar flow The differences in current speed and force in a stream or channel.

Low-head dam A small dam that creates a hydraulic immediately downstream.

Movable control point (MCP) A system of pulleys that serves as the heart of a tethered boat highline system; also known as a trolley.

Packaging The process of preparing a victim for movement as a unit, often accomplished with a long spine board or similar device.

Parbuckling A water rescue technique that uses lengths of rope, webbing, or netting to roll a victim from the water's surface into a small boat.

Personal floatation device (PFD) Formerly known as a life jacket.

Pluff mud Soft, semi-liquid material that resembles quicksand. It is found in many tidal creeks and salt marshes.

Point last seen (PLS) The last location where someone actually saw a missing person.

"Reading the river" A set of learned skills that involves the ability to assess a variety of river features and determine hazards, safe places to navigate, and places of relative safety in swiftwater or floodwater.

Ring buoy A Type IV personal floatation device that resembles a floating ring.

Riverboard A larger, two-person version of a boogie board.

Squeeze A barotrauma episode characterized by painful pressure in the sinuses and ear canals.

Strainers Porous objects such as trees and fences that allow water to pass through but trap people, boats, equipment, and debris.

Suspended load A neutrally buoyant object moving between the surface and the bottom of a water channel.

Swim aids Items designed to assist rescuers and/or victims with floatation, vision, breathing, or propulsion.

Telfer system A highline system that uses technical rope rescue techniques to control and position a boat in strong river current.

Tension diagonal A lowline system stretched at an approximately 45-degree angle across a stream, anchored at both ends, and then tightened by means of a mechanical advantage pulley system.

Top load A positively buoyant object floating on top of water.

Vertical helical flow Water moving perpendicular to the laminar flow or vertical exaggeration of a typical eddy.

V-lower A swiftwater technique whereby a boat or a tethered swimmer is moved downstream by means of an upstream rope anchored on each side of the stream.

Water recovery A water emergency that meets the water rescue definition, plus the victim is known to be dead, has been trapped underwater for more than 90 minutes, or is trapped in water from which there is no reasonably safe way to accomplish a rescue.

Water rescue The act of accessing and removing a person who is endangered in a water environment, with due regard for any injuries and/or contamination that may be present.

X-lower A swiftwater technique whereby a boat or a tethered swimmer is moved downstream by means of an upstream rope and a downstream rope anchored on each side of the stream.

Rescue Responder *in Action*

A major hurricane has passed over the coast and is now over your city. It has been raining and has become increasingly windy for several hours. The water is rising fast in the local creeks and in the two rivers that bisect your city. It is two hours before sunset when your engine company is dispatched to investigate a report of rising water at an occupied apartment complex.

You arrive to find the apartment complex next to the river with 2 feet of rushing flood water covering the parking lot and entering the first floor of the three-story apartment buildings next to the river. At least 20 people are trapped. As you and your engine company officer get off the rig and start to walk up to the scene, you realize that your company has no water rescue equipment other than PFDs and helmets for the crew, two throw bag ropes, a ring buoy, and a few carabiners. The water is still rising quickly, and the weather report calls for the river to crest at midnight at a level 10 feet higher than the present water level. A young mother screams that her two older children and a baby are trapped with her, and that her neighbor is trapped in her electric wheelchair on the first floor of a nearby apartment.

1. After the engine officer establishes command, what should his first action be?
   A. Send the crew to perform a Wade rescue for the most endangered occupants.
   B. Request that the dive team respond.
   C. Drive the engine through the floodwater to effect multiple rescues.
   D. Request at least one swiftwater rescue team and flood-capable boats.

2. What is the next action that the members of the engine company should take at this incident?

   A. Secure access to the scene and don't let anyone enter the floodwater.

   B. Don full turnout gear so that fire fighters are readily identifiable.

   C. Borrow a neighbor's canoe and paddle it to rescue the victim in the wheelchair.

   D. Use a bullhorn to tell the mother to grab the baby and other children and wade toward the engine company's location.

3. To improve scene safety, the engine officer should:

   A. assign a fire fighter as an upstream lookout.

   B. have every fire fighter don a PFD and a water rescue helmet.

   C. assign at least one fire fighter to set up with a throw bag/rope downstream of the incident.

   D. All of the above.

4. When the swiftwater rescue team arrives, what should the incident commander's next action be?

   A. Direct the swiftwater rescue team to immediately attempt Go rescues.

   B. Appoint the swiftwater rescue team officer to supervise the incident's logistics.

   C. Meet with the swiftwater rescue team leader to determine the team's capabilities and the likely best course of action.

   D. Have the swiftwater rescue team officer assume command of the incident.

5. A fire fighter falls into the water and is being swept downstream. What is the safest way to rescue this fire fighter?

   A. Have another fire fighter swim out and make a Go rescue.

   B. Have several fire fighters perform a group Wade rescue.

   C. Have a fire fighter make a Reach rescue with a throw bag/rope.

   D. Wait for the Coast Guard to make a Helo rescue.

NFPA 1670 Standard

10.1 General Requirements.

Organizations operating at wilderness search and rescue incidents shall meet the requirements specified in Chapter 4. [p.235]

10.1.1* The AHJ, as part of its hazard identification and risk assessment (*see 4.4.2*), shall identify all locations in the jurisdiction that meet the definition of *wilderness*. [p.234]

10.2 Awareness Level.

10.2.1 Organizations operating at the awareness level at wilderness search and rescue incidents shall meet the requirements specified in Section 10.2. [p.234–246]

10.2.2 Members of organizations at the awareness level shall be permitted to assist in support functions on a wilderness search and rescue operation but shall not be deployed into the wilderness. [p.235]

10.2.3 Organizations operating at the awareness level at any wilderness search and rescue incident shall have the following capabilities: [p.234–246]

(1) Recognizing the need for a wilderness search and rescue-type response

(2)* Initiating the emergency response system for wilderness search and rescue

(3)* Initiating site control and scene management

(4)* Recognizing the general hazards associated with wilderness search and rescue incidents

(5) Recognizing the type of terrain involved in wilderness search and rescue incidents

(6)* Recognizing the limitations of conventional emergency response skills and equipment in various wilderness environments

(7)* Initiating the collection and recording of information necessary to assist operational personnel in a wilderness search and rescue

(8)* Identifying and isolating any reporting parties and witnesses

Additional NFPA Standards

NFPA 1001, *Standard for Fire Fighter Professional Qualifications*

NFPA 1006, *Standard for Technical Rescuer Professional Qualifications*

NFPA 1983, *Standard on Line Safety Rope and Equipment for Emergency Services*

Knowledge Objectives

After studying this chapter, you will be able to:

- Identify the need for wilderness search and rescue (SAR) operations.
- Describe various types of wilderness SAR emergencies.
- List general hazards associated with a wilderness SAR incident and understand how to reduce those hazards.
- Describe the resources needed to conduct wilderness SAR operations.
- Describe response planning and incident management requirements related to a wilderness SAR incident.
- Describe initial actions that will expedite operations and lay the groundwork for specialized rescue units.

Skill Objectives

There are no skill objectives for this chapter.

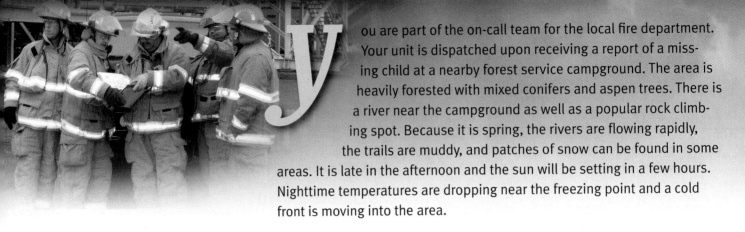

ou are part of the on-call team for the local fire department. Your unit is dispatched upon receiving a report of a missing child at a nearby forest service campground. The area is heavily forested with mixed conifers and aspen trees. There is a river near the campground as well as a popular rock climbing spot. Because it is spring, the rivers are flowing rapidly, the trails are muddy, and patches of snow can be found in some areas. It is late in the afternoon and the sun will be setting in a few hours. Nighttime temperatures are dropping near the freezing point and a cold front is moving into the area.

As the first crew jumps into the rescue truck, you throw in area maps and ensure that the rope rescue equipment is ready. Because you do not know the medical status of the child, you ensure EMS personnel are on standby. You arrive on scene to find frantic parents who tell you that they just looked away for a moment.

1. Are there any scene safety issues?
2. Which other details should be gathered about the missing child?
3. Which resources might be needed?
4. What are the most likely search areas to deploy a hasty team immediately?
5. Does the topography limit movement?

Introduction

Search and rescue (SAR) personnel respond to lost, overdue, injured, and stranded persons. These individuals may be part of the fire service, law enforcement, emergency medical services (EMS), government agencies, or even private organizations. In most cases, SAR personnel are volunteers. The public often does not understand which agency SAR falls under; however, many people expect fire fighters to take care of all emergencies from stranded cats to mass-casualty situations. SAR operations can happen at any time of the year and in any location—from urban areas to remote, backcountry wilderness areas.

As the population in the United States has steadily increased, so has the recreational use of our public lands. These public lands include national forests, national parks, Bureau of Land Management lands, tribal lands, and state lands, to name a few. Their topographic and vegetative characteristics vary widely, from low-lying deserts and grasslands to dense coniferous forests and jagged mountain peaks.

The definition of **wilderness** can also vary greatly from person to person. The Wilderness Act of 1964 defines "wilderness" as follows: "a wilderness, in contrast with those areas where man and his own works dominate the landscape, is hereby recognized as an area where the earth and its community of life are untrammeled by man, where man himself is a visitor who does not remain." The act continues with the definition of "wilderness": "an area of undeveloped Federal land retaining its primeval character and influence, without permanent improvements or human habitation, which is protected and managed so as to preserve its natural conditions." Perhaps more relevant to rescuers is the NFPA 1670 definition of "wilderness" as "an uncultivated, uninhabited, and natural area usually, but not necessarily, far from human civilization and trappings."

If you are dealing with an SAR event in a federally designated wilderness, check with the U.S. Department of Agriculture's (USDA) Forest Service office to understand the restrictions within the area (e.g., no motorized equipment). The authority having jurisdiction (AHJ) determines which areas in a jurisdiction are to be considered wilderness.

Some SAR events happen in or near the **wildland–urban interface (WUI)**, which is the area where human-made improvements intermix with wildland. As more people build homes and pursue leisure activities in the WUI, the chance of fire personnel being involved in incidents in these areas also increases. In addition, non-firefighting SAR personnel are often mobilized during wildland fires to help with evacuation of citizens in the WUI.

Just as a structure fire requires a fast and efficient response, so too does an SAR operation. Timeliness may mean the difference between life and death in such an event, whether you are searching for a lost person or rescuing an injured or stranded person. The most complex SAR events may involve locating the missing person and then rescuing him or her from a precarious situation. In SAR, a mnemonic commonly used to summarize this process is LAST **Table 11-1 ▶**.

Table 11-1 LAST

L	Locate
A	Access
S	Stabilize
T	Transport

First, the victim(s) must be located. This task may be easy if dispatch calls in a known location. At other times, it can involve an extended search that can last for several days.

Sometimes you know where the victim is—you may even be able to see the person—but reaching him or her is a challenge. Specialized equipment such as ice axes, crampons, and ropes may be needed. As a rescue responder, you must determine the most effective, timely, and safe way to reach the victim. Be sure to size up the situation for potential hazards before heading in. Remember that whatever happened to the victim can happen to a rescuer as well.

Once you reach the victim, you must assess his or her condition. Identify any life threats to the victim and treat any conditions based on the level of your medical training. Stabilization can involve bandaging wounds, splinting, treating hypothermia, or immobilizing the victim's spine if damage is suspected.

Finally, you need to extricate the victim from his or her predicament and transport the victim to safety. You must properly package the victim so that you can transport the individual in a safe and efficient manner while allowing for any medical care that may be needed. Transportation options can range from the victim walking out under his or her own power to use of a helicopter. The type of transportation that is appropriate for the victim is determined by his or her medical condition, the condition and training of your personnel, the topography, the weather conditions, available resources, and other factors.

At search and rescue incidents, the awareness-level responder's role is to support rescuers with a higher level of training. Responders at the awareness level are not to be deployed into the wilderness.

Applicable Standards

Wilderness search and rescue responders can find guidance by referring to two National Fire Protection Association (NFPA) standards: NFPA 1006, *Standard for Technical Rescuer Professional Qualifications*; and NFPA 1670, *Standard on Operations and Training for Technical Search and Rescue Incidents*.

NFPA 1006 establishes the job performance requirements (JPRs) needed for rescuers to perform Level I (operations-level) and Level II (technician-level) rescue skills. These skills include those related to communications, navigation, survival, and victim management. NFPA 1006 does not include requirements for awareness-level responders but can offer these responders insight into wilderness search and rescue operations.

NFPA 1670 establishes organizational requirements to operate safely and effectively at wilderness search and rescue events. This standard identifies the requirements for training, docu-

mentation, written procedures, hazard identification and risk assessment, response planning, equipment, and safety protocols. The general requirements in Chapter 4 of NFPA 1670 must be met by any organization operating at a wilderness search and rescue incident.

Types of Wilderness Emergencies and Response

The initial SAR size-up is classified into one of four categories: known location, known condition; known location, unknown condition; unknown location, known condition; and unknown location, unknown condition **Figure 11-1 ▾**. The most common type of SAR dispatch involves missing or overdue people. In these cases, you do not know the location or condition of the individuals. Are they safe at a nearby tavern or have they fallen into a crevasse far from their planned travel route?

Recognizing the need for a wilderness search and rescue response starts with knowing your local area, your department's capabilities, and the location of the SAR operation. The topography, vegetation, and access of the landscape will be the first consideration when determining whether a wilderness search and rescue operation is needed. For example, a larger number of personnel will be required to search areas with thick vegetative cover quickly and efficiently. The next step is identifying the most likely area to search for the missing or stranded person. If the search area is located several miles from the nearest road, a wilderness search and rescue operation will most likely be required.

Search

Some wilderness SAR events focus primarily on the search part of the operation. Conducting a search may or may not require specialized teams. If the missing person is a hiker who planned to stay on marked trails, you can send out a **hasty team** to cover the route quickly. A hasty team is a quick response team that searches the areas where the missing person is most likely to be.

The degree and urgency of your response will depend to some extent on the situation. Consider the following two missing person scenarios:

- A backpacker is 6 hours overdue. The missing person is a 35-year-old male with years of backcountry experience. He has extra food, water, and clothing. He has a history of being somewhat unreliable, but usually comes out of

Known Location Known Condition	Unknown Location Known Condition
Known Location Unknown Condition	Unknown Location Unknown Condition

Figure 11-1 Initial SAR size-up grid.

potentially dangerous situations with no problems. The sun is setting, but the weather is mild with moderate temperatures.

- A child has wandered away from a campsite. The 5-year-old girl has not been to this area before and rarely goes out of view of her parents. She does not have any extra clothing or food, and nighttime temperatures are in the 40s. The sun is setting, and there look to be storm clouds on the horizon.

Which scenario is more urgent? Do you want to send a search party out at night for the missing backpacker? What about the missing girl? You will have many questions that need to be answered in the initial part of an SAR operation. Some questions you will ask yourself, while other questions may be asked by frantic parents or even the media.

A National Association for Search and Rescue (NASAR) form provides some guidance in determining the urgency of response Figure 11-2 ▶ . Similar to a fire call, the SAR emergency response system is generally initiated by a phone call to the 911 dispatch center. Once the fire service has been notified of a potential SAR operation, then several questions need to be asked:

1. Where will this operation take place? Where should the incident command post and staging areas be located?
2. Which resources will be needed? Which resources do we currently have available, and which resources will need to be ordered?
3. How urgent is the situation?

Once some of these questions have been answered, site control and scene management can be considered. Scene management for a swift water rescue will vary considerably from a missing person in a vast wilderness.

The object of your search is ultimately the missing person(s), but initially it could be a vehicle, a boat, or an aircraft. These items may be referred to as search objects, whereas the missing person would be called the search subject. Search tactics vary based on the situation and available resources. General tactics are classified as passive or active.

Passive Tactics

Passive tactics include attraction, containment, and investigation. These search techniques do not involve physically entering the search area (the area in which responders will search for the missing person).

Attraction techniques attempt to draw the subject toward a desired location where rescuers are waiting. Examples of sound attraction include using noise makers such as sirens, horns, whistles, yelling, and bullhorns. When using sound attraction, make sure to take time to listen as well: The missing person may be yelling back. Radio communications or establishment of set time intervals can allow search teams to coordinate noise and silence periods effectively. It is also important for all search teams to have a general idea where each team is and which noise makers they will be using. Visual attraction can include strobes, beacons, high-power spotlights, flares, fires, and smoke. Use of such techniques will largely depend on the terrain, the forest type, and the weather. Most of the visual techniques work best in low light conditions (e.g., dusk, dawn, and night), with the exception of smoke.

Containment techniques are employed to keep the search area as small as possible and possibly intercept the missing person as he or she is trying to find his or her way to safety; they are often combined with attraction techniques. In the early stages of the search, the search management team needs to identify likely exit points from the search area—for example, roads, trail junctions, trailheads, or drainages. Personnel should be staged at these likely egress points while other personnel patrol roads. Moving patrols need to keep their eyes open for signs that the missing person may have crossed the road or trail. Another containment technique is to run flagging, with arrows and signs, to direct the missing person should he or she come upon that information.

To focus the search, it is important to obtain as many details about the incident as possible. Use of the last passive technique discussed here should actually begin as soon as dispatch receives the missing person call. Dispatch personnel should have a form available to help focus their questions Figure 11-3 ▶ . Interviewing friends, family members, and co-workers can also add clues to the search puzzle. Other pertinent questions are listed here:

- What sort of backcountry experience does the subject have?
- What was the subject's plan (e.g., to hike in Rapid Creek on Saturday, camp at Buffalo Meadow, and hike out Baker Draw on Sunday)?
- Which equipment did the subject have upon leaving?
- What is the general health and fitness level of the subject?
- Does the subject have any significant medical problems (e.g., past medical issues, allergies, medications)?
- What are some of the subject's personality attributes (e.g., organized and reliable or flighty and unreliable)?

Because the incident commander (IC) will be busy coordinating the SAR mission, someone should be assigned to gather this information. Sometimes an information officer will take over this task. If available, law enforcement personnel are effective resources for gathering information. All contacts should be instructed to call if they hear from the missing person or get new information on the person's possible condition and whereabouts.

Active Tactics

Active tactics include all methods used inside the search area to locate the search subject or object. Three general types of active search tactics are distinguished:

- Type 1: Rapid or hasty search
- Type 2: Quick area search
- Type 3: Thorough area search

The type of search that is performed is often determined by the amount and type of resources available. For example, even if a situation dictates a Type 3 search, the resources might be limited to those supporting a Type 1 search capability. As more search resources become available, the IC may elect to change the search type or have several search types performed simultaneously.

Type 1 searches use hasty teams. A hasty team should be a highly mobile, well-trained, physically fit, and self-sufficient unit consisting of 2–4 personnel. This team searches the most likely locations where the subject or clues might be found. Examples of

Search Urgency

Remember the lower the number the more urgent the response!!!

Date Completed: _____
Time Completed: _____
Initials: _____

A. SUBJECT PROFILE ....   _____

Age
       Very Young 1
       Very Old 1
       Other 2-3
Medical Condition
       Known or suspected injury or illness 1-2
       Healthy 3
       Known fatality 3
Number of Subjects
       One alone 1
       More than one (unless separation suspected) 2-3

B. WEATHER PROFILE ....   _____

Existing hazardous weather 1
Predicted hazardous weather (8 hours or less) 1-2
Predicted hazardous weather (more than 8 hours) 2
No hazardous weather predicted 3

C. EQUIPMENT PROFILE ....   _____

Inadequate for environment 1
Questionable for environment 1-2
Adequate for environment 3

D. SUBJECT EXPERIENCE PROFILE ....   _____

Not experienced, not familiar with the area 1
Not experienced, knows the area 1-2
Experienced, not familiar with the area 2
Experienced, knows the area 3

E. TERRAIN & HAZARDS PROFILE ....   _____

Known hazardous terrain or other hazards 1
Few or no hazards 2-3

        TOTAL ....   _____

If any of the seven categories above are rated as a one (1), regardless of the total, the search could require an emergency response.

●●● THE TOTAL SHOULD RANGE FROM **7** TO **21** WITH **7** BEING THE MOST URGENT. ●●●

8-11 Emergency Response   *12-16 Measured Response*   *17-21 Evaluate & Investigate*

DCC - revised 10/23/96

Figure 11-2 Search Urgency Score Sheet.

CARBON COUNTY SEARCH AND RESCUE

Form # 1 **Initial Report Form—To Be Completed by Dispatch**		
Person Receiving Report:	**Date Received:**	**Time Received:**
Name, Contact Information, and Current Location of Informant:		
Report Received by (telephone, person, etc.):		
Name of Missing Person:		
Date and Time Last Seen:		
Activity or Event that led to Needed SAR Response:		
Point Last Seen (PLS) or Last Known Position (LKP):		
What Does Informant Think Happened:		
What Does Informant Want Done, and Instructions from Dispatch to Informant:		
Time of First Page and Name of Responding SAR personnel:		

Form modified from ERI International, Inc. CCSAR Form 1, April 07

Figure 11-3 Initial report form.

hasty team search locations include trailheads, trails, campsites, and cabins. Depending on the search area and available SAR personnel, command may decide to send out multiple hasty teams. Hasty teams must remain vigilant for clues as to the location of the subject. As clues are located, the team should use a global positioning system (GPS) to determine their exact locations and pass this information back to the IC and the mission planners. Any clues should be marked with flagging.

Some search incidents lend themselves well to operations by hasty teams, such as topographic and vegetative features that funnel human travel **Figure 11-4 ▶**. Other incidents may have less prominent landscape features, such as a cornfield or a relatively

A.

B.

Figure 11-4 The features of a search area can help determine the type of search used. **A.** Areas that naturally guide human travel lend themselves to hasty teams. **B.** Areas with fewer distinguishing features are better suited to Type 2 and 3 searches.

flat pine forest. When this is the case, area searches (Type 2 and 3) become more effective. Unlike a hasty search, an area search has defined boundaries for each team. The best way to organize an area search is to use some type of line or grid search, wherein each individual searcher walks a line parallel to the next searcher.

While the Type 1 search is in progress, the mission planners should be organizing Type 2 teams. The Type 2 area search will most likely require more personnel than are included in Type 1 teams. The goal of a Type 2 search is to cover a large area fairly quickly and systematically. Overall, this type of search will be more thorough than a Type 1 search, but less thorough than a Type 3 search. When a very large area needs to be searched and no clues are available to focus the rescue responders' efforts, a Type 2 search is the way to go. Ideally, search team members should be able to maintain visual and voice contact, although this is not strictly necessary with a Type 2 search. Type 2 searches are typically conducted in the early stages of the operation and often key into areas where the hasty team found clues. Because Type 2 searches are looser in formation, it is better to have experienced searchers on these teams. These experienced personnel will have a better idea where to look and what to look for.

A Type 3 area search is often used when the other search tactics have failed to produce acceptable results. This kind of search is designed to ensure that the search object is not in the search area at the time of the search. The tight spacing employed by searchers does not allow for the object to be missed by rescuers. Unfortunately, this type of search has several weaknesses: It takes a great deal of effort, it is very slow, and it can end up destroying clues that are missed by less skilled searchers.

Clue and Sign Awareness

There are always more clues than there are missing persons. Consequently, clue detection is a major component of the field personnel's job. Every footstep is a clue for those who are trained

to see them. Clue seekers must practice these skills frequently to maintain proficiency. On some search incidents, professional trackers may be called in to assist in the search. Often these professionals are called only after many other people have been searching and, therefore, have contaminated the area. Effective clue detection can resolve a search quickly. Some commonly encountered clues include trash, depressions in vegetation where the person may have rested, broken branches, and tracks.

Safety Tips

Seeking clues requires practice and great attention to detail. Professional trackers can even be called in to assist with the search.

Important Search Concepts

Incident planners need to define the search area (the area in which responders will search for the missing person). In most cases, the planners can say there is a 99.9 percent chance that the missing person is on planet Earth. However, a search area of that magnitude is unreasonable. The planners must, therefore, reduce the search area to the smallest practical size that will have a high probability of containing the missing person or object.

The **point last seen (PLS)** is determined by information gathered from someone who actually saw the person and can identify that location on a map. This information is useful if the location is somewhere along the intended route of travel and within the search area. A PLS of the grocery store before the person started his or her trip would not be helpful.

The **last known point (LKP)** can be more useful in some cases than a PLS. It might be identified as the subject's car parked at the trailhead, a campsite, a backpack abandoned in the woods, the subject's signature in a summit log book, or similar informa-

Voices of Experience

The Beartooth Highway winds 65 miles from Red Lodge, Montana, briefly through Wyoming, and on to the northeast entrance of Yellowstone National Park. This route covers some of the most beautiful and rugged country in North America. Along this highway you will find the Twin Lakes and Twin Lakes Headwall near the Montana–Wyoming border. The Headwall is a favorite late spring/early summer ski area. At an altitude of 11,000 feet, the area is truly wilderness. No humans live there; they only visit. There are no phones or cell phone service. Even radio communication is limited to the range of simplex systems.

> **"The patient status required rapid transport, effective communication, and coordination among responding units."**

On June 20, 2007, a young male snowboarder lost control and fell approximately 1000 feet down the side of the mountain. During the fall, he struck a rock outcropping. His companion flagged down a passerby, who then drove to where there was cell service to request help. The first 911 call came in to the sheriff's office in Red Lodge. Carbon County SAR (CCSAR) and Red Lodge EMS (RLEMS) were immediately dispatched to the incident. The incident was inside Wyoming; however, CCSAR and RLEMS were the closest agencies by at least 60 miles and nearly 2 hours. (Keep in mind that in a mountainous setting, travel is slower.) RLEMS immediately responded with a three-person ALS ambulance, as CCSAR prepared rescue equipment. As the IC, I ensured that coordination with Wyoming's Park County Sheriff and SAR took place. Bystander reports of the patient status justified dispatching Billings, Montana, HELP Flight.

A paramedic with RLEMS arrived on scene and hiked down to the patient, who was being attended to by his companion. The paramedic reported that the patient was unresponsive with a serious head injury. The patient status required rapid transport, effective communication, and coordination among responding units. CCSAR arrived on scene moments after RLEMS, when patient packaging was under way. The steep, snow-covered slopes and high altitude forced HELP Flight to land about 200 yards below the patient. At this point, the Park County SAR team arrived on scene to aid with transport. The rescue teams loaded the patient onto the helicopter and it departed quickly for the trauma center in Billings.

The IC of an SAR event needs to be prepared to deal with multiple agencies at once. At one point I was coordinating personnel from RLEMS, CCSAR, Park County SAR, and HELP Flight. After almost 20 years in the fire/EMS service, I am still impressed by the way people can come together to get things done in an emergency.

Aaron McDowell, BS, REMT-Paramedic
Deputy Chief
Red Lodge Fire Rescue
Red Lodge, Montana

tion. With the LKP, no one actually saw the subject there, but evidence and clues reveal that the person was there. There may be only one PLS but several LKPs may be discovered as clues are found and verified.

Several key points should be kept in mind when planning a search. First, any search should be considered an emergency because the condition of the subject is unknown. The missing person might be injured or unprepared for inclement weather. Second, if the subject is on the move, every minute that passes could equate to a larger search area. Third, search resources need to be deployed effectively, efficiently, and in a timely manner. Search has been called a "classic mystery" because the clues are out there—if the searchers know where to look and what to look for. For this reason, rescue responders should implement confinement as soon as feasible; searching for a person who is no longer in the search area is a waste of time and resources. Use a tight grid search (Type 3) as a last resort.

Figure 11-5 Specialized litters with wheeled attachments can aid in transporting victims who cannot walk.

Rescue Tips

All searches should be considered emergencies because the condition of the missing person is unknown.

The United States offers a wide variety of terrain, ranging from low-lying deserts to alpine mountain ranges. The barren deserts of the Southwest may provide little restriction in direction of travel, whereas a series of jagged north–south ridges in the Rocky Mountains will provide a significant discouragement to a lost hiker moving in an east–west direction. Large lakes and fast-moving rivers can also affect movement across the landscape. Given these factors, knowledge of local terrain is invaluable during search operations. It is not uncommon for more than one lost person to travel or end up in the same place as other lost persons.

Every tool that helps to focus the search better is worthwhile. In his book *Analysis of Lost Person Behavior*, W. G. Syrotuck details the findings on his research of 229 case studies of people lost in wilderness settings. Certain categories of lost persons were defined based the activity, age, and mental capacity of the lost persons. Categories listed in the book include small children (1–6 years old), children (6–12 years old), hunters, hikers, elderly persons, mentally challenged individuals, despondent persons, and miscellaneous lost persons. While this book and others like it won't tell you exactly where to look, they can help the mission planners streamline the initial search areas.

■ Rescue

Although *locating* the victim may be the most time-consuming part of the SAR process, *accessing*, *stabilizing*, and *transporting* the victim are just as critical. Getting the victim to safety is the focus of rescue. Sometimes that may involve transport on an all-terrain vehicle (ATV) or snowmobile in a wilderness event, while other situations may require the use of technical rope equipment (see Chapter 5) or water rescue equipment (see Chapter 10). In addition, SAR mission planners need to identify the need for

specialized rescue teams early. For example, if two ice climbers are long overdue from their trip, you should have a good idea that ice climbing rescuers may be needed. You must also consider the possibility that the missing ice climbers may have gotten lost far from the ice wall.

Keep in mind that most people require rescue because they are unable to take care of themselves, especially if they have suffered an injury. If the person is unable to walk, the SAR team may have to carry the person out. Specialized litters with wheeled attachments can facilitate this process **Figure 11-5 ▲**. In other cases, the terrain or the urgency of the victim's medical condition may warrant the use of helicopter rescue. In such a situation, it is best to notify the IC early about the potential need for a helicopter. In other cases, the person may be uninjured, but other factors such as poor weather may have stranded him or her. SAR teams will determine which equipment is needed to access the stranded person.

The diversity of special rescue teams available will depend mostly on the area and the time of year. For example, agencies on the Island of Kauai would probably not need a proficient ice climbing rescue team, but they would need personnel experienced with rope rescue. By comparison, SAR teams in northern Michigan would need personnel experienced with ice rescue, and teams near Mount Whitney would need experienced mountaineers. Other examples of specialized SAR teams include those focusing on swiftwater, ice, open ocean, and helicopter rescue. A list detailing these resources and their contact information should be kept and updated regularly.

▌Resource Requirements

The resource requirements for a successful SAR operation can vary from a light daypack to a high-angle rope rescue team to a Sikorsky S-70 Black Hawk helicopter. Furthermore, any of this equipment is only as good as the person using it. SAR personnel must be able to endure environmental extremes, have good physical strength and endurance, and keep a level head during stressful situations. They must also be qualified to complete the tasks assigned to them.

Search and Rescue Equipment

An SAR event may be cancelled before team members even leave the house, or it could last for days. A rapid response on responders' part can make the difference between life and death. For this reason, as an SAR team member you must prepare yourself and your equipment as best as possible. Many SAR personnel maintain a "ready pack," which is loaded with the essentials at all times **Table 11-2 ▾**. When the call comes in, they simply grab their packs and head to the staging area. Of course, the contents of the pack will vary based on the location and the season.

Other specialized equipment can include the following items:

- Rope rescue equipment
- Swiftwater equipment
- Avalanche beacons, probes, and shovels
- Litters (collapsible, rigid, wheeled)
- ATVs, snowmobiles, and four-wheel-drive trucks

Table 11-2 Contents of an SAR Ready Pack

- Backpack—approximately 2500–3000 cubic inches, or 40–50 liters
- Hydration—water bottles/bladder, minimum 2 liters
- Water purification tablets
- All-weather boots—insulated in colder climates
- Gaiters
- Synthetic socks and spare pair
- GORE-TEX–type jacket and pants for rain, wind, or snow
- Insulating layers such as wool, down, fleece, or synthetic
- Hat—with brim for sunny climates, insulated for colder climates)
- Gloves—waterproof and leather
- Headlamp with spare batteries and spare bulb
- Compass, maps, and GPS
- Sunglasses, safety glasses, and hearing protection
- Multipurpose tool and knife
- Whistle, signal mirror, and flagging tape
- Personal first-aid kit—including sunscreen, insect repellant, and lip balm
- Notepad and pencil or pen
- Toilet paper
- Space blanket
- Fire-starting materials
- Lightweight tarp for shelter
- Lightweight synthetic sleeping bag
- Foam pad
- Small stove
- Compact cookware
- Meals Ready to Eat (MRE) or similar trail meal
- High-energy foods such as trail mix and energy bars
- Insect repellant

Rescue Team Resources

An effective SAR organization requires some type of administrative structure. Depending on the organization, the head SAR person may be a captain or the SAR director. The lead SAR officer helps to set the focus and mission of the organization through coordination of the activities of other response agencies. He or she should also oversee logistics, training, and other issues related to search. Once an operation has begun, the standard incident command structure should be established.

Team equipment needs to be maintained and kept in good working order. It is extremely helpful for an SAR organization to have a vehicle (preferably a 4 × 4) that can get personnel to the staging area. Good choices for SAR vehicles can both transport personnel and carry a variety of rescue equipment **Figure 11-6 ▾**. Examples of items kept in the SAR vehicle may include technical rope gear, helmets, litters, first-aid kits, flashlights, spare batteries, radios, satellite phones, GPS units, food, and water. Other important vehicles for SAR organizations may include snowmobiles, ATVs, boats, and mountain bikes. Because many organizations that run SAR operations have limited funding and available equipment, their vehicles and equipment are often supplemented by searchers' own personal vehicles and equipment.

As technology becomes more advanced and widely available, many SAR teams are moving into the digital era. Units with adequate funding are outfitting their vehicles with laptops, printers, satellite phones, and wireless Internet connections **Figure 11-7 ▸**. New technology is allowing for use of a GPS–map real-time interface, which means that incident commanders can watch SAR teams as they move across the landscape. Additionally, many organizations are making use of thermal imaging cameras, night vision cameras, and forward-looking infrared radar (FLIR) as part of their search capabilities. Larger organizations may have a dedicated command vehicle.

Individual Skills and Training

An SAR event can come when you least expect it. Mission success may be hindered by SAR personnel who slow the response due to their poor physical condition. Core components of physical con-

Figure 11-6 A good SAR vehicle can transport not only rescue personnel, but also a variety of rescue equipment.

Figure 11-7 Some rescue units are outfitted with modern technology to facilitate search operations.

Figure 11-8 Clothing should be layered with a wicking, synthetic material closest to the skin, a middle insulating layer, and a water- and wind-resistant outer layer.

ditioning include strength, flexibility, and endurance. Endurance may be improved through cardiovascular training such as biking, running, or swimming several times per week. Strength should be maintained or improved through weight training. Responders should also remember that it is important to stretch before and after workouts.

When responding in a wilderness area, rescue responders must understand that they may be outdoors for an undetermined length of time and that weather and temperatures could change dramatically over the course of the incident. Layering clothing is an effective way of efficiently regulating body temperature **Figure 11-8 ▶**. A wicking, synthetic material should be placed closest to the skin, followed by a middle insulating layer (preferably synthetic or wool), and then a water- and wind-resistant outer layer. SAR personnel should wear reflective stripping and bright colors when possible. Cotton clothing should be avoided. Boots should be durable, at least ankle high, broken in, waterproof, and breathable. And don't forget extra socks!

Being properly outfitted will get you only so far without adequate training. SAR members need a strong understanding of backcountry travel and navigation. Achieving this level of expertise requires training and field time using a map, compass, and GPS. Of course, a map can blow away, a compass can break, and the batteries in a GPS can run out, so responders must have a good understanding of the general landscape before hiking in. If all else fails, have a backup plan (e.g., hike west to hit the dirt road).

Another skill set required for SAR personnel comprises basic survival skills. SAR personnel must be prepared to take care of not only the victim, but also themselves. Essential survival techniques include the ability to obtain water, find shelter, and build a fire. Although food is important, the human body can last for days without it. In some parts of North America, water can be found around every turn. In other places, you could hike many miles without finding a drop. Know the area where you are hiking, and either bring plenty of water or be prepared to locate and treat it. Effective shelter and fire building can be a life saver in cold and wet environments. A fire can be used for warming people and drying clothes and equipment, as well as for cooking food and signaling. Always carry waterproof fire-starting equipment.

Backcountry first aid requires a slightly different skill set than administration of care in more accessible locations. As a rescuer in the wilderness, you do not have ready access to an ambulance full of medical supplies; you have only what you carry in and what is around you. You may have to improvise a litter, a splint, or a pelvic support. **Table 11-3 ▶** lists common supplies in a backcountry first-aid kit.

Additionally, you must have good assessment skills. Although you will not be able to reduce a victim's increasing intracranial pressure or provide similar treatments in the wilderness, you should be able to identify the severity of the victim's condition. Your rapid assessment of the victim and request for helicopter evacuation may mean the difference between a happy ending and a bad one. On the same note, you should be able to accurately identify stable victims who can be evacuated without a helicopter. Certifications such as Wilderness First Responder and Wilderness Emergency Medical Technician are strongly suggested for at least one member on each field team.

All personnel involved in SAR operations should understand the basics of the Incident Command System (ICS) as outlined by NIMS, search theory, and communication protocols. Establishing an effective chain of command, skilled searching, and good communication will all increase the likelihood of a positive outcome. As noted earlier, another desirable skill for SAR personnel is the ability to detect signs and clues. When moving down trails and dirt roads, you must be on the lookout for evidence that the missing person was there. Candy wrappers, items of clothing, notes, and boot tracks could be important clues. For example, a confirmed boot track could update the search subject's LKP and help to focus the search area.

Table 11-3 Backcountry First Aid Kit Supplies

Medical information	Wilderness first-aid book
	Patient assessment forms
	Pencils
Medication	Pain medications (ibuprofen, acetaminophen, aspirin)
	Gastrointestinal medications (antidiarrheal, antacid, laxative)
	Cold medicine
	Glucose paste
	Electrolyte powder
	Emergency dental (temporary filling mixture, clove oil)
	Povidone-iodine
	Sting/itch relief wipes or lotion
Personal protection	CPR pocket mask
	Nitrite gloves
	Antiseptic hand wipes
Trauma	Trauma pads, dressings, bandages, gauze, and tape
Tools	EMT shears
	Splinter/tick remover
	Thermometer
	Magil forceps
	Disposable scalpels
	Safety pins
	Permanent marker
	Duct tape
	Stethoscope and blood pressure cuff
Miscellaneous	Blister care (moleskin, second skin)
	Antibiotic ointment
	Sunscreen
	Adhesive bandages
	SAM splint

Figure 11-9 EMS resources can include a helicopter.

Emergency Medical Services Resources

SAR missions often require close communication with EMS. It is the responsibility of the IC to determine the appropriate EMS response. Often an ambulance will be waiting at the trailhead where SAR members are bringing out the victim. At other times, the EMS response may be in the form of a helicopter with a flight nurse and/or paramedic **Figure 11-9 ▶**. Some fire/EMS organizations have specially trained members who make up a wilderness medical team. When the potential for injury exists, it is a good idea to put EMS on standby, both for the victims and for the rescue responders.

Response Planning

An effective ICS needs to work for incidents of all sizes and must be standardized to ensure seamless integration of outside resources. When you are dispatched to an SAR event, initial response planning involves evaluating the situation and allocating resources appropriately to resolve the situation effectively.

Although most SAR events are resolved within 24 hours, responders must be prepared to handle a situation that stretches out over several days. The ICS helps to manage emergency situations of varying sizes and complexity and is discussed in more detail in Chapter 3.

Command

The IC is responsible for overall management of the incident. The IC must be aware of his or her own limitations and be prepared to ask for help and delegate duties when needed. In the initial phase of an SAR operation, the IC assumes responsibilities in all ICS sections. Several support positions may also be established within the command section, including a liaison officer to coordinate with other agencies (e.g., forest service, state land managers, and law enforcement), a safety officer to oversee the safety of involved personnel, and an information officer to deal with the media and possibly assist the IC in interviewing friends, family, and bystanders.

Operations

The operations section is responsible for carrying out the incident action plan (IAP). In the initial phases of an SAR event, the IC will likely perform the duties of the operations section. When the incident grows in complexity, however, the operations sections chief is often the first position assigned. As more resources become available and the span of control increases, the operations section may add new positions to coordinate the actions of specialized teams (e.g., dog teams, air operations, and medical teams).

Here are some examples of what the operations chief will be doing in the initial phase of operations:

- Receive initial report and guidance from the IC
- Coordinate with other agencies that may be involved in the operation

- Coordinate with landowners to set up access where needed
- Obtain initial search planning information (i.e., time and location of PLS or LKP, the subject's travel plans and level of experience, maps and topographic data, a weather report)
- Gather critical information to give to the search team, such as the subject's name, shoe size, clothing, medical history, and any habits that searchers may be able to pick up on
- Evaluate available and en-route resources
- Determine the immediate needs of the operations team and inform the IC
- Supervise and manage operations team members
- Determine the communications protocol to be used
- Track the field team's current location and their findings
- Maintain a field team tracking board with status updates (could be accomplished in electronic format as well)

Planning

The planning section is responsible for incident planning. These duties include developing the IAP, collecting and evaluating incident situation status, displaying support information on the event, and tracking overall resources. One of the most important roles of the planning section is to look beyond the current events and identify resources that may be needed in the future and potential problems—for example, the fact that snowshoes or cross-country skis may be needed given an incoming storm front. This section keeps other sections up-to-date with daily briefings, strategies, maps and topographical data, and weather tracking.

Logistics

The logistics section, which is generally established as part of the IMS for larger incidents, is responsible for supporting all needs related to the incident, including obtaining personnel, facilities, equipment, transportation, food, and supplies. The coordination and organization of logistics become increasingly difficult if the SAR event lasts more than 24 hours and outside resources start arriving. The logistics section is responsible for identifying and maintaining staging areas. In an extended search operation, staging areas may be set up in the backcountry, but more often they are set up near roads, campgrounds, or trailheads. A staging area manager should be assigned to keep the area organized and operating smoothly. The logistics chief focuses on keeping the field teams well equipped, fueled, fed, hydrated, and rested.

Finance/Administration

SAR events can quickly increase in complexity, such that they require more specialized resources. In larger incidents, the finance/administration chief tracks payroll information, investigates and processes claims of damage or injury to equipment or personnel, and maintains records of money spent.

Hazards and Hazard Assessment

SAR incidents are often very hazardous situations, not just because factors such as geography or weather may complicate matters, but also because emotions can run high. Whether you are running into a burning building to save a victim or heading off into the wilderness to save someone, it is important to assess the potential hazards. Once hazards are identified, a plan to avoid or mitigate them should be made.

Scene Assessment and Hazard Mitigation

One of the greatest challenges for responding SAR personnel is to know their limits. People who require wilderness rescue can often get themselves into extremely dangerous predicaments. Rescuers want to help victims, but they must ensure their personal safety and the team's safety first.

When performing a scene size-up, review these three criteria: safety, mechanism of injury, and numbers. When assessing safety of the situation, always ensure your individual safety first, followed by your partner's safety, the public's safety, and finally the patient's safety. Remember that you have done the victim no good if you become another victim by being unsafe. Because the victim is already injured, he or she is the last priority in evaluating safety.

Evaluate the mechanism of injury (i.e., the cause of the injury) by looking around and gathering clues. Did the person fall? Was he or she struck by a falling rock? You will want to avoid the same fate befalling you and your team.

Consider the numbers involved. How many victims are there? How many rescuers do you have, and how many will you need to safely and efficiently evacuate the victim(s)? It is also important to be sure you have accounted for any other possible victims who may not be seen immediately.

Safety Tips

At a wilderness SAR incident, your own safety is the most important consideration, followed by your partner's safety, the public's safety, and finally the victim's safety.

When operating in wilderness settings, the luxuries of civilization can be hours or even days away. As a consequence, the responders need to be prepared mentally and physically for the incident, and to have the proper equipment and supplies to complete the mission safely. Ideally, SAR personnel will be well rested and in top physical condition. In reality, some members may not be in optimal condition for a mission. Before sending personnel into the field, the IC or operations chief should ensure that each individual is adequately equipped, mentally ready, and physically capable to handle the task at hand.

Environmental hazards should be considered before teams leave the staging area. Here are a few examples of appropriate questions to ask:

- What is the current temperature? What are the expected highs and lows for the next 12 hours? Relative humidity should be evaluated as well.
- What are the current weather conditions? What is the forecasted weather? A sunny, clear day can turn into a cold, overcast day with poor visibility in just a short time.
- What are the landscape concerns? Are there impassable rock faces or rivers in the area? Is there a chance of personnel falling into a crevasse? Will topographic features prevent radio communication?

- Are water sources available in the area? Do rescue personnel need to carry all the water they need, or are there creeks and lakes in the area?
- At what altitudes will the rescuers operate? Is there a chance that altitude sickness could be a factor? Will high altitude hinder a helicopter evacuation?
- Are there poisonous or dangerous animals in the area? What can be done to minimize chances of human–wildlife conflict?

Wrap-Up

Chief Concepts

- SAR personnel respond to lost, overdue, injured, and stranded persons in both frontcountry and backcountry settings. Wilderness operations create unique search and rescue situations that require thorough planning, specialized equipment, and trained personnel.
- Remember the guiding acronym for SAR operations: LAST—Locate, Access, Stabilize, and Transport.
- Search tactics may be either active or passive. Passive tactics take place outside the search area; they include attraction, containment, and investigation. Active tactics take place within the search area and include three types of searches: Type 1 (hasty search), Type 2 (quick area search), and Type 3 (thorough area search).
- SAR callouts can happen any time of day and any time of year. Personnel need to be physically and mentally fit to respond effectively. SAR team members should keep a "ready pack" loaded with the essentials and ready to go at any time.

- SAR personnel require specialized training so that they can mount an effective response. Specifically, they need a strong understanding of backcountry navigation, map and compass, and GPS skills. Basic survival skills such as water procurement, fire building, and emergency shelter construction are required as well. SAR personnel also need to understand and be proficient at delivering backcountry first aid.
- The ICS is the ideal structure for running an SAR operation. It allows for flexibility, good communications, clear chain of command, and defined span of control. It also allows for near-seamless integration of other assets that understand and use the ICS.
- SAR incidents can be very hazardous and have the potential of endangering response personnel. Incident planners and field team leaders should constantly reevaluate the situation for new and ongoing hazards. Safety, mechanism of injury, and numbers issues should all be considered as part of the scene size-up.

■ Hot Terms

Attraction A method in which audio and visual signals are used to attract a missing person.

Containment A method to limit movement of a missing person from inside to outside the search area.

Hasty team A small, self-sufficient, well-trained, physically fit team that responds quickly to search those areas in which there is the highest probability of locating the missing person or object.

Last known point (LKP) The last location where a missing person was believed to be (i.e., his or her vehicle or campsite).

Point last seen (PLS) The last location where someone actually saw a missing person.

Search area The area that is determined by the IC or planners to most likely contain the missing person or object.

Wilderness An uncultivated, uninhabited, and natural area that tends to be far from human civilization.

Wildland–urban interface (WUI) The area where homes, businesses, and other human-made structures are mixed with, or located right next to, wildland.

Rescue Responder *in Action*

Your department is dispatched to a local trailhead for a report of two missing boys. Your crew grabs hasty packs and jumps into the rescue truck. While en route to the location, dispatch informs you that the boys are 14 years old and are part of a Boy Scout group. The Boy Scout leader and other Scouts will be waiting for you at the trailhead. One of the fire fighters on your truck says he heard there was some bad weather moving in.

When you arrive at the trailhead, you are met by a worried Scout leader. He tells your crew leader that the boys did not come back to the campsite last night after they were out looking for firewood.

"I told them to stay close," he says worriedly. The Scout leader says they called and looked for the boys all last night. When the boys weren't back by mid-morning, the Scouts hiked back to the trailhead to get help. He tells you that the campsite is approximately four miles from the trailhead, right along the trail.

1. Which additional question(s) should you ask the Scout leader?
   A. What is the level of experience of the boys?
   B. Which additional supplies did they have with them?
   C. Is anyone waiting at the campsite for the boys?
   D. All of the above

2. Which additional resources are needed for this incident?
   A. More trained personnel
   B. Rope rescue team
   C. Search dogs
   D. Both A and C

3. Which type of search makes sense in this situation?
   A. Type 1
   B. Type 2
   C. Type 3
   D. Both A and C

Chapter 4 General Requirements

4.1 General.

4.1.1* The authority having jurisdiction (AHJ) shall establish levels of operational capability needed to conduct operations at technical search and rescue incidents safely and effectively, based on hazard identification, risk assessment, training level of personnel, and availability of internal and external resources.

4.1.2 At a minimum, all technical search and rescue organizations shall meet the awareness level for each type of search and rescue incident for which the AHJ has identified a potential hazard (see 4.2.1).

4.1.3* In jurisdictions where identified hazards might require a search and rescue capability at a level higher than awareness, a plan to address this situation shall be written.

4.1.3.1 The AHJ shall determine distribution of roles and responsibilities in order to focus training and resources at the designated level to maintain proficiency.

4.1.3.2 Where an advanced level of search and rescue capability is required in a given area, organizations shall have a system in place to utilize the most appropriate resource(s) available, through the use of local experts, agreements with specialized resources, and mutual aid.

4.1.4 The AHJ shall establish written standard operating procedures (SOPs) consistent with one of the following operational levels for each of the disciplines defined in this document:

(1)* *Awareness Level.* This level represents the minimum capability of organizations that provide response to technical search and rescue incidents.

(2)* *Operations Level.* This level represents the capability of organizations to respond to technical search and rescue incidents and to identify hazards, use equipment, and apply limited techniques specified in this standard to support and participate in technical search and rescue incidents.

(3) *Technician Level.* This level represents the capability of organizations to respond to technical search and rescue incidents and to identify hazards, use equipment, and apply advanced techniques specified in this standard necessary to coordinate, perform, and supervise technical search and rescue incidents.

4.1.5* It is not the intent of this document to have an organization deem itself capable of an advanced skill level in any of the disciplines defined herein simply by training or adhering to the requirements set forth. Maintaining an operations- or technician-level capability in any discipline shall require a combination of study, training, skill, and frequency of operations in that discipline.

4.1.6 The AHJ shall establish operational procedures consistent with the identified level of operational capability to ensure that technical search and rescue operations are performed in a manner that minimizes threats to rescuers and others.

4.1.7 The same techniques used in a search and rescue operation shall be considered equally useful for training, body recovery, evidence search, and other operations with a level of urgency commensurate with the risk/benefit analysis.

4.1.8 Operational procedures shall not exceed the identified level of capability established in 4.1.4.

4.1.9* At a minimum, medical care at the basic life support (BLS) level shall be provided by the organization at technical search and rescue incidents.

4.1.10 Training.

4.1.10.1 The AHJ shall provide for training in the responsibilities that are commensurate with the operational capability of the organization.

4.1.10.1.1 The minimum training for an organization shall be at the awareness level.

4.1.10.1.2 Organizations expected to perform at a higher operational level shall be trained to that level.

4.1.10.2* The AHJ shall provide for the continuing education necessary to maintain all requirements of the organization's identified level of capability.

4.1.10.3 An annual performance evaluation of the organization based on requirements of this standard shall be performed.

4.1.10.4* The AHJ shall evaluate its training program to determine whether the current training has prepared the organization to function at the established operational level under abnormal weather conditions, extremely hazardous operational conditions, and other difficult situations.

4.1.10.5* Documentation.

4.1.10.5.1 The AHJ shall be responsible for the documentation of all required training.

4.1.10.5.2 This documentation shall be maintained and available for inspection by individual team members and their authorized representatives.

4.1.11 Prior to operating at a technical search and rescue incident, an organization shall meet the requirements specified in Chapter 4 as well as all relevant requirements of Chapters 6 through 15 for the specific technical rescue incident.

4.1.12 Standard Operating Procedure.

4.1.12.1 The AHJ shall ensure that there is a standard operating procedure to evacuate members from an area and to account for their safety when an imminent hazard condition is discovered.

4.1.12.2 This procedure shall include a method to notify all members in the affected area immediately by any effective means, including audible warning devices, visual signals, and radio signals.

4.1.13* The AHJ shall comply with all applicable local, state, and federal laws.

4.1.14* The AHJ shall train responsible personnel in procedures for invoking, accessing, and using relevant components of the *U.S. National Search and Rescue Plan, the U.S. National Response Framework*, and other national, state, and local response plans, as applicable.

4.2 Hazard Identification and Risk Assessment.

4.2.1* The AHJ shall conduct a hazard identification and risk assessment of the response area and shall determine the feasibility of conducting technical search and rescue operations.

4.2.2 The hazard identification and risk assessment shall include an evaluation of the environmental, physical, social, and cultural factors influencing the scope, frequency, and magnitude of a potential technical search and rescue incident and the impact they might have on the ability of the AHJ to respond to and to operate while minimizing threats to rescuers at those incidents.

4.2.3* The AHJ shall identify the type and availability of internal resources needed for technical search and rescue incidents and shall maintain a list of those resources.

4.2.4* The AHJ shall identify the type and availability of external resources needed to augment existing capabilities for technical search and rescue incidents and shall maintain a list of these resources, which shall be updated at least once a year.

4.2.5* The AHJ shall establish procedures for the acquisition of those external resources needed for technical search and rescue incidents.

4.2.6 The hazard identification and risk assessment shall be documented.

4.2.7 The hazard identification and risk assessment shall be reviewed and updated on a scheduled basis and as operational or organizational changes occur.

4.2.8 At intervals determined by the AHJ, the AHJ shall conduct surveys in the organization's response area for the purpose of identifying the types of technical search and rescue incidents that are most likely to occur.

4.3 Incident Response Planning.

4.3.1 The procedures for a technical search and rescue emergency response shall be documented in the special operations incident response plan.

4.3.1.1 The plan shall be a formal, written document.

4.3.1.2 Where external resources are required to achieve a desired level of operational capability, mutual aid agreements shall be developed with other organizations.

4.3.2 Copies of the technical search and rescue incident response plan shall be distributed to agencies, departments, and employees having responsibilities designated in the plan.

4.3.3 A record shall be kept of all holders of the technical search and rescue incident response plan, and a system shall be implemented for issuing all changes or revisions.

4.3.4 The technical search and rescue incident response plan shall be approved by the AHJ through a formal, documented approval process and shall be coordinated with participating agencies and organizations.

4.4 Equipment.

4.4.1 Operational Equipment.

4.4.1.1* The AHJ shall ensure that equipment commensurate with the respective operational capabilities for operations at technical search and rescue incidents and training exercises is provided.

4.4.1.2 Training shall be provided to ensure that all equipment is used and maintained in accordance with the manufacturers' instructions.

4.4.1.3 Procedures for the inventory and accountability of all equipment shall be developed and used.

4.4.2 Personal Protective Equipment (PPE).

4.4.2.1* The AHJ shall ensure that the protective clothing and equipment are supplied to provide protection from those hazards to which personnel are exposed or could be exposed.

4.4.2.2 Personnel shall be trained in the care, use, inspection, maintenance, and limitations of the protective clothing and equipment assigned or available for their use.

4.4.2.3 The AHJ shall ensure that all personnel wear and use PPE while working in known or suspected hazardous areas during technical search and rescue incidents and training exercises.

4.4.2.4 The AHJ shall ensure that atmosphere-supplying respirators in the form of supplied air respirators (SAR) or self contained breathing apparatus (SCBA) are available when required for technical search and rescue operations and that they meet the requirements specified in Chapter 7 of NFPA 1500, *Standard on Fire Department Occupational Safety and Health Program*.

4.4.2.4.1 Breathing apparatus shall be worn in accordance with the manufacturer's recommendations.

4.4.2.4.2 A supply source of breathing air meeting the requirements of ANSI/CGA G7.1, *Commodity Specification for Air*, with a minimum air quality of Grade D shall be provided for all atmosphere-supplying respirators.

4.4.2.4.3 A supply source of breathing air meeting the requirements of ANSI/CGA G7.1, *Commodity Specification for Air*, with a minimum air quality of Grade E shall be provided for all atmosphere-supplying respirators used for dive operations.

4.4.2.4.4 Supplied air respirators shall be used in conjunction with a self-contained breathing air supply capable of providing enough air for egress in the event of a primary air supply failure.

4.5 Safety.

4.5.1 General.

4.5.1.1 All personnel shall receive training related to the hazards and risks associated with technical search and rescue operations.

4.5.1.2 All personnel shall receive training for conducting search and rescue operations while minimizing threats to rescuers and using PPE.

4.5.1.3 The AHJ shall ensure that members assigned duties and functions at technical search and rescue incidents and training exercises meet the relevant requirements of the following chapters and sections of NFPA 1500, *Standard on Fire Department Occupational Safety and Health Program*:
(1) Section 5.4, Special Operations Training
(2) Chapter 7, Protective Clothing and Protective Equipment
(3) Chapter 8, Emergency Operations

4.5.1.4* Where members are operating in positions or performing functions at an incident or training exercise that pose a high potential risk for injury, members qualified in BLS shall be standing by.

4.5.1.5* Rescuers shall not be armed except when it is required to meet the objectives of the incident as determined by the AHJ.

4.5.2 Safety Officer.

4.5.2.1 At technical search and rescue training exercises and in actual operations, the incident commander shall assign a safety officer with the specific knowledge and responsibility for the identification, evaluation, and, where possible, correction of hazardous conditions and unsafe practices.

4.5.2.2 The assigned safety officer shall meet the requirements specified in Chapter 6, Functions of the Incident Safety Officer, of NFPA 1521, *Standard for Fire Department Safety Officer*.

4.5.3 Incident Management System.

4.5.3.1* The AHJ shall provide for and utilize training on the implementation of an incident management system that meets the requirements of NFPA 1561, *Standard on Emergency Services Incident Management System*, with written SOPs applying to all members involved in emergency operations. All members involved in emergency operations shall be familiar with the system.

4.5.3.2 The AHJ shall provide for training on the implementation of an incident accountability system that meets the requirements of NFPA 1561, *Standard on Emergency Services Incident Management System*.

4.5.3.3 The incident commander shall ensure rotation of personnel to reduce stress and fatigue.

4.5.3.4 The incident commander shall ensure that all personnel are aware of the potential impact of their operations on the safety and welfare of rescuers and others, as well as on other activities at the incident site.

4.5.3.5 At all technical search and rescue incidents, the organization shall provide supervisors who possess skills and knowledge commensurate with the operational level identified in 4.1.4.

4.5.4* Fitness. The AHJ shall ensure that members are psychologically, physically, and medically capable to perform assigned duties and functions at technical search and rescue incidents and to perform training exercises in accordance with Chapter 10 of NFPA 1500, *Standard on Fire Department Occupational Safety and Health Program*.

4.5.5 Nuclear, Biological, and Chemical Response.

4.5.5.1* The AHJ, as part of its hazard identification and risk assessment, shall determine the potential to respond to technical search and rescue incidents that might involve nuclear or biological weapons, chemical agents, or weapons of mass destruction, including those with the potential for secondary devices.

4.5.5.2 If the AHJ determines that a valid risk exists for technical search and rescue response into a nuclear, biological, and/or chemical environment, it shall provide training and equipment for response personnel.

Chapter 5 Rope Rescue

5.1 General Requirements.

5.1.1 Organizations operating at rope rescue incidents shall meet the requirements specified in Chapter 4.

5.1.2* The AHJ shall evaluate the need for missing person search where rope rescues might occur within its response area and shall provide a search capability commensurate with the identified needs.

5.2 Awareness Level.

5.2.1 Organizations operating at the awareness level for rope rescue incidents shall meet the requirements specified in Section 5.2.

5.2.2 Organizations operating at the awareness level for rope rescue incidents shall develop and implement procedures for the following:
- (1) Recognizing the need for a rope rescue
- (2)* Identifying resources necessary to conduct rope rescue operations
- (3)* Carrying out the emergency response system where rope rescue is required
- (4)* Carrying out site control and scene management
- (5)* Recognizing general hazards associated with rope rescue and the procedures necessary to mitigate these hazards
- (6)* Identifying and utilizing PPE assigned for use at a rope rescue incident

Chapter 6 Structural Collapse Search and Rescue

6.1 General Requirements. Organizations operating at structural collapse incidents shall meet the requirements specified in Chapter 4.

6.2 Awareness Level.

6.2.1 Organizations operating at the awareness level for structural collapse incidents shall meet the requirements specified in Sections 6.2 and 7.2 (awareness level for confined space search and rescue).

6.2.2 Organizations operating at the awareness level for structural collapse incidents shall implement procedures for the following:
- (1) Recognizing the need for structural collapse search and rescue
- (2)* Identifying the resources necessary to conduct structural collapse search and rescue operations
- (3)* Initiating the emergency response system for structural collapse incidents
- (4)* Initiating site control and scene management
- (5)* Recognizing the general hazards associated with structural collapse incidents, including the recognition of applicable construction types and categories and the expected behaviors of components and materials in a structural collapse
- (6)* Identifying the five types of collapse patterns and potential victim locations
- (7)* Recognizing the potential for secondary collapse
- (8)* Conducting visual and verbal searches at structural collapse incidents, while using approved methods for the specific type of collapse
- (9)* Recognizing and implementing a search and rescue/ search assessment marking system, building marking system (structure/hazard evaluation), victim location marking system, and structure marking system (structure identification within a geographic area), such as the ones used by the FEMA Urban Search and Rescue System
- (10) Removing readily accessible victims from structural collapse incidents
- (11)* Identifying and establishing a collapse safety zone
- (12)* Conducting reconnaissance (recon) of the structure(s) and surrounding area

Chapter 7 Confined Space Search and Rescue

7.1 General Requirements.

7.1.1 Organizations operating at confined space incidents shall meet the requirements specified in Chapter 4.

7.1.2* The requirements of this chapter shall apply to organizations that provide varying degrees of response to confined space emergencies.

7.1.3 All confined space rescue services shall meet the requirements defined in 7.1.3.1 through 7.1.3.12.

7.1.3.1 Each member of the rescue service shall be provided with, and trained to use properly, the PPE and rescue equipment necessary for making rescues from confined spaces according to his or her designated level of competency.

7.1.3.2 Each member of the rescue service shall be trained to perform the assigned rescue duties corresponding to his or her designated level of competency.

7.1.3.3 Each member of the rescue service shall also receive the training required of authorized rescue entrants.

7.1.3.4* Each member of the rescue service shall practice making confined space rescues once every 12 months, in accordance with the requirements of 4.1.10 of this document, by means of simulated rescue operations in which he or she removes dummies, mannequins, or persons from actual confined spaces or from representative confined spaces resembling all those to which the rescue service could be required to respond in an emergency within their jurisdiction.

7.1.3.5 Representative confined spaces should—with respect to opening size, configuration, and accessibility—simulate the types of confined spaces from which rescue is to be performed.

7.1.3.6 Each member of the rescue service shall be certified to the level of first responder or equivalent according to U.S. Department of Transportation (DOT) *First Responder Guidelines.*

7.1.3.7 Each member of the rescue service shall successfully complete a course in cardiopulmonary resuscitation (CPR) taught through the American Heart Association (AHA) to the level of a "Health Care Provider," through the American Red Cross (ARC) to the "CPR for the Professional Rescuer" level, or through the National Safety Council's equivalent course of study.

7.1.3.8* The rescue service shall be capable of responding in a timely manner to rescue summons.

7.1.3.9 Each member of the rescue service shall be equipped, trained, and capable of functioning to perform confined space rescues within the area for which they are responsible at their designated level of competency.

7.1.3.10 The requirements of 7.1.3.9 shall be confirmed by an annual evaluation of the rescue service's capabilities to perform confined space rescues in terms of overall timeliness, training, and equipment and to perform safe and effective rescue in those types of spaces to which the team must respond.

7.1.3.11 Each member of the rescue service shall be aware of the hazards he or she could confront when called on to perform rescue within confined spaces for which the service is responsible.

7.1.3.12 If required to provide confined space rescue within U.S. federally regulated industrial facilities, the rescue service shall have access to all confined spaces from which rescue could be necessary so that they can develop rescue plans and practice rescue operations according to their designated level of competency.

7.1.4 A confined space rescue team shall be made up of a minimum of six individuals for organizations operating at the technician level, and a minimum of four individuals for organizations operating at the operations level.

7.2 Awareness Level.

7.2.1 Organizations operating at the awareness level for confined space search and rescue incidents shall meet the requirements specified in Sections 7.2 and 5.2 (awareness level for rope rescue).

7.2.2 The organization shall have an appropriate number of personnel meeting the requirements of Chapter 4 of NFPA 472, *Standard for Competence of Responders to Hazardous Materials/ Weapons of Mass Destruction Incidents,* commensurate with the organization's needs.

7.2.3 Organizations at the awareness level shall be responsible for performing certain nonentry rescue (retrieval) operations.

7.2.4 Organizations operating at the awareness level for confined space search and rescue incidents shall implement procedures for the following:

(1)　Recognizing the need for confined space search and rescue

(2)　Initiating contact and establishing communications with victims where possible

(3)*　Recognizing and identifying the hazards associated with nonentry confined space emergencies

(4)*　Recognizing confined spaces

(5)* Performing a nonentry retrieval
(6)* Implementing the emergency response system for confined space emergencies
(7)* Implementing site control and scene management

Chapter 8 Vehicle Search and Rescue

8.1* General Requirements. Organizations operating at vehicle search and rescue incidents shall meet the requirements specified in Chapter 4.

8.2 Awareness Level.

8.2.1 Organizations operating at the awareness level for vehicle emergencies shall meet the requirements specified in Section 8.2.

8.2.2 All members of the organization shall meet the requirements specified in Chapter 4 of NFPA 472, *Standard for Competence of Responders to Hazardous Materials/Weapons of Mass Destruction Incidents*, commensurate with the organization's needs.

8.2.3 Organizations operating at the awareness level for vehicle emergencies shall implement procedures for the following:
(1) Recognizing the need for a vehicle search and rescue
(2)* Identifying the resources necessary to conduct operations
(3)* Initiating the emergency response system for vehicle search and rescue incidents
(4)* Initiating site control and scene management
(5)* Recognizing general hazards associated with vehicle search and rescue incidents
(6) Initiating traffic control

Chapter 9 Water Search and Rescue

9.1 General Requirements. Organizations operating at water search and rescue incidents shall meet the requirements specified in Chapter 4.

9.2 Awareness Level.

9.2.1 Organizations operating at the awareness level at water search and rescue incidents shall meet the requirements specified in Section 9.2.

9.2.2 Each member of an organization operating at the awareness level shall be a competent person as defined in 3.3.24.

9.2.3 Organizations operating at the awareness level at water search and rescue incidents shall implement procedures for the following:
(1) Recognizing the need for water search and rescue
(2)* Implementing the assessment phase
(3)* Identifying the resources necessary to conduct safe and effective water operations
(4)* Implementing the emergency response system for water incidents
(5)* Implementing site control and scene management
(6)* Recognizing general hazards associated with water incidents and the procedures necessary to mitigate these hazards within the general search and rescue area
(7) Determining rescue versus body recovery

Chapter 10 Wilderness Search and Rescue

10.1 General Requirements. Organizations operating at wilderness search and rescue incidents shall meet the requirements specified in Chapter 4.

10.1.1* The AHJ, as part of its hazard identification and risk assessment (*see 4.2.2*), shall identify all locations in the jurisdiction that meet the definition of *wilderness*.

10.2 Awareness Level.

10.2.1 Organizations operating at the awareness level at wilderness search and rescue incidents shall meet the requirements specified in Section 10.2.

10.2.2 Members of organizations at the awareness level shall be permitted to assist in support functions on a wilderness search and rescue operation but shall not be deployed into the wilderness.

10.2.3 Organizations operating at the awareness level at any wilderness search and rescue incident shall have the following capabilities:
(1) Recognizing the need for a wilderness search and rescue–type response
(2)* Initiating the emergency response system for wilderness search and rescue
(3)* Initiating site control and scene management
(4)* Recognizing the general hazards associated with wilderness search and rescue incidents
(5) Recognizing the type of terrain involved in wilderness search and rescue incidents
(6)* Recognizing the limitations of conventional emergency response skills and equipment in various wilderness environments

(7)* Initiating the collection and recording of information necessary to assist operational personnel in a wilderness search and rescue

(8)* Identifying and isolating any reporting parties and witnesses

Chapter 11 Trench and Excavation Search and Rescue

11.1 General Requirements. Organizations operating at trench and excavation search and rescue incidents shall meet the requirements specified in Chapter 4.

11.2 Awareness Level.

11.2.1 Organizations operating at the awareness level at trench and excavation emergencies shall meet the requirements specified in Sections 11.2 and 7.2 (awareness level for confined space search and rescue).

11.2.2 Each member of the organization shall meet the requirements specified in Chapter 4 of NFPA 472, *Standard for Competence of Responders to Hazardous Materials/Weapons of Mass Destruction Incidents*, and shall be a competent person as defined in 3.3.24.

11.2.3 Organizations operating at the awareness level at trench and excavation emergencies shall implement procedures for the following:

(1) Recognizing the need for a trench and excavation rescue

(2)* Identifying the resources necessary to conduct safe and effective trench and excavation emergency operations

(3)* Initiating the emergency response system for trenches and excavations

(4)* Initiating site control and scene management

(5)* Recognizing general hazards associated with trench and excavation emergency incidents and the procedures necessary to mitigate these hazards within the general rescue area

(6)* Recognizing typical trench and excavation collapse patterns, the reasons trenches and excavations collapse, and the potential for secondary collapse

(7)* Initiating a rapid, nonentry extrication of noninjured or minimally injured victim(s)

(8)* Recognizing the unique hazards associated with the weight of soil and its associated entrapping characteristics

Chapter 12 Machinery Search and Rescue

12.1* General Requirements. Organizations operating at machinery search and rescue incidents shall meet the requirements specified in Chapter 4.

12.2 Awareness Level.

12.2.1 Organizations operating at the awareness level for machinery emergencies shall meet the requirements specified in Section 12.2.

12.2.2 All members of the organization shall meet the requirements specified in Chapter 4 of NFPA 472, *Standard for Competence of Responders to Hazardous Materials/Weapons of Mass Destruction Incidents*, commensurate with the organization's needs.

12.2.3 Organizations operating at the awareness level for machinery emergencies shall implement procedures for the following:

(1) Recognizing the need for a machinery search and rescue
(2)* Identifying the resources necessary to conduct operations
(3)* Initiating the emergency response system for machinery search and rescue incidents
(4)* Initiating site control and scene management
(5)* Recognizing general hazards associated with machinery search and rescue incidents

Correlation Guide: NFPA® 1670, *Standard on Operations and Training for Technical Search and Rescue Incidents*, 2009 Edition
© National Fire Protection Association

Chapter 4 General Requirements

NFPA 1670, *Standard on Operations and Training for Technical Search and Rescue Incidents*, 2009 Edition	Corresponding Chapter(s)	Corresponding Page(s)
4.1	General	General
4.1.1	3, 9	70–85, 183–196
4.1.2	1	7
4.1.3	1	7
4.1.3.1	1	2
4.1.3.2	1	2
4.1.4	3, 9	84, 186–191
4.1.5	1	2
4.1.6	3	83–84
4.1.7	3	77–84
4.1.8	3	84
4.1.9	3, 4, 9	77, 94–95, 98, 104, 186
4.1.10	Training	Training
4.1.10.1	3	76
4.1.10.1.1	3	76–77, 84
4.1.10.1.2	3	76–77
4.1.10.2	3	77
4.1.10.3	3	84
4.1.10.4	3	73–84
4.1.10.5	Documentation	Documentation
4.1.10.5.1	3	84
4.1.10.5.2	3	84
4.1.11	3	77
4.1.12	Standard Operating Procedure	Standard Operating Procedure
4.1.12.1	3	84
4.1.12.2	3	84
4.1.13	3, 4	84, 98
4.1.14	3	84
4.2	Hazard Identification and Risk Assessment	Hazard Identification and Risk Assessment
4.2.1	3	75–76
4.2.2	3	74–79
4.2.3	3	76
4.2.4	3	76–78
4.2.5	3	76–78
4.2.6	3	78–79
4.2.7	3	78
4.2.8	3	78
4.3	Incident Response Planning	Incident Response Planning
4.3.1	3	73
4.3.1.1	3	73
4.3.1.2	3	76, 83
4.3.2	3	83
4.3.3	3	73
4.3.4	3	83–84
4.4	Equipment	Equipment
4.4.1	Operational Equipment	Operational Equipment
4.4.1.1	2, 3	12–43, 76–78
4.4.1.2	2, 3	14–43, 81
4.4.1.3	2, 3	12–43, 70
4.4.2	Personal Protective Equipment (PPE)	Personal Protective Equipment (PPE)
4.4.2.1	2	18–24
4.4.2.2	2	18–26
4.4.2.3	2	18–25
4.4.2.4	2	23–24
4.4.2.4.1	2	23–24
4.4.2.4.2	2	13
4.4.2.4.3	2	13
4.4.2.4.4	2	13–14, 23–24
4.5	Safety	Safety
4.5.1	General	General
4.5.1.1	3	74–77
4.5.1.2	3	76
4.5.1.3	3	83
4.5.1.4	3	77
4.5.1.5	3	85
4.5.2	Safety Officer	Safety Officer
4.5.2.1	3	72–73
4.5.2.2	3	72–73
4.5.3	Incident Management System	Incident Management System
4.5.3.1	3	70
4.5.3.2	3	83
4.5.3.3	3	72
4.5.3.4	3	71–74
4.5.3.5	3	70
4.5.4	3	82–83
4.5.5	Nuclear, Biological, and Chemical Response	Nuclear, Biological, and Chemical Response
4.5.5.1	3	76
4.5.5.2	3	76

Chapter 5 Rope Rescue

NFPA 1670, *Standard on Operations and Training for Technical Search and Rescue Incidents,* 2009 Edition	Corresponding Chapter(s)	Corresponding Page(s)
5.1	General Requirements	General Requirements
5.1.1	5	112
5.1.2	5	114–115, 118
5.2	Awareness Level	Awareness Level
5.2.1	5	110–112
5.2.2	5	110–112, 115

Chapter 6 Structural Collapse Search and Rescue

NFPA 1670, *Standard on Operations and Training for Technical Search and Rescue Incidents,* 2009 Edition	Corresponding Chapter(s)	Corresponding Page(s)
6.1	6	126
6.2	Awareness Level	Awareness Level
6.2.1	6	136
6.2.2	6	125–141

Chapter 7 Confined-Space Search and Rescue

NFPA 1670, *Standard on Operations and Training for Technical Search and Rescue Incidents,* 2009 Edition	Corresponding Chapter(s)	Corresponding Page(s)
7.1	General Requirements	General Requirements
7.1.1	7	149
7.1.2	7	149–151
7.1.3	7	147–157
7.1.3.1	7	149–150
7.1.3.2	7	149–150
7.1.3.3	7	149
7.1.3.4	7	149
7.1.3.5	7	148–149

NFPA 1670, *Standard on Operations and Training for Technical Search and Rescue Incidents,* 2009 Edition	Corresponding Chapter(s)	Corresponding Page(s)
7.1.3.6	7	147
7.1.3.7	7	147
7.1.3.8	7	147
7.1.3.9	7	149–150
7.1.3.10	7	149
7.1.3.11	7	151–157
7.1.3.12	7	151
7.1.4	7	149–150
7.2	Awareness Level	Awareness Level
7.2.1	7	150
7.2.2	7	151
7.2.3	7	158
7.2.4	7	147–159

Chapter 8 Vehicle Search and Rescue

NFPA 1670, *Standard on Operations and Training for Technical Search and Rescue Incidents,* 2009 Edition	Corresponding Chapter(s)	Corresponding Page(s)
8.1	9	184
8.2	Awareness Level	Awareness Level
8.2.1	9	184
8.2.2	9	191
8.2.3	9	185–196

Chapter 9 Water Search and Rescue

NFPA 1670, *Standard on Operations and Training for Technical Search and Rescue Incidents,* 2009 Edition	Corresponding Chapter(s)	Corresponding Page(s)
9.1	10	205
9.2	Awareness Level	Awareness Level
9.2.1	10	203–226
9.2.2	10	205
9.2.3	10	203–226

Chapter 10 Water Search and Rescue

NFPA 1670, *Standard on Operations and Training for Technical Search and Rescue Incidents,* 2009 Edition	Corresponding Chapter(s)	Corresponding Page(s)
10.1	11	235
10.1.1	11	234
10.2	Awareness Level	Awareness Level
10.2.1	11	234–246
10.2.2	11	235
10.2.3	11	234–246

Chapter 11 Trench and Excavation Search and Rescue

NFPA 1670, *Standard on Operations and Training for Technical Search and Rescue Incidents,* 2009 Edition	Corresponding Chapter(s)	Corresponding Page(s)
11.1	8	169
11.2	Awareness Level	Awareness Level
11.2.1	8	169
11.2.2	8	174
11.2.3	8	164–177

Chapter 12 Machinery Search and Rescue

NFPA 1670, *Standard on Operations and Training for Technical Search and Rescue Incidents,* 2009 Edition	Corresponding Chapter(s)	Corresponding Page(s)
12.1	9	184
12.2	Awareness Level	Awareness Level
12.2.1	9	184
12.2.2	9	191
12.2.3	9	185–196

Glossary

Access and egress equipment Equipment used by confined-space rescue teams to enter and exit a confined space. This equipment may include ladders, tripods, davit arms, and rope rescue equipment.

Advanced life support (ALS) Advanced life-saving procedures, such as cardiac monitoring, administration of IV fluids and medications, and use of advanced airway adjuncts.

A-frame A piece of equipment designed to provide vertical lifting capabilities at a point away from a rescue vehicle. It consists of two poles that are connected at the working end, with the base of each pole being connected to the vehicle, and is stabilized by one or more cables attached to the vehicle.

Approach hazards Hazards that rescuers may encounter as they travel from their current location, such as a fire station, to the actual collapse site.

Ascending A means of traveling up a fixed rope safely with the use of an ascent device.

Attraction A method in which audio and visual signals are used to attract a missing person.

Authority having jurisdiction (AHJ) The organization, office, or individual responsible for approving equipment, materials, installation, or a procedure.

Awareness level The first level of rescue training provided to all responders, which emphasizes recognizing hazards, securing the scene, and calling for appropriate assistance. There is no actual use of rescue skills at the awareness level.

Basic life support (BLS) Noninvasive emergency life-saving care that is used to treat airway obstruction, respiratory arrest, or cardiac arrest.

Benching (benching system) A method of protecting employees from cave-ins in which the side of a trench is excavated to form one or a series of horizontal levels or steps, usually with vertical or near-vertical surfaces between levels.

Bend A knot that joins two ropes or webbing pieces together.

Bends A type of barotrauma emergency that occurs when a diver ascends from depth too quickly without adequate decompression.

Block creel construction A type of rope without knots or splices in the yarns, ply yarns, strands, or braids.

Body substance isolation (BSI) An infection control concept that treats all bodily fluids as potentially infectious.

Boom A piece of equipment designed to provide vertical lifting capabilities at a point away from a rescue vehicle. It consists of a telescoping beam that can be rotated 360 degrees.

Bottom load A heavy object with negative buoyancy that sinks to the bottom of a water channel.

Branch A segment with the ICS that may be functional or geographic in nature. Branches are established when the number of divisions or groups exceeds the recommended span of control for the operations section chief or for geographic reasons.

Canine search team Specially trained dogs, dog handlers, and, in some cases, personnel who direct the canine search teams.

Carabiner A piece of metal hardware used in rope rescue operations. It is generally an oval-shaped device with a spring-loaded gate that can be used to connect pieces of rope, webbing, and other rope hardware.

Cascade system A compressed air system used for refilling SCBA bottles and/or regulating for direct supply to supplied-air respirators.

Cave-in The separation of a mass of soil or rock material from the side of an excavation or trench, or the loss of soil from under a trench shield or support system. The sudden movement of a sufficient quantity of this material into the excavation, either by falling or sliding, could entrap, bury, or otherwise injure and immobilize a person.

Compartment syndrome A medical condition in which nerves and blood vessels are compressed within an enclosed space.

Competent person An individual who is capable of identifying existing and predictable hazards and is authorized to take measures to eliminate them.

Confined space A space large enough and configured so that a person can enter and perform assigned work, that has limited or restricted means for entry or exit (e.g., tanks, vessels, silos, storage bins, hoppers, vaults, and pits), and that is not designed for continuous human occupancy.

Containment A method to limit movement of a missing person from inside to outside the search area.

Coxswain The responder responsible for steering a powered boat.

Crew A group of personnel working without apparatus and led by a leader or boss.

Crush syndrome A medical condition in which a large body part is released after a prolonged period of compression.

Cut-back operation Use of heavy digging equipment to dig a parallel trench or a hole to create a void.

Descending A means of traveling down a fixed rope safely with the use of a descent control device.

Disentanglement The removal and/or manipulation of vehicle/machinery components to allow for proper victim removal.

Division A segment within the ICS established to divide an incident into physical or geographical areas of operation.

Dynamic rope A rope that is designed to be elastic and will stretch when it is loaded.

Eddy A horizontal current reversal along the riverbank and/or immediately downstream of an obstacle that penetrates the water's surface.

Edge protection A device that prevents damage to rope or other rope software from sharp or jagged edges or from friction.

Emergency Management Assistance Compact (EMAC) An organization established by the U.S. Congress that provides form and structure to interstate mutual aid. Through EMAC, a state that has experienced a disaster can request and receive assistance from other member states.

Engulfment A type of victim entrapment that may occur when a liquid product or a granular solid product flows into or moves within a confined space and then surrounds and/or buries a victim.

Entrapment A condition in which a victim is trapped by debris, soil, or other material and is unable to perform self-extrication.

Excavation Any human-made cut, cavity, trench, or depression made in an earth surface.

Extrication The removal of trapped victims from an entrapment.

Fall protection A system of associated rope hardware and software used to protect workers from falling from an elevated position. Fall protection also refers to equipment to which workers may be attached while working at elevations or near a fall hazard, which is meant to capture the individuals if they fall.

Finance/administration section Staff function responsible for the accounting and financial aspects of an incident, as well as any legal issues that may arise.

Gin pole A piece of equipment designed to provide vertical lifting capabilities at a point away from a rescue vehicle. It consists of a single pole that attaches to a rescue vehicle, with stabilizing cables being attached to the vehicle on either side.

Global positioning system (GPS) An electronic device that pinpoints a location on the earth's surface by triangulation with two or more special satellites designed for this purpose.

Group A segment within the ICS established to divide an incident into functional areas of operation.

Hard line communications systems Wire cables used for communication purposes that run between a control panel on the outside of the space and are attached to the rescuer.

Harness A piece of safety equipment that can be worn by a rescuer and attached to a rope rescue system, thereby allowing the rescuer to work safely in a high-angle environment during rope rescue operations.

Hasty team A small, self-sufficient, well-trained, physically fit team that responds quickly to search those areas in which there is the highest probability of locating the missing person or object.

Hazard analysis The process of identifying situations or conditions that have the potential to cause injury to people, damage to property, or damage to the environment.

Hazardous materials Materials or substances that pose an unreasonable risk of damage or injury to persons, property, or the environment if not properly controlled during handling, storage, manufacture, processing, packaging, use and disposal, or transportation.

High-angle rope rescue A rope rescue operation where the angle of the slope is greater than 45 degrees. Rescuers depend on the rope for the primary support mechanism rather than a fixed support surface such as the ground.

Hitch A knot that attaches to or wraps around an object such that when the object is removed, the knot will fall apart.

Horizontal helical flow Water moving in a circular motion, parallel with the river's surface.

Hydraulic (water flow) A form of vertical helical flow (current reversal) that occurs in rivers and streams and that has trapped and drowned many people, including water rescuers.

Immediate danger to life and health (IDLH) An atmospheric concentration of any toxic, corrosive, or asphyxiant substance that poses an immediate threat to life or could cause irreversible or delayed adverse health effects.

Incident action plan (IAP) An oral or written plan containing general objectives reflecting the overall strategy for managing an incident. It may include the identification of operational resources and assignments. It may also include attachments that provide direction and important information for management of the incident during one or more operational periods.

Incident commander (IC) The individual with overall responsibility for the management of all incident operations.

Incident Command System (ICS) A management structure that provides a standard approach and structure to managing operations, thereby ensuring that operations are coordinated, safe, and effective, especially when multiple agencies must work together.

Infection control plan A written plan that provides a comprehensive infection control system to maximize protection against communicable diseases.

Kern The center or core of a rope; it provides approximately 70 percent of the strength of the rope.

Knot A fastening made by tying together lengths of rope or webbing in a prescribed way.

Laminar flow The differences in current speed and force in a stream or channel.

Last known point (LKP) The last location where a missing person was believed to be (i.e., his or her vehicle or campsite).

Liaison officer The incident commander's point of contact for outside agencies. This officer coordinates information and resources among cooperating and assisting agencies and establishes contacts with agencies that may be capable or available to provide support.

Life safety harness A system component that is an arrangement of materials secured about the body and used to support a person during rescue.

Life safety rope A compact but flexible, torsionally balanced, continuous structure of fibers produced from strands that are twisted, plaited, or braided together. It serves primarily to support a load or transmit a force from the point of origin to the point of application.

Lockout/tagout system Method of ensuring that the electricity, gases, solids, or liquids have been shut down and that switches and valves are locked and cannot be turned on at an incident scene.

Logistics section Staff function responsible for all support requirements needed to facilitate effective and efficient incident management, including providing supplies, services, facilities, and materials during the incident.

Low-angle rope rescue A rope rescue operation where the slope of the ground is less than 45 degrees. Rescuers depend on the ground for their primary support, and the rope system becomes the secondary means of support.

Low-head dam A small dam that creates a hydraulic immediately downstream.

Mantle The sheath of a rope; the braided covering that protects the core of the rope from dirt and abrasion. It provides 30 percent of the strength of the rope.

Mass-casualty incident An emergency situation that involves more than one victim and that can place such great demand on equipment or personnel that the system is stretched to its limit (or beyond).

Mechanical advantage system A system that creates a leverage force using levers, pulleys, or gearing and is described in terms of a ratio of output force to input force.

Movable control point (MCP) A system of pulleys that serves as the heart of a tethered boat highline system; also known as a trolley.

Operations level The level of rescue training that allows for limited entry into a confined space with very few hazards, in such a way that a person could enter with self-contained breathing apparatus (SCBA) with no obstructions, and where the victim can be seen from the entryway of the confined space.

Operations section Staff function responsible for development, direction, and coordination of all tactical operations conducted in accordance with an incident action plan.

Organizational analysis A process to determine if it is possible for an organization to establish and maintain a given capability.

Packaging The process of preparing a victim for movement as a unit, often accomplished with a long spine board or similar device.

Parbuckling A water rescue technique that uses lengths of rope, webbing, or netting to roll a victim from the water's surface into a small boat.

Patient packaging The process of preparing a victim for movement as a unit, often accomplished with a long spine board or similar device.

Patient transfer The process of removing a patient from a hazard zone and the subsequent hand-off of the patient to a transport vehicle.

Personal floatation device (PFD) Formerly known as a life jacket.

Personal protective equipment (PPE) The equipment provided to shield or isolate personnel from infectious, chemical, physical, and thermal hazards.

Physical search team Personnel who have been trained to perform a physical search in a collapse environment and do not require specialized equipment.

Pinch points Areas where rescuers' fingers, hands, feet, or legs may become entrapped.

Pipe string The line of concrete pipe positioned along the length of a trench, ready for installation.

Planning section Staff function responsible for the collection, evaluation, dissemination, and use of information and intelligence critical to the incident.

Pluff mud Soft, semi-liquid material that resembles quicksand. It is found in many tidal creeks and salt marshes.

Point last seen (PLS) The last location where someone actually saw a missing person.

Pre-incident plan (preplan) A document containing detailed information about a facility or site that allows staff or first responders (e.g., police, fire fighters, emergency medical staff) to respond to any crisis situation at that location quickly and effectively.

Primary search The deployment of search resources to initiate a rapid search of an area that is thought likely to contain survivors.

Public information officer (PIO) Staff position that interfaces with the media and provides a single point of contact for information related to an incident.

Pulley A metal sheave (wheel) mounted on a bearing or metal bushing used during rope rescue to change direction, provide a mechanical advantage, reduce friction over an edge, or provide rope tension.

"Reading the river" A set of learned skills that involves the ability to assess a variety of river features and determine hazards, safe places to navigate, and places of relative safety in swiftwater or floodwater.

Reconnaissance (recon) An initial survey or rapid visual check of a rescue environment to obtain information that can be used to develop a search plan and direct search and rescue operations.

Recovery The removal of a body from a trapped location to a location where it can be examined and identified.

Regulation A rule, ordinance, or law requiring specific conduct of individuals and organizations.

Rescue Those activities directed at locating endangered persons at an emergency incident, removing those persons from danger, treating the injured, and providing for transport to an appropriate healthcare facility.

Rescue area An area that surrounds the incident site and whose size is proportional to the hazards that exist there.

Resource management A standard system of assigning and tracking the resources involved in an incident.

Resource typing Categorization of the resources requested, deployed, and used in incidents according to their capability to perform given tasks.

Respiratory protection A protective device to provide safe breathing air to a user in a hostile or dangerous atmosphere.

Respiratory protection program A written program, with workplace-specific procedures addressing the major elements of the program, established whenever respirators are necessary to protect the health of the employee.

Response planning (pre-incident planning) The process of compiling information that will assist the organization should an incident occur.

Ring buoy A Type IV personal floatation device that resembles a floating ring.

Risk/benefit analysis An assessment of the risk to the rescuers versus the benefits that can be derived from their intended actions.

Riverboard A larger, two-person version of a boogie board.

Rope hardware Rigid mechanical auxiliary equipment that can include, but is not limited to, anchor plates, carabiners, and mechanical ascent and descent control devices.

Rope rescue hardware Rigid mechanical auxiliary equipment that can include, but is not limited to, anchor plates, carabiners, and mechanical ascent and descent control devices.

Rope rescue incident A situation where rescuing an injured or trapped patient requires rope rescue equipment and techniques.

Rope rescue software A flexible fabric component of rope rescue equipment that can include, but is not limited to, anchor straps, pick-off straps, and rigging slings.

Rope rescue system A system consisting of rope rescue equipment and an appropriate anchor system intended for use in the rescue of a subject.

Rope software A flexible fabric component of rope rescue equipment that can include, but is not limited to, anchor straps, pick-off straps, and rigging slings.

Running end The part of a rope used for lifting or hoisting a load.

Safety knot A type of knot used to secure the ends of ropes to prevent them from coming untied.

Safety officer Staff position responsible for enforcing general safety rules and developing measures for ensuring personnel safety.

Search area The area that is determined by the IC or planners to most likely contain the missing person or object.

Search phase The phase of a rescue incident in which the victims are located.

Secondary collapse A partial or total failure of structural members from a building that had been damaged from an earlier collapse event.

Secondary search A slower, more methodical, and detailed search of a rescue environment to ensure that no victims were overlooked during the initial (primary) search operation.

Shield A structure that is able to withstand the forces imposed on it by a cave-in and, therefore, protects workers within the structure. When used in a trench, these structures are known as trench shields or trench boxes.

Shoring Tools and equipment used to support damaged buildings and damaged structural members, so as to stabilize the area involved and to avoid further (secondary) collapse.

Shoring (shoring system) A structure such as a metal hydraulic, pneumatic/mechanical, or timber shoring system designed to support the sides of an excavation and prevent cave-ins.

Single resource An individual vehicle and its assigned personnel.

Sloping system A protection system that uses inclined excavating to form sides that incline away from the excavation so as to prevent a cave-in.

Spoil pile A pile of excavated soil placed next to the excavation or trench.

Squeeze A barotrauma episode characterized by painful pressure in the sinuses and ear canals.

Stable rock Natural mineral matter that, when excavating the vertical sides of a trench, remains stable when exposed to a lack of support.

Staging area manager Staff position responsible for ensuring that all resources in the staging area are available and ready for assignment.

Standard A model or example commonly or generally accepted for the conduct of individuals and organizations.

Standard operating guideline (SOG) An organizational directive that is similar to a standard operating procedure but allows more flexibility in the decision-making process.

Standard operating procedure (SOP) A written organizational directive that establishes or prescribes specific operational or administrative methods to be followed routinely for the performance of designated operations or actions.

Standing part The part of a rope between the working end and the running end when tying a knot.

Static rope A rope that is designed not to stretch when it is loaded.

Strainers Porous objects such as trees and fences that allow water to pass through but trap people, boats, equipment, and debris.

Strike team A group of five units of the same type working on a common task or function.

Surcharge load Any weight in the proximity of a trench or excavation that increases its instability or likelihood of collapse. Also called overpressure load or superimposed load.

Surface hazards Unstable debris piles, loose and uneven rubble, and sharp materials found at a collapse site.

Surface victims The victims of a structural collapse who may be found on top of the collapse debris pile or along the perimeter of a collapsed building.

Suspended load A neutrally buoyant object moving between the surface and the bottom of a water channel.

Swim aids Items designed to assist rescuers and/or victims with floatation, vision, breathing, or propulsion.

Task force A group of up to five single resources of any type.

Technical rescue The application of special knowledge, skills, and equipment to safely resolve unique and/or complex rescue situations.

Technical search The use of mechanical search equipment, including visual, thermal, and acoustic devices, to locate victims trapped within a collapsed structure.

Technical specialists Advisors who have the special skills required at a rescue incident.

Technician level The level of rescue training that allows the rescuer to be directly involved in the rescue operation itself, including difficult and hazardous confined-space entries.

Telfer system A highline system that uses technical rope rescue techniques to control and position a boat in strong river current.

Tension diagonal A lowline system stretched at an approximately 45-degree angle across a stream, anchored at both ends, and then tightened by means of a mechanical advantage pulley system.

Top load A positively buoyant object floating on top of water.

Trench An excavation that is relatively narrow in comparison to its width. In general, the depth of the trench is greater than its width, but the width of a trench is not greater than 15 feet.

Triage The process of sorting victims based on the severity of their injuries and medical needs, with the goal of establishing treatment and transport priorities.

Type A soil Cohesive soils and cemented soils such as clay, silty clay, sandy clay, and hardpan.

Type B soil Cohesive soil with less compressive strength than Type A soils, such as angular gravel, silt, and silt loam.

Type C soil Sandy soils, sandy loam, and gravel soils, submerged soils, or soils where water is freely flowing from the trench walls.

Unified command An incident management tool and process that allows agencies with different legal, geographic, and functional responsibilities to coordinate, plan, and interact effectively to manage emergencies. In a unified command structure, multiple agency representatives make command decisions instead of just a single incident commander.

Urban search and rescue (USAR) Those activities directed at locating and rescuing victims in the collapse of a structure.

Utility rope A rope used for securing objects, for hoisting equipment, or for securing a scene to prevent bystanders from being injured. Utility rope is not used for life safety purposes or to support a person's weight in any way.

Vacuum excavation A two-phase process (reduction and removal) in which mechanical vacuum devices are used to remove soil from around a victim.

Vehicle/machinery stabilization The process of securing a vehicle or machine to prevent unexpected movement during a rescue operation, thereby avoiding injury to rescuers and preventing further injuries to the entrapped victims.

Ventilation A hazard control method involving the use of ventilation fans or blowers to introduce clean air into a confined space in an attempt to remove or reduce hazardous atmospheres.

Vertical helical flow Water moving perpendicular to the laminar flow or vertical exaggeration of a typical eddy.

Victim management All aspects of identifying hazards to which a victim may be exposed, gaining access to a victim, performing triage of multiple victims, assessing patient injuries, stabilizing victim injuries, communicating with victims, and moving and transferring those victims who require technical rescue assistance.

Victim packaging The process of preparing a victim for movement as a unit, often accomplished with a long spine board or similar device.

Victim packaging and removal equipment Equipment used in the process of securing a victim in a transfer device, with regard to existing and potential injuries or illnesses, so as to prevent further harm to the victim during movement.

V-lower A swiftwater technique whereby a boat or a tethered swimmer is moved downstream by means of an upstream rope anchored on each side of the stream.

Water recovery A water emergency that meets the water rescue definition, plus the victim is known to be dead, has been trapped underwater for more than 90 minutes, or is trapped in water from which there is no reasonably safe way to accomplish a rescue.

Water rescue The act of accessing and removing a person who is endangered in a water environment, with due regard for any injures and/or contamination that may be present.

Weapons of mass destruction (WMD) Hazards that include, but are not limited to, four main categories of threats: chemical, biological, radiological/nuclear, and explosive (CBRNE).

Wildland–urban interface (WUI) The area where homes, businesses, and other human-made structures are mixed with, or located right next to, wildland.

Wilderness An uncultivated, uninhabited, and natural area that tends to be far from human civilization.

Working end The part of the rope used for forming a knot.

X-lower A swiftwater technique whereby a boat or a tethered swimmer is moved downstream by means of an upstream rope and a downstream rope anchored on each side of the stream.

Index

Photo Credits

Voices of Experience © fred goldstein/ShutterStock, Inc.; **You Are the Rescue Responder** Courtesy of Scott Dornan, ConocoPhillips Alaska; **cloud of smoke** © Greg Henry/ShutterStock, Inc.

Chapter 1

1-1 © Kelley McCall/AP Photos; **1-2** Courtesy of Captain David Jackson, Saginaw Township Fire Department

Chapter 2

2-1A, 2-1C © Steve Redick; **2-4A** Courtesy of Eagle Compressors, Inc.; **2-4B** Courtesy of BelAire Compressors; **2-5A** Courtesy of AMKUS; **2-5B** Courtesy of Holmatro, Inc.; **2-6A** Courtesy of American Honda Motor Co., Inc.; **2-6B** Courtesy of the Berwyn Heights Volunteer Fire Department & Rescue Squad, Berwyn Heights, Maryland; **2-7, 2-8** Courtesy of Akron Brass Company; **2-9** © Glen E. Ellman; **2-11A** Courtesy of M&T Fire and Safety, Inc.; **2-11B** © Tim James/The Gray Gallery/Alamy Images; **2-11C** Courtesy of Tru-Spec by Atlanco; **2-11D** Courtesy of Mustang Survival; **2-12B, 2-12C** Courtesy of PMI West; **2-13** Courtesy of Captain David Jackson, Saginaw Township Fire Department; **2-15A** © Byron Moore/Alamy Images; **2-15B** © Andrew Ammendolia/Alamy Images; **2-16B** © 0neuser/ShutterStock, Inc.; **2-18** Courtesy of Sperian Respiratory Protection; **2-21** © Courtesy of Biomarine Ntron, Inc.; **2-22** Courtesy of Dräeger Safety, Inc.; **2-30** Courtesy of Skedco, Inc.; **2-31** © Alaettin Yildirim/ShutterStock, Inc.; **2-32** © TerryM/ShutterStock, Inc.; **2-33** Copyrighted image courtesy of CMC Rescue, Inc.; **2-35** Courtesy of Ingersoll Rand Industrial Technologies; **2-37A** Courtesy of Holmatro, Inc.; **2-41** Courtesy of Hot-Stick USA, Inc.; **2-42** Courtesy of Robert Bosch Tool Corporation; **2-43** © 2003, Berta A. Daniels; **2-45** Courtesy of CON-SPACE Communications; **2-46A** Courtesy of Junkin Safety; **2-46B** Courtesy of Ferno Washington, Inc.; **2-46C** Courtesy of Junkin Safety; **2-47A, 2-47B** Courtesy of Ferno Washington, Inc.; **2-48** Courtesy of Skedco, Inc.; **2-52** Courtesy of Fire Hooks Unlimited; **2-53** Courtesy of Unirope Limited; **2-55** Courtesy of Skedco, Inc.; **2-57A, 2-57B, 2-58** Courtesy of SMC—Seattle Manufacturing Corporation; **2-59** Courtesy of Gibbs Products, Inc.; **2-60** Courtesy of Donald M. Colarusso, allhandsfire.com

Chapter 3

3-1 © Steven Townsend/Code 3 Images; **3-3** Courtesy of Captain David Jackson, Saginaw Township Fire Department

Chapter 4

4-1 Courtesy of Yellowstone National Park/NPS; **4-2** Courtesy of Jocelyn Augustino/FEMA; **4-3** © Dennis Wetherhold, Jr.; **4-5** © Scott Downs/Dreamstime.com; **4-7A** Courtesy of Junkin Safety; **4-7B, 4-7C, 4-7D** Courtesy of Ferno Washington, Inc.; **4-9** Courtesy of the Prince George's County Fire/EMS Department; **4-10** © Fred L. Isom/ShutterStock, Inc.

Chapter 5

Opener © Dale A. Stork/ShutterStock, Inc.; **5-1** © Keith D. Cullom; **5-3** Courtesy of Robert Womer/Rock-N-Rescue; **5-4** © Greg Epperson/ShutterStock, Inc.; **5-5A, 5-5B** Courtesy of Donald Carek/Karl Kuemmerling, Inc.; **5-6** Courtesy of Skedco, Inc.; **5-7** Courtesy of Spelean Pty Ltd, Artarmon, Australia; **5-8** Courtesy of Mike Moore/FEMA

Chapter 6

Opener Courtesy of Jocelyn Augustino/FEMA; **6-1** © Baloncici/ShutterStock, Inc.; **6-2** Courtesy of CON-SPACE Communications; **6-4** © Glen E. Ellman; **6-7** © Sue Ashe/ShutterStock, Inc.; **6-8** © wheatley/ShutterStock, Inc.; **6-9** Courtesy of J. Arn Womble, Ph.D., P.E.; **6-17** Courtesy of Lara Shane/FEMA; **6-18** © Steve Allen/Brand X Pictures/Alamy Images; **6-19** © A.S. Zain/ShutterStock, Inc.; **6-20** Courtesy of FEMA

Chapter 7

7-1B Courtesy of Jack B. Kelley, Inc.; **7-4B** © Joan Stabnaw/ShutterStock, Inc.; **7-4C** © Richard Thornton/ShutterStock, Inc.

Chapter 8

Opener © Samuel Hoffman, *The Journal Gazette*/AP Photos; **8-8, 8-9** Courtesy of Cecil "Buddy" Martinette; **8-11** Courtesy of Captain David Jackson, Saginaw Township Fire Department

Chapter 9

Opener © Corbis; **9-1** © Jack Dagley/ShutterStock, Inc.; **9-2** © Jim Parkin/ShutterStock, Inc.; **9-3** © Corbis; **9-4** © Stan Rohrer/Alamy Images; **9-5** © Bob Pardue/Alamy Images; **9-6** © Kristian Gonyea, The Press of Atlantic City/AP Photos; **9-8** Courtesy of Captain David Jackson, Saginaw Township Fire Department; **9-9** © Louis Brems, *The Herald-Dispatch*/AP Photos; **9-10** Courtesy of Robert Kaufmann/FEMA; **9-11** © 1125089601/ShutterStock, Inc.; **9-12** Courtesy of Mark Woolcock

Chapter 10

Opener Courtesy of Jocelyn Augustino/FEMA; **10-1A** © Willem Dijkstra/ShutterStock, Inc.; **10-1B, 10-1C** © Photodisc; **10-1D** © Simon Krzic/ShutterStock, Inc.; **10-3A** Courtesy of Carol Read; **10-3B** Courtesy of Oceanid—Water Rescue Craft; **10-4A** Courtesy of The Coleman Company, Inc.; **10-4B, 10-4C, 10-4D, 10-4E** Courtesy of Johnson Outdoors Watercraft, Inc.; **10-5** Courtesy of Life Safer, Inc. www.life-safer.com; **10-6B** Courtesy of Carol Read; **10-7** Courtesy of C.J. Johnson; **10-8** Courtesy of Carol Read; **10-9** © Etienne Oosthuizen/ShutterStock, Inc.; **10-10A** Courtesy of CMC Rescue, Inc. www.cmcrescue.com; **10-10B** © jeff gynane/ShutterStock, Inc.; **10-11A** Photo of Zodiac Minuteman 420 RAD, courtesy of Zodiac of North America, Inc.; **10-11B** Courtesy of Oceanid—Water Rescue Craft; **10-11C** Courtesy of SeaArk Boats; **10-12A** Courtesy of Applied Combustion Technology, Inc.; **10-12C** © Purestock/age fotostock; **10-13** Courtesy of Garmin International; **10-14** Courtesy of ISG/INFRASYS; **10-17** Courtesy of Marvin Nauman/FEMA; **10-18** Courtesy of Innovative Scuba Concepts, Inc.; **10-19** Courtesy of The Coleman Company, Inc.; **10-20** Courtesy of Andrea Booher/FEMA; **10-22** © Stacey Lynn Payne/Dreamstime.com; **10-23** Courtesy of T.M. Smalley—MN DNR; **10-25** © Michael Zysman/ShutterStock, Inc.; **10-26A** © R. Filip/ShutterStock, Inc.; **10-26B** Courtesy of Marvin Nauman/FEMA; **10-26C** Courtesy of David Savile/FEMA; **10-27** © Chris James/Peter Arnold, Inc.

Chapter 11

Opener © Greg Epperson/ShutterStock, Inc.; **11-4A** © Corbis; **11-4B** © Caleb Foster/ShutterStock, Inc.; **11-5** Courtesy of Cascade Rescue Company

Unless otherwise indicated, all photographs are under copyright of Jones and Bartlett Publishers, LLC, photographed by Glen E. Ellman, or have been provided by the authors.

Reprinted with permission from NFPA 1670, *Operations and Training for Technical Search and Rescue Incidents*, Copyright © 2009, National Fire Protection Association, Quincy, MA 02169. This reprinted material is not the complete and official position of the NFPA on the referenced subject, which is represented only by the standard in its entirety.

Additional Resources
from Jones and Bartlett Publishers

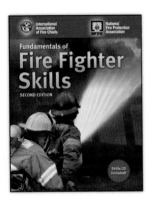

Fundamentals of Fire Fighter Skills

SECOND EDITION

International Association of Fire Chiefs, National Fire Protection Association
ISBN-13: 978-0-7637-7145-4
Paperback • 1068 Pages • © 2009

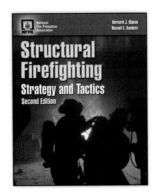

Structural Firefighting

Strategy and Tactics

SECOND EDITION

National Fire Protection Association, Bernard J. Klaene, Russell E. Sanders
ISBN-13: 978-0-7637-5168-5
Hardcover • 379 Pages • © 2008

Engine Company Fireground Operations

THIRD EDITION

National Fire Protection Association, Harold Richman, Steve Persson
ISBN-13: 978-0-7637-4495-3
Paperback • 192 Pages • © 2008

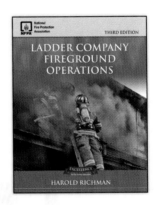

Ladder Company Fireground Operations

THIRD EDITION

National Fire Protection Association, Harold Richman, Steve Persson
ISBN-13: 978-0-7637-4496-0
Paperback • 251 Pages • © 2008

JONES AND BARTLETT
PUBLISHERS
BOSTON   TORONTO   LONDON   SINGAPORE

www.jbpub.com/Fire

Jones and Bartlett Publishers has developed a complete line of resources to help you advance your career within the fire service.

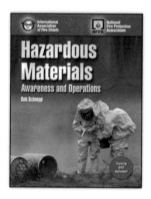

Hazardous Materials Awareness and Operations

International Association of Fire Chiefs, National Fire Protection Association, Rob Schnepp
ISBN-13: 978-0-7637-3872-3
Paperback • 432 Pages • © 2010

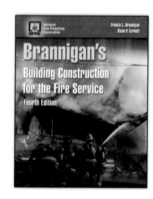

Brannigan's Building Construction for the Fire Service

FOURTH EDITION

National Fire Protection Association, Francis L. Brannigan, Glenn P. Corbett
ISBN-13: 978-0-7637-4494-6
Hardcover • 348 Pages • © 2008

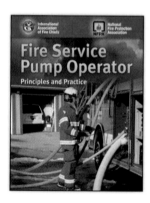

Fire Service Pump Operator

Principles and Practice

International Association of Fire Chiefs, National Fire Protection Association
ISBN-13: 978-0-7637-3908-9
Paperback • 400 Pages • © 2010

Fire Service Instructor

Principles and Practice

International Association of Fire Chiefs, National Fire Protection Association, International Society of Fire Service Instructors
ISBN-13: 978-0-7637-4910-1
Paperback • 296 Pages • © 2009

For more information on these resources, or to place your risk-free order, call 1-800-832-0034 or visit: www.jbpub.com/Fire.

1-800-832-0034